Introductory Plant Biology

Introductory Plant Biology

edition seven

KINGSLEY R. STERN

California State University ~ Chico

Boston, Massachusetts   Burr Ridge, Illinios   Dubuque, Iowa
Madison, Wisconsin   New York, New York   San Francisco, California   St. Louis, Missouri

WCB/McGraw-Hill

A Division of The **McGraw·Hill** *Companies*

Project Team

Editor *Marge Kemp*
Developmental Editor *Kathy Loewenberg*
Production Editor *Kay J. Brimeyer*
Marketing Manager *Tom Lyon*
Designer *Kaye Farmer*
Art Editor *Jodi K. Banowetz*
Photo Editor *Carrie Burger*
Advertising Coordinator *Heather Wagner*

President and Chief Executive Officer *Beverly Kolz*
Vice President, Director of Editorial *Kevin Kane*
Vice President, Sales and Market Expansion *Virginia S. Moffat*
Vice President, Director of Production *Colleen A. Yonda*
Director of Marketing *Craig S. Marty*
National Sales Manager *Douglas J. DiNardo*
Executive Editor *Michael Lange*
Advertising Manager *Janelle Keeffer*
Production Editorial Manager *Renée Menne*
Publishing Services Manager *Karen J. Slaght*
Royalty/Permissions Manager *Connie Allendorf*

Copyedited by *Sarah Lane*
Interior & cover design by *Jamie O'Neal*
Cover photo: © Lois Ellen Frank/West Light

To the memory of Franklin Charles Lane

1928–1971

Botanist, Teacher, Friend

Table of Contents

Preface

*I*ntroductory Plant Biology is designed as an introductory text in botany. It assumes little knowledge of the sciences on the part of the student. The text includes sufficient information for some shorter introductory botany courses open to both majors and nonmajors, but it is arranged so that certain sections—for example, "Early History and Development of Plant Study," "Soils," "Division Psilotophyta"—can be omitted without disrupting the overall continuity of the course.

Botany instructors vary greatly in their opinions concerning the depth of coverage needed for the topics of photosynthesis and respiration in a text of this type. Some feel that nonmajors, in particular, should have a brief introduction only, while others consider a more detailed discussion essential. In this text, photosynthesis and respiration are discussed at three levels. Some may find one or two levels sufficient, and others may wish their students to become familiar with the processes at all three levels.

Despite eye-catching chapter titles and headings, many texts for majors and nonmajors relegate the current interests of a significant number of students to comparative obscurity. This text emphasizes current interests without giving short shrift to botanical principles. Present interests of students include subjects such as global warming, ozone layer depletion, acid rain, genetic engineering, organic gardening, Native American and pioneer uses of plants, pollution and recycling, houseplants, backyard vegetable gardens, natural dye plants, poisonous and hallucinogenic plants, and the nutritional values of edible plants. The rather perfunctory coverage or absence of such topics in many botany texts has occurred partly because botanists previously have tended to believe that some of the topics are more appropriately covered in anthropology and horticulture courses. I have found, however, that both majors and nonmajors in botany who may be initially disinterested in the subject matter of a required course frequently become engrossed if the material is repeatedly related to such topics. Accordingly, a considerable amount of ecological and ethnobotanical material has been included with traditional botany throughout the book—without, however, resorting to excessive use of technical terms.

ORGANIZATION OF THE TEXT

A relatively conventional sequence of botanical subjects is included. Chapters 1 and 2 cover introductory and back-ground information; chapters 3 through 11 deal with structure and function; chapters 12 and 13 introduce meiosis and genetics. Chapter 14 discusses plant biotechnology; chapter 15 introduces evolution. Chapter 16 presents a five-kingdom system of classification; chapters 17 through 23 stress, in phylogenetic sequence, the diversity of organisms traditionally regarded as plants, and chapter 24 deals with ethnobotanical aspects and information of general interest pertaining to sixteen major families of flowering plants. Chapter 25 is an overview of the vast topic of ecology, although ecological topics and applied botany are included in most of the preceding chapters as well. Some of these subjects are broached in anecdotes that introduce the chapters, while others are mentioned in the human and ecological relevance sections (with which most of the chapters in the latter half of the book conclude).

AIDS TO THE READER

Review questions, discussion questions, and additional reading lists are provided for each chapter. New terms are defined as they are introduced, and those used more than once are boldfaced and included in a pronunciation glossary. The use of scientific names throughout the body of the text has been held to a minimum, but a list of the scientific names of all organisms mentioned is given in Appendix 1. Appendix 2 deals with biological controls and companion planting; Appendix 3 lists wild edible plants, poisonous plants, hallucinogenic plants, spices, and natural dye plants. Appendix 4 discusses pruning and grafting, and gives horticultural information on houseplants; information on the cultivation and nutritional value of vegetables is included. Appendix 5 gives some metric equivalents.

NEW TO THIS EDITION

More than 100 new or revised illustrations have been added to this edition, and information throughout the text has been updated or augmented, particularly in the area of plant physiology. Boxed inserts by Dr. Daniel Scheirer about interesting recent specific events and discoveries have been added to about half of the chapters; his contributions are gratefully acknowledged. Discussion of major types of grafting has been moved from the Plant Biotechnology chapter to Appendix 4. The metric conversion table in Appendix 5 has been expanded.

ADDITIONAL LEARNING AIDS

Instructor's Manual/Test Item File

The Instructor's Manual/Test Item File is available with *Introductory Plant Biology* and offers a variety of course schedules while providing overviews, goals, suggested answers, film sources, and examination questions for each text chapter.

Laboratory Manual

The Laboratory Manual that accompanies *Introductory Plant Biology* has been revised for the seventh edition. It is written for the student entering the study of botany for the first time. The exercises utilize plants to introduce biological principles and the scientific method, and are written to allow for maximum flexibility in sequencing.

Student Study Guide

A Student Study Guide, prepared by Daniel Scheirer, Northeastern University, is also available. The guide provides students an opportunity to study at their own pace. It contains learning objectives, chapter outlines, key terms/concepts (referenced to the text), and a set of objective questions for each chapter.

Transparencies

The transparency package includes 100 two- and four-color acetate overlays that are available free to adopters. These figures represent key illustrations from the text that merit additional visual review and discussion.

Micro Test III

Micro Test III is a computerized testing program that offers you the most effective and flexible software to date. And there are several convenient format options available to all adopters: Use your MacIntosh or IBM PC to pick and choose your test questions, edit them, and add your own questions.

   If you do not have a microcomputer, you can still pick and choose your questions via our call-in/mail-in service. Within two working days of your request, we'll put a test master, a student answer sheet, and an answer key in the mail to you. Call-in hours are 8:30-5:00 CST, Monday through Friday.

Readings

Critical Thinking: A Collection of Readings, by David J. Stroup and Robert D. Allen is available to instructors who are working to integrate critical thinking into their curriculum. This inexpensive text will help in the planning and implementation of programs intended to develop students' abilities to think logically and analytically. The reader is a collection of articles that provide instruction and examples of current programs. The authors have included descriptions and evaluations of their personal experience with incorporating critical thinking study in their coursework. (ISBN 14556)

   In addition to helping students learn about botany through writing, *Writing to Learn Botany,* by Randy Moore, will help improve communication skills—crucial to any profession. (ISBN 17455)

Videotapes

Tapes One, Two, and Five in the *Life Science Animations Videotape Series* will provide you with more than 30 animations of the most difficult-to-learn concepts found in a botany course.

ACKNOWLEDGMENTS

The contributions of many individuals to the development of this book are gratefully acknowledged. Critical reviewers, who provided many valuable suggestions for improving and updating the text, include:

Reviewers

Holly Adrian, *St. Johns University*
Steve K. Alexander, *University of Mary Hardin-Baylor*
Dale Benham, *Nebraska Wesleyan University*
George S. Ellmore, *Tufts University*
John Green, *Nicholls State University*
Helen G. Kiss, *Miami University, Oxford, Ohio*
John Z. Kiss, *Miami University, Oxford, Ohio*
Jerry McClure, *Miami University, Oxford, Ohio*
H. Gordon Morris, *University of Tennessee—Martin*
Alison M. Mostrom, *Philadelphia College of Pharmacy and Science*
Jim Nelson, *Southwest Texas State University*
Daniel C. Scheirer, *Northeastern University*
C. Gerald Van Dyke, *North Carolina State University*

Additional persons who read parts of the manuscripts of various editions and made many helpful criticisms and suggestions include Robert I. Ediger, Richard S. Demaree, Jr., Robert B. McNairn, Donald T. Kowalski, Larry Hanne, Patricia Parker, and Robert A. Schlising. Others whose encouragement and contributions are deeply appreciated include Timothy Devine, W. T. Stearn, Lorraine Wiley, Isabella A. Abbott, Paul C. Silva, Donald E. Brink, Jr., William F. Derr, Beverly Marcum, Robert McNulty, the faculty and staff of the Department of Biological Sciences, California State University, Chico, my many inspiring students, the Lyon Arboretum of the University of Hawaii, the editorial, production, and design staffs of the Wm. C. Brown Company Publishers, and most of all my wife, Janet, and my children, Kevin and Sharon. Special thanks are due to artists Denise Robertson Devine, Janet Monelo, and Sharon Stern.

Kingsley R. Stern
Chico, California

Steershead (Dicentra uniflora), a diminutive relative of bleeding hearts, native to the mountains of the western United States and Canada.

The Development of Plant Study

FIGURE 1.3A   Ripening coffee berries. They are picked by hand when they are red. The seeds are extracted for roasting after the berries have been fermented.

FIGURE 1.3B   Coffee beans being roasted.

of plant habitats caused primarily by the huge increase in the number of earth inhabitants. This subject and related matters are further discussed in Chapter 25.

Some have suggested that the odds are against humanity saving itself from itself and have even indicated that it might become necessary to emigrate to another planet. If so, microscopic algae could play a vital role in space exploration. Experiments with portable oxygen generators have been in progress for many years. Tanks of water teeming with tiny green algae are taken aboard a spacecraft and installed so that they are exposed to light for at least part of the time. The algae not only produce oxygen, which the spacecraft inhabitants can breathe, but they also utilize the waste carbon dioxide end product of respiration. In addition, as they multiply, excess algae can be fed to a special kind of shrimp, which in turn multiply and become food for the space travelers. Other wastes are recycled by different microscopic organisms. When this self-supporting arrangement, called a *closed system,* is perfected, the range of spacecraft should greatly increase because heavy oxygen tanks will not be necessary, and fewer food reserves will be needed.

Inhabitants of undeveloped areas still use plants not only for food, shelter, clothing, and medicine but also in hunting and fishing. For example, plant poisons are applied to the tips of blow gun darts used in hunting, and the bulbs and seeds of certain plants, when thrown in dammed streams, stupefy fish so that they float to the surface. Today, small teams of botanists, anthropologists, and medical doctors are interviewing medical practitioners and herbal healers in remote tropical regions about various uses of plants by primitive peoples. These scientists are doing so in the hope of preserving at least some plants with potential uses for modern civilization before disruption of their habitats results in their extinction.

EARLY HISTORY AND DEVELOPMENT OF PLANT STUDY

On a number of occasions, I have received visits from anxious mothers or from pediatricians wanting to know if part of backyard plants that young children have consumed are poisonous. I have also been consulted by various law enforcement officials, landscape architects, pollen collectors, students interested in wildflowers, vegetarians, organic and traditional gardeners, and a variety of other professional and amateur persons with a wide range of interest in plant life. Some have wanted plants identified. Others have wanted suggestions for treating diseased plants. Still others have wanted to know about grafting techniques, the effect of "gray water" (water that has been used for bathing or washing dishes) on plants, the suitability of plants for specific locations, the preservation of plants, the edibility of wild plants, and a host of other plant-related subjects.

FIGURE 1.4 Cotton plants. The white fibers, in which seeds are embedded, are the source of textiles and fabrics. The seeds are the source of vegetable oils used in margarine and shortening. After the oils have been extracted, the remaining "cotton cake" is used for cattle feed.

FIGURE 1.5 A polluted waterway in an urban area.

Knowledge about plant life throughout the world has now become so vast that it is impossible for any person to be an authority on more than a tiny fraction of it. Our libraries contain thousands of books dealing with virtually every facet of botanical investigation, and research journals publish the latest discoveries from around the world on an almost daily basis. Why and how has all this knowledge accumulated? To answer this question we need to take a brief look at the early history and development of plant study.

Plants and Primitive Peoples

Archaeological evidence indicates that between 15,000 and 35,000 years ago humans migrated to the Americas and Australia. Migrations to the Americas evidently occurred between Siberia and Alaska across the Bering Strait, which was a land bridge during parts of the Pleistocene era. These early humans were primarily hunters. Indeed, the extinction of many large land mammals in North America coincides with the appearance and activities of humans between 13,000 and 9000 B.C., although major changes in climate and vegetation also occurred at that time. If you are a hunter or a fisherman, however, you are well aware that success in hunting or fishing can vary considerably, depending on various environmental and other circumstances.

By 8000 B.C., our ancestors had begun to develop more reliable sources of food through primitive forms of agriculture. Archaeological evidence obtained from the walls of tombs, mummy wrappings, hieroglyphics, cave paintings, and carvings indicates the cultivation of grains legumes and certain fruits (e.g., figs, olives, pomegranates, dates) was well established in the Near East by 6501 B.C. The Near Eastern Center and other major centers of origin of cultivated plants are discussed in Chapter 24.

By 4000 B.C., the date had become one of the most important crops to the Assyrians and Egyptians. It was eaten fresh or dried, and the sap of date palms was fermented for wine. Although the Assyrians knew nothing of the details, they were aware of sexuality in the date palm and pollinated the female trees by hand (Fig. 1.6). By the seventh century B.C., they had produced a systematically arranged list of medicinal plants, which suggests that the physicians and pharmacists of the day had a noteworthy knowledge of plants and their uses.

The Chinese have been cultivating medicinal and other useful plants for at least 4,500 years. Records from that far back in history are fragmentary, and it is often impossible to separate fact from legend. Some authorities agree, however, that the founder of Chinese agriculture was an emperor by the name of Shen Nung, who was born in 2737 B.C. Shen Nung is said to have invented the plow. He is also believed to have established an annual ceremony during which seeds of soybeans, wheat, rice, millet, and sorghum were sown or planted by royalty. He appears to have been an authority on poisons and antidotes. He also wrote a book on drugs and medicines that was incorporated into the Pun-tsao, a Chinese *pharmacopoeia* (an officially recognized book describing

FIGURE 1.6   A date palm.

and how they were put together. This inquisitiveness led to plant study becoming a **science,** which broadly defined is simply "a search for knowledge of the natural world." A science may be distinguished from other fields of intellectual endeavor by several features. It involves the observation, recording, organization, and classification of facts, and more importantly, it involves what is done with the facts. Scientific procedure involves experimentation, observation, and the verifying or discarding of information, chiefly through inductive reasoning from known samples. There is no universal agreement on the precise details of the process. A few decades ago, scientific procedure was considered to involve a routine series of steps. This series of steps came to be known as the *scientific method,* and there are still instances where such a structured approach works well. In general, however, the scientific method now describes the procedures of assuming and testing *hypotheses.*

Hypotheses

A **hypothesis** is simply a tentative, unproven explanation for something that has been observed. It may not be the correct explanation—testing will determine whether it is correct or incorrect. To be accepted by scientists, the results of any experiments designed to test a hypothesis must be repeatable and capable of being duplicated by others. The nature of the testing will vary according to the circumstances and materials. We may, for example, *observe* that apples are red fruits that taste sweet. We may then *hypothesize* that all red fruits taste sweet. We may *test* the hypothesis by tasting red fruits, and as a result of our testing (since red crab apples are bitter), we may *modify* the *hypothesis* to state that only some red fruits are sweet.

When a hypothesis is tested, *data* (bits of information) are accumulated and may lead to the formulation of a useful generalization called a *principle.* Several related principles may lend themselves to grouping into a *theory,* which is not simply a guess. A theory is a group of generalizations (principles) that help us understand something. We reject or modify theories only when new principles increase our understanding of a phenomenon.

Plant Science and Ancient Greece

However one defines *science,* it is clear that plant science existed in ancient Greece. As in even older cultures, the study of plants in Greece started when people developed a practical interest in food and drug plants, and it grew as they also became curious about the structure and function of plants. As the physicians and pharmacists of the era gathered and used medicinal plants, they studied the variations and forms and came to recognize apparent relationships among them.

One of these Greek herbal physicians, in 384 B.C., had a son who became one of the most renowned philosophers of all time—Aristotle. Although Aristotle is perhaps better known for his philosophical works, he was an accomplished mathematician and he also acquired extensive knowledge in

drugs and medicines) of 40 volumes, published during the 17th century. During the Han dynasties, which lasted from about 200 B.C. until the birth of Christ, gardens became very extensive in China and many ornamental plants were cultivated. Plants such as primroses, poppies, and chrysanthemums were brought to the Western world from China over 2,000 years ago.

The Egyptians cultivated primitive forms of wheat and barley from about 5000 to 3400 B.C., although some authorities place the cultivation of these two cereals as far back as 10,000 to 15,000 B.C. More modern forms of cereals, such as six-rowed barley, may have been under cultivation by 2000 B.C. By the fifth century B.C., the Egyptians apparently were brewing *booza,* a beer, from barley.

Botany as a Science

The study of plants, called **botany**—from the French word *botanique* (botanical) and the three Greek words *botanikos* (botanical), *botane* (plant or herb), and *boskein* (to feed)—appears to have had its origins with Stone Age peoples who sought to modify their surroundings and feed themselves. Initially, the primary interest in plants was practical, centering around how plants might provide food, fibers, fuel, and medicine. Eventually, however, an intellectual interest arose. Individuals became curious about how plants reproduced nearly all aspects of natural history. In fact, he combined philosophical and scientific interests as few other philosophers have done.

At the age of 17, Aristotle went to Athens, where he met and became a pupil of Plato. He left Athens after Plato's death in 347 B.C. and studied marine animals at a coastal area for several years, eventually returning to Athens to found the first botanical garden of which there is any record.

When Aristotle died, he willed the botanical garden and its associated library to his pupil and assistant, Theophrastus (Fig. 1.7). Theophrastus was an extraordinary man who not only learned virtually all Aristotle knew about plants but also added prodigiously to Aristotle's knowledge from his own observations. It is said that he had 2,000 disciples and wrote 200 treatises. The most important of the latter to have survived are two books entitled *History of Plants* and *Causes of Plants*. So great were his contributions to botany as a science that the famous 18th-century Swedish botanist Linnaeus gave him the title "Father of Botany." Few, if any, dispute his right to the honor.

Herbals Appear

Two books that had a significant influence on botanical studies appeared during the second century A.D. Pliny's *Historia Naturalis* contained lists of food or medicinal plants. Dioscorides' *Materia Medica* was the first book to contain illustrations of plants, all laboriously copied by hand. Many of the common names used by Dioscorides are still used today. European scholars who followed Dioscorides continued, by hand, to copy these books, which became known as **herbals,** and held them in such high esteem that it was considered heresy to question anything in their contents; consequently, few new ideas were added during the Dark and Middle Ages that lasted from 400 to 1400 A.D.

When the printing press appeared in the middle of the 15th century, the number of herbals mushroomed, and the period from about 1500 to 1700 A.D. became known as the Age of Herbals. These botanical works were primarily the products of German botanists, although some Italian and English botanists made their own contributions between 1470 and 1670.

The *herbalists,* as they were called, were mostly concerned with medicinal plants, which they studied in the botanical gardens that had become numerous and extensive in Europe by this time. They produced elaborate and intriguing illustrations for the herbals, occasionally accompanied by outlandish stories and descriptions. Some of the stories became legends and developed into the *Doctrine of Signatures*. According to this doctrine, if a part of a plant had the shape of a part of the human body, it would be useful in treating a disease of the human part it most closely resembled. For example, the meat of a walnut, which somewhat resembles a miniature brain, was used in treating brain diseases, and hepatica leaves, which have lobes reminiscent of those of the liver, were used to treat ailments of that organ (Fig. 1.8). One of the more famous herbalists was Otto Brunfels,

FIGURE 1.7 Theophrastus.
(Courtesy National Library of Medicine)

FIGURE 1.8 *Hepatica* plants.

who published a three-volume herbal in 1530. His work had excellent illustrations and is considered to be a link between ancient and modern botany (Fig. 1.9).

THE FIRST MICROSCOPES

The microscope had and continues to have a profound effect not only on plant studies but also on the biological sciences and related fields as a whole. In 1590, Zacharias and Francis

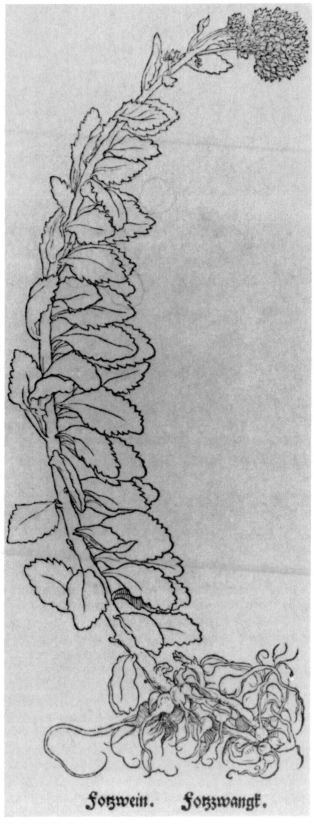

FIGURE 1.9 An illustration from Otto Brunfel's herbal (1530 A.D.).

(Courtesy National Library of Medicine)

Janssen, Dutch brothers who were spectacle makers, drew on the experience of their father, Hans, who was famous for his optical work. They discovered how to combine two convex lenses in the interior of a tube and produced the first instrument for magnifying minute objects. Because of this work, Zacharias Janssen, in particular, is often referred to as the inventor of the compound microscope, although it was Faber of Bamberg, a physician serving Pope Urban VIII, who originally applied the term microscope to the instrument during the first half of the 17th century. A Dutch draper by the name of Anton van Leeuwenhoek (1632–1723), who ground lenses and made microscopes in his spare time, is best known for his development of primitive microscopes. Leeuwenhoek was the first to describe bacteria, sperms, and other tiny cells he observed with his microscopes, some of which could magnify as much as 200 times. In his will, he left 26 of his 400 handmade microscopes to the Royal Society of London. Pictures of both primitive and modern microscopes are found on pages 28 and 30.

Before the invention of the microscope, plant study had been dominated by investigations based primarily on the external features of plants. The magnification of the early microscopes was not very great by present standards, but these instruments nevertheless led to the discovery of *cells* (discussed in Chapter 3) and opened up whole new areas of study.

DIVERSIFICATION OF PLANT STUDY

Plant anatomy, which is concerned chiefly with the internal structure of plants, was established through the efforts of several scientific pioneers. Early plant anatomists of note included Marcello Malpighi (1628–1694) of Italy, who discovered various tissues in stems and roots, and Nehemiah Grew (1628–1711) of England, who described the structure of wood more precisely than any of his predecessors (Fig. 1.10).

Plant physiology, which is concerned with plant function, was established by J. B. van Helmont (1577–1644), a Flemish physician and chemist, who was the first to demonstrate that plants do not have the same nutritional needs as animals. In a classic experiment, van Helmont planted a willow branch weighing 5 pounds in an earthenware tub filled with 74.4 kilograms (200 pounds) of dry soil. He covered the soil to prevent dust settling on it from the air, and after five years he reweighed the willow and the soil. He found that the soil weighed 56.7 grams (2 ounces) less than it had at the beginning of the experiment, but that the willow had gained 164 pounds. He concluded that the tree had added to its bulk and size from the water it had absorbed. We know now that most of the weight came as a result of photosynthetic activity (discussed in Chapter 10), but van Helmont deserves credit for landmark experimentation in plant physiology.

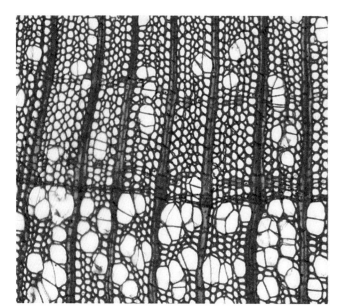

FIGURE 1.10 A thin section of wood as seen through a microscope.

FIGURE 1.11 Ecologists, geographers, and other biologists recognize large communities of plants and animals that occur in areas with distinctive combinations of environmental features. These areas, called *biomes,* are represented here by the Tropical Rain Forest, which, although occupying less than 5% of the earth's surface, is home to more than half of the world's species of organisms.

In the 15th century when Columbus visited Cuba, he found local Indian tribes cultivating corn (maize). This important food plant had apparently been in use by the pre-Incas of Peru some 5,000 years earlier. By the time explorers ventured into the Americas in the 1600s, they found that maize culture had spread from Argentina in the South to the St. Lawrence River area in the North. American Indians had also domesticated the white potato, and Indians were cultivating flowers and medicinal plants in Mexico.

The 17th century saw a marked increase in botanical explorations to various parts of the globe. The explorers took large numbers of plants back to Europe with them, and it soon became clear to those working with the plants that some sort of formalized system was necessary just to keep the collections straight. Several *plant taxonomists* (botanists who specialize in the identifying, naming, and classifying of plants) proposed ways of accomplishing this, but we owe our present system of naming and classifying plants to the Swedish botanist Carolus Linnaeus (1707–1778) (see Fig. 16.2).

Plant taxonomy (also called *plant systematics*), which is the oldest branch of plant study, began in antiquity, but Linnaeus did more for the field than any other person in history. Thousands of plant names in use today are those originally recorded in Linnaeus's book *Species Plantarum,* published in 1753. An expanded account of Linnaeus and his system of classification is given in Chapter 16.

Theophrastus (fourth century B.C.) was the first person on record to have systematically discussed the relationship of plants to their surroundings, but the discipline of **plant geography,** the study of how and why plants are distributed where they are, did not develop until the 19th century (Fig. 1.11). The allied field of **plant ecology,** which is the study of the interaction of plants with one another and with their environment, also developed in the 19th century.

Public awareness of the field of ecology as a whole increased considerably after 1962, following publication of a book entitled *Silent Spring* by Rachel Carson. In this bestseller, based on more than four years of literature research, the author called attention to the fact that more than 500 new toxic chemicals annually are put to use as pesticides in the United States alone, and she detailed the insidious impact of these chemicals and other pollutants on all facets of human life and the environment.

Among the most noteworthy of the early plant geographers were two natives of Berlin, Germany, Carl Willdenow (1765–1812) and Alexander von Humboldt (1769–1859), who published books on the relationship of seed dispersal to plant distribution and on the associations of various plants with one another in tropical and temperate climates. These studies were brought to a climax by Sir Joseph D. Hooker (1817–1911), who eventually became director of the Royal Botanic Gardens in Kew, England. Hooker traveled widely, studying plant life in both the Northern and Southern Hemispheres, and he also published floras (accounts of the plants of a specific region) of India and Antarctica. Charles Darwin, whose books (particularly his *On the Origin of Species*) revolutionized some basic biological concepts of the adaptation of organisms to their environment, said of Hooker's *Flora Antarctica,* "It is by far the grandest and most interesting essay on subjects of nature I have ever read."

The study of the form and structure of plants, **plant morphology,** was developed during the 19th century, and by the beginning of the 20th century, much of the basic information incorporated in the plant sciences today had been

Plant Biology and the Web

The **World Wide Web** (WWW) is a rich area of cyberspace that contains formatted text documents, color graphics, maps, audio clips, video images, and other interesting items. It contains a virtual storehouse of scientific knowledge just waiting to be explored. If you have never experienced the **Internet** and the **World Wide Web,** you are about to experience virtual biology.

What is the Internet? A technical answer would include a description of the historical origins of the Internet in national defense, research, and education, as well as the physical connection of computers to one another. However, the Internet has come to mean much more than this. It is frequently described as the *Information Superhighway,* the *Infobahn,* or *Cyberspace.* What do people do on the Internet? There are several components to the Internet such as exchanging e-mail, following newsgroups, and downloading data files, images, and sound files.

One aspect of the Internet is that it is international. Because it is a global network, one minute you may be retrieving a file from France or Japan, and the next minute you are tapped into a computer at your local university or college. The interesting part about this is that you frequently do not even know that you have crossed national boundaries.

Access to the Internet begins with a connection that can be supplied by numerous Internet providers or some large commercial on-line services like America Online or CompuServe. These services charge a fee for access to their computer, but once connected, you have the full global capabilities of exploring the vast amounts of information and entertainment features on the Internet.

The *client* and *server* are terms that are used to explain the information flow from remote computers (server) to your computer (client). Your personal computer has software that controls what you see on the screen (user interface menus) and responds to your interactions. This is the *client.* When you request a file, the client software program sends a message to a *server* (on another computer) to retrieve the file. The server then returns the file to the client software, which interprets and displays the information in the file. The diagram below summarizes this interaction.

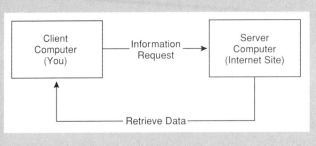

The Internet has several information servers that provide different ways to access information. They range from the easy-to-use to the more complex and arcane. The World Wide Web is similar to the other information servers (FTP, Gopher, WAIS, Veronica) but has several distinct advantages that make it a very popular way of browsing for information. First, it offers formatted text and graphics in the form of pages instead of menu lists. These pages begin with a *home page* (a central navigational point) and are read much like the pages of a book. Additionally, documents are linked together using hypertext formatting that allows users to browse from one linked document to another, not in an hierarchical tree, but in a true web of interrelated topics.

The first thing you need to start browsing on the web is web browser software like *Netscape Navigator* or *Mosaic.* You are then ready to type in an address, called a URL (universal resource locator) and start exploring. If you type in a URL address, the client software interprets the URL and initiates communication with the specified server. The parts of a URL consists of the following:

protocol://server/directory/filename

This is the URL for the web site at Wm. C. Brown Publishers (the publishers of this text).

---

discovered and elucidated. The number of scientists engaged in investigating plants had also increased conspicuously. **Genetics,** the science of heredity, had been founded by the Austrian monk Gregor Mendel (1822–1884) through his classic experiments with peas. **Cytology,** the science of cell structure and function (now called **cell biology**), had received great impetus from the discovery of how cells multiply and function in sexual reproduction. The mid-20th-century development of *electron microscopes* (see Chapter 3) further spurred cell research and led to vast new insights into cells and new forms of cell research that continue to the present. Many other significant 20th-century discoveries are discussed at appropriate junctures throughout the chapters to follow.

http://www.wcbp.com/wcbprintf.html#plants

1. **http** is the protocol called hypertext transfer protocol and is used by the client and server to communicate with each other.
2. **www.wcbp.com** is the name of the server
3. **wcbprintf.html#plants** is the directory and filename (pathname)

What if you want to search the Web for a specific topic? There are several *search engines* available that allow you to search by key word(s). The search software scans the numerous Web servers for your key word(s) and returns to you any number of hits, or positive matches. You can go directly to any of the matches by clicking the mouse pointer on the hyperlinked search results. One widely used search engine is called *WebCrawler (http://www.Webcrawler.com/)*. Another one that offers several different search engines is **Quarter-deck** *(http://www.qdeck.com/qdeck/search/)*.

What botanical information is available on the Web? You will be surprised at the variety and amount of information accessible. The following are some interesting web sites that I have explored and their URLs. Try them out sometime! Maybe you'll find a good idea for a research paper.

Some plant biology Web sites:

Australian National Botanic Gardens WWW Server provides a wealth of botanical and biological information about Australia.

http://www.anbg.gov.au/

California Flora Database contains geographic and ecological distribution information for 6,717 California vascular plant taxa, as well as additional habitat information for rare taxa and species of the Sierra Nevada.

http://s27w007.pswfs.gov/calflora/index.html

Carnivorous Plants Database includes over 3,000 entries giving an exhaustive nomenclatural synopsis of all carnivorous plants.

http://www.hpl.hp.com/bot/cp_home

Common Conifers of the Pacific Northwest provides information about the conifers of Oregon, including a dichotomous key for their identification.

http://www.orst.edu/instruct/for241/con/

Florida Wildflower Page provides hundreds of photos of Florida wildflowers.

http://www-wane-leon.scri.fsu.edu/~mikems

GardenNet is an information center for garden and gardening enthusiasts.

http://trine.com/gardennet/

National Wildflower Research Center is a nonprofit research and educational organization committed to the conservation and reestablishment of native wildflowers, grasses, shrubs, and trees.

http://www.onr.com/wildflower.html

Orchid Greenhouse provides information, photographs, and methods of growing orchids.

http://yakko.cs.wmich.edu/~charles/orchids/

Poisonous Plant Database is a set of working files of scientific information about the animal and human toxicology of vascular plants and herbal products of the world.

http://vm.cfsan.fda.gov/~djw/readme.html

Tropical Rainforest in Surinam provides a virtual tour of the rain forests of Surinam, complete with many fine photographs.

http://www.euronet.nl/users/mbleeker/suri_eng.html

PLANT SCIENCES AND THE FUTURE

There is still a vast amount of information to be discovered, and new discoveries continue to be made daily. Back in 1938, for example, 11,000 papers on botanical subjects were published in that year alone; the number in recent years is many times greater. Further, it appears probable that at least one-third of all the organisms traditionally regarded as plants (particularly algae and fungi) have yet to be named, let alone thoroughly investigated.

Wild plants and animals are becoming extinct at a rapidly accelerating rate as their natural habitats are destroyed through development and pollution; in fact, it is evident that many undescribed organisms are becoming extinct before we have learned anything about them. Efforts must be upgraded to educate the general public on the necessity of preserving natural habitats so that the numerous tangible and aesthetic benefits of doing so will be available to succeeding generations. Also, both basic and applied research in botany need additional support if the earth's burgeoning human population is to continue to be fed, clothed, and housed.

Summary

1. Genetic engineering, which involves the introduction of desirable genes from one organism into another, holds much potential for the future in many areas including crop improvement, frost damage control, pollution control, weed inhibition, insect repulsion, and soil reclamation and binding.

2. We are totally dependent on green plants because they alone can convert the sun's energy into forms that are usable by and vital to the very existence of animal life.

3. We largely take plants and plant products for granted. Animals, animal products, many luxuries and condiments, and other useful substances, such as fibers, lumber, coal, medicines, and drugs, either depend on plants or are produced by them.

4. To ensure human survival, all persons soon may need to acquire some knowledge of plants and how to use them. Plants will undoubtedly play a vital role in space exploration as portable oxygen generators.

5. Teams of scientists are interviewing medical practitioners and herbal healers in the tropics to locate little-known plants used by primitive peoples before the plants become extinct.

6. There are thousands of books dealing with botanical subjects, the knowledge having accumulated over the span of human existence.

7. Humans migrated to the Americas between 10,000 and 35,000 years ago; they originally were hunters, but gradually they learned how to cultivate crops. The Assyrians and Egyptians were cultivating fruits and cereals by 4000 B.C., and the Chinese practiced primitive agriculture at least 4,500 years ago.

8. Botany, the study of plants, apparently began with Stone Age peoples' practical uses of plants. Eventually, botany became a science, as intellectual curiosity about plants arose.

9. A science involves observation, recording, organization, and classification of facts. The verifying or discarding of facts is done chiefly from known samples through inductive reasoning. The scientific method involves specifically following a routine series of steps and generally assuming and testing hypotheses.

10. Plant science existed in ancient Greece. A fourth-century B.C. Greek, Theophrastus, contributed much to the field of botany.

11. During the second century A.D., Pliny and Dioscorides produced books on food or medicinal plants. These books were copied by hand and became known as *herbals*. During the 15th century, the number of herbals increased greatly with the advent of the printing press. The herbals were illustrated and sometimes contained bizarre stories about plants; these stories resulted in the development of the *Doctrine of Signatures*.

12. The compound microscope had a profound effect on studies in the biological sciences and led to the discovery of cells.

13. Plant anatomy and plant physiology developed during the 17th century. J. B. van Helmont was the first to demonstrate that plants have nutritional needs different from those of animals. During the 17th century, Europeans engaged in botanical exploration on other continents and took plants back to Europe.

14. During the 18th century, Linnaeus produced the elements of our present system of naming and classifying plants.

15. During the 19th century, plant ecology, plant geography, and plant morphology developed, and by the beginning of the 20th century, genetics and cell biology became established. Much remains yet to be discovered and investigated.

Review Questions

1. Briefly indicate contributions to plant science made by the following: Shen Nung, the ancient Egyptians, Theophrastus, Grew, and Linnaeus.

2. What is meant by the *scientific method?*

3. Distinguish among *herb, herbal,* and *herbalist.*

4. What was the significance of van Helmont's experiment with the willow tree?

5. What is the oldest branch of botany, and why did it precede other branches?

6. What is the thrust of each of the other branches of botany?

7. What is the *Doctrine of Signatures?*

Discussion Questions

1. Since humans survived on wild plants for thousands of years, might it be desirable to return to that practice?

2. On the basis of what you have read, would you single out any one individual as having contributed the most to the development of plant study? Why?

3. How would you guess that Stone Age peoples discovered medicinal uses for plants?

4. Many of the early botanists were also doctors. Why do you suppose this is no longer so?

5. Consider the following hypothesis: "The majority of mushrooms that grow in grassy areas are not poisonous." How could you go about testing this hypothesis scientifically?

Additional Reading

Anderson, F. J. 1985. *An illustrated history of the herbals.* New York: Columbia University Press.

Carson, R. 1994. *Silent spring.* Boston: Houghton Mifflin Co.

Ewan, J. (Ed.). 1969. *A short history of botany in the United States.* Forestburgh, NY: Lubrecht & Cramer.

Greene, E. L. 1983. In F. N. Egerton (Ed.), *Landmarks of botanical history* (2 vols.). Palo Alto, CA: Stanford University Press.

Harvey-Gibson, R. J. 1981. In I. B. Cohen (Ed.), *Outlines of the history of botany.* Salem, NH: Ayer Company Publishers.

Morton, A. G. 1981. *History of botanical science: An account of the development of botany from the ancient time to the present.* San Diego, CA: Academic Press.

Nordenskiold, E. 1988. *The history of biology.* Irvine, CA: Reprint Services Corp. (Reprint of 1935 edition)

Swift, L. H. 1974. *Botanical bibliographies: A guide to bibliographic materials applicable to botany.* Ann Arbor, MI: University Microfilms International. (Reprint of 1970 edition)

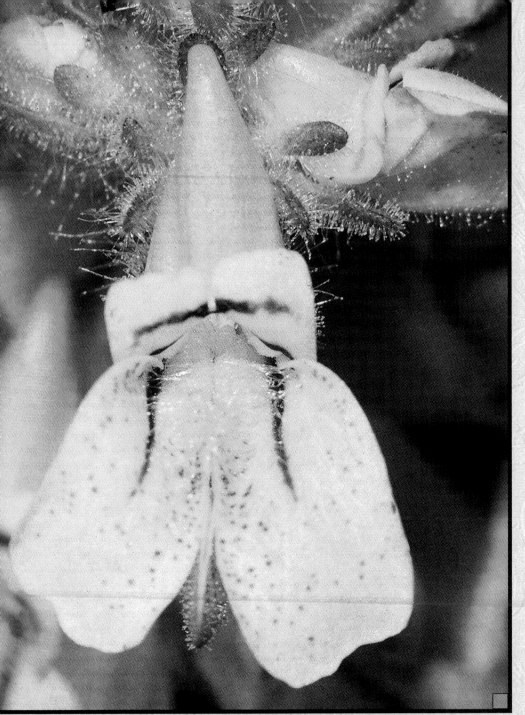

A flower of brownstain collinsia (Collinsia tinctoria). *The clear sap of this plant, on contact with human skin, produces a yellowish-brown stain; hence the plant's name.*

The Nature of Life

O v e r v i e w ────────────────────────

This chapter begins with a discussion of the attributes of living organisms. These include growth, reproduction, response to stimuli, metabolism, movement, complexity of organization, and adaptation to the environment. Then the chapter examines the chemical and physical bases of life. A brief look at the elements and their atoms is followed by a discussion of compounds, molecules, bonds, ions, valence, mixtures, acids, bases, and salts. Forms of energy and the chemical components of protoplasm are examined next. The chapter concludes with an introduction to macromolecules: carbohydrates, lipids, proteins, and nucleic acids.

S o m e   L e a r n i n g   G o a l s

1. Learn the attributes of living organisms.
2. Define *matter*; describe its basic state.
3. Differentiate compounds from mixtures and describe acids, bases, and salts.
4. Know the various forms of energy.
5. Learn the elements found in protoplasm.
6. Understand the nature of carbohydrates, lipids, and proteins.

Have you ever dropped a pellet of dry ice (frozen carbon dioxide) into a pan of water and watched what happens? As the solid pellet is rapidly converted to a gas due to its contact with the warmer water, it darts randomly about the surface looking like a highly energetic bug waterskiing. Does all that motion make the dry ice alive? Hardly; yet one of the attributes of living things is movement. But if living things move, what about plants? Is a tree not alive because it does not crawl down the sidewalk? Again the answer is no, but these questions do serve to point out some of the difficulties encountered in defining *life*. In fact, some contend that there is no such thing as life—only living organisms—and that life is a concept based on the collective attributes of living organisms.

ATTRIBUTES OF LIVING ORGANISMS

Composition

The activities of living organisms emanate from *protoplasm*, the physical basis or the "stuff of life," discussed in the next chapter.

Structure

The protoplasm is contained in tiny structural units called *cells*, which are unique to living things. Cells are discussed in the next chapter.

Growth

The complex phenomenon of **growth** has been described simply as an increase in mass (a body of matter), which is usually also correlated with an increase in volume. Growth, which results primarily from the production of new protoplasm, includes variations in *form*—some the result of inheritance, some the result of environmental response. As an example of environmental response, consider what might happen if you were to plant two apple seeds of the same variety in poor soil and subsequently give them unequal treatment. If you were to give one just barely enough water to allow it to germinate and grow, while you not only gave the other an ample water supply but also worked fertilizers and conditioners into the soil around it, you might expect the second one to grow larger and be more productive than the first. In other words, although your two apple trees grew from the same variety of apple seed, they would differ in form, following patterns of growth dictated by the protoplasm and the environment. Various aspects of growth are discussed in Chapter 11.

Reproduction

Dinosaurs were abundant 160 million years ago, but none exist today. Numerous mammals, birds, reptiles, plants, and other organisms are now on lists of endangered or threatened species, and many species are doomed to extinction within the next decade or two. All these once-living or living things have one feature in common: It became impossible or it has become difficult for them to reproduce. **Reproduction** is such an obvious feature of living organisms that we take it for granted—until it is lost.

When reproduction occurs, the offspring are always similar to the parents: Guppies never have puppies—just more guppies—and a petunia seed, when planted, will not develop into a pineapple plant. In addition, offspring of one kind tend to resemble their parents more than they do other individuals of the same kind. The laws governing these aspects of inheritance are discussed in Chapter 13.

Response to Stimuli

If you were to stick a pin into a pillow, you certainly would not expect any reaction from the pillow, but if you were to stick the same pin into a friend, you know the reaction would be instantaneous (assuming the friend was conscious) because response to stimuli is a major characteristic of all living things. You might argue, however, that when you stuck a pin into your houseplant nothing happened, even though you were fairly certain the plant was alive. What you might not have been aware of is that the houseplant did indeed respond but in a manner very different from that of a human. Plant responses to stimuli are generally much slower than those of animals and usually are of a different nature. If the houseplant's food-conducting tissue was pierced, the plant probably responded by producing a plugging substance called

callose in the affected cells. Some studies have shown that callose may form within as little as 5 seconds after wounding. In addition, an unorganized tissue called **callus,** which forms much more slowly, may be produced at the site of the wound. Responses of plants to injury and to other stimuli, such as light, temperature, and gravity, are discussed in Chapters 9 through 11.

Metabolism

Metabolism has been defined as the "collective product of all the biochemical reactions taking place within an organism." All living organisms undergo various metabolic activities, which include the production of new protoplasm, the repair of damage, and normal maintenance. The most important activities include **respiration,** an energy-releasing process that takes place in all living things; **photosynthesis,** an energy-harnessing process in green cells that is, in turn, associated with energy storage; **digestion,** the conversion of large or insoluble food molecules to smaller soluble ones; and **assimilation,** the conversion of raw materials into protoplasm and other cell substances. These topics are discussed in Chapters 9 through 11.

Movement

As observed at the beginning of this chapter, plants generally do not move from one place to another (although their reproductive cells may do so). This does not mean, however, that plants do not exhibit movement, a universal characteristic of living things. The leaves of sensitive plants (*Mimosa pudica*) fold within a few seconds after being disturbed or subjected to sudden environmental changes, and the tiny underwater traps of bladderworts (*Utricularia*) snap shut in less than one 100th of a second. But most plant movements, when compared with those of animals, are slow and imperceptible and are primarily related to growth phenomena. They become obvious only when demonstrated experimentally or when shown by time-lapse photography. The latter often reveals many types and directions of motion, particularly in young organs. Movement is not confined to the organism as a whole but occurs down to the cellular level. *Cyclosis,* a streaming motion of protoplasm, occurs constantly in living cells. The streaming tends to resemble a river flowing clockwise or counterclockwise within the boundaries of each cell, but movement may actually be in various directions.

Complexity of Organization

The cells of living organisms are composed of large numbers of **molecules** (the smallest unit of an element or compound retaining its own identity). There are typically more than 1 trillion molecules in a single cell. The molecules are not simply mixed, like the ingredients of a cake or the concrete in a sidewalk, but are organized into "compartments," membranes, and other structures within cells and tissues. Even the most complex nonliving object has only a tiny fraction of the types of molecules of the simplest living organism, and in the living organism, the arrangements of these molecules are highly structured and complex. Bacteria, for example, are considered to have the simplest cells known, yet each cell contains a minimum of 600 different kinds of protein in addition to hundreds of other substances, and each component has a specific place or structure within the cell. When larger living objects, such as flowering plants, are examined, the complexity of organization is overwhelming, and the number of molecule types can run into the millions.

Adaptation to the Environment

Assume that you skip a flat stone across a body of water and it lands on the opposite shore. The stone is not affected by the change from air to water to land during its quick journey; it does not respond to its environment. Living organisms, however, do respond to the air, light, water, and soil of their environment, as will be explained in later chapters. In addition, they are in many subtle ways genetically adapted to their environment, after countless generations of natural selection (as discussed in Chapter 15). Some weeds (e.g., dandelions) can thrive in a wide variety of soils and climates, while many species now threatened with extinction have adaptations to their environment that are so specific they cannot tolerate even relatively minor changes.

CHEMICAL AND PHYSICAL BASES OF LIFE

The Elements: Units of Matter

The basic "stuff of the universe" is called matter. On earth, matter occurs in three states: *solid, liquid,* and *gas.* In simple terms, matter's characteristics are as follows:

1. It occupies space.

2. It has mass (with which we commonly associate weight).

3. It is composed of **elements,** of which there are at present 109 known (92 occur naturally on our planet; the others have been produced artificially). Only a few of the natural elements (e.g., nitrogen, oxygen, gold, silver, copper) occur in pure form; the others are found combined together chemically in various ways. Each element has a designated symbol, often derived from its Latin name. The symbol for copper, for example, is *Cu* (from the Latin *cuprum*); and for sodium it is *Na* (from the Latin *natrium*). The symbols for carbon, hydrogen, and oxygen are *C, H,* and *O,* respectively.

The smallest stable subdivision of an element that can exist is called an **atom.** Atoms are so minute that individual atoms were not rendered directly visible to us until the mid-1980s

Table 2.1

ATOMIC NUMBERS AND MASSES OF SOME ELEMENTS FOUND IN PLANTS

ELEMENT	ATOMIC NUMBER	USUAL ATOMIC MASS
Hydrogen (H)	1	1
Boron (B)	5	11
Carbon (C)	6	12
Nitrogen (N)	7	14
Oxygen (O)	8	16
Magnesium (Mg)	12	24
Phosphorus (P)	15	31
Sulphur (S)	16	32
Chlorine (Cl)	17	35
Potassium (K)	19	39
Calcium (Ca)	20	40
Iron (Fe)	26	56

by even the most powerful electron microscopes. We do know from experimental evidence, however, that each atom has a tiny **nucleus** consisting of particles called **protons,** which have positive electrical charges, and other particles called **neutrons,** which have no electrical charges. If the nucleus, which contains nearly all of the atom's mass, were enlarged so that it were as big as a beach ball, the atom, which is mostly space, would be larger than an average-sized professional football stadium. Each kind of atom has a specific number of protons in its nucleus, ranging from 1 in the lightest element, hydrogen, to 92 in the heaviest element, uranium. Each element has an *atomic number* that is based on the number of protons present in a single atom. The atomic mass of an element is determined by the number of protons and neutrons present in a single atom (Table 2.1). An atom's protons and neutrons equal each other in mass.

Whirling around the nucleus are associated negative electric charges called **electrons**. Electron masses are about 1,840 times lighter than those of both protons and neutrons and are so infinitesimal that they are generally disregarded.

The region in which electrons whirl around the nucleus is called an **orbital.** Orbitals each have an imaginary axis and are somewhat cloudlike but are without precise boundaries, so we cannot be certain of an electron's position within an orbital at any time. This has led to an orbital being defined as a *volume of space in which a given electron occurs 90% of the time.*

An important feature of orbitals is that each is limited to two electrons; an orbital with only one electron can attract another electron to fill the available space. The innermost orbital is more or less spherical and is so close to the nucleus that it is usually not shown on diagrams of atoms. The one to several additional orbitals, which are shaped mostly like the tips of cotton swabs, generally occupy much more space.

The electrons of each orbital tend to repel those of other orbitals, so the axes of all the orbitals of an atom are oriented as far apart from each other as possible, although the outer parts of the orbitals actually overlap more than shown in diagrams of them. Orbitals usually have diameters thousands of times more extensive than that of an atomic nucleus (Fig. 2.1).

Electrons usually equal the protons in number, so the positive electric charges of the protons balance the negative charges of the electrons, making the atom electrically neutral. The number of neutrons in the atoms of an element can vary slightly, so the element may occur in forms having different weights but with all forms behaving alike chemically. Such variations of an element are called **isotopes.** The element oxygen (Fig. 2.2), for example, has seven known isotopes. The nucleus of one of these isotopes contains eight protons and eight neutrons; the nucleus of another isotope holds eight protons and ten neutrons, and the nucleus of a third isotope consists of eight protons and nine neutrons. If the number of neutrons in an isotope of a particular element varies too greatly from the average number of neutrons for its atoms, the isotope may be unstable and split into smaller parts, with the release of a great deal of energy. Such an isotope is said to be *radioactive.*

Molecules: Combinations of Elements

The atoms of most elements have the property of binding to other atoms of the same or different elements and forming new combinations; in fact, most elements do not exist independently as single atoms. When two or more elements are united in a definite ratio by chemical bonds, the substance is

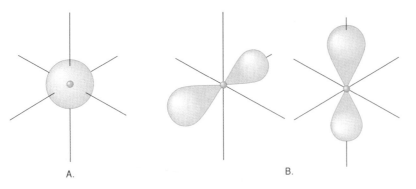

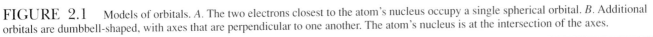

FIGURE 2.1   Models of orbitals. *A.* The two electrons closest to the atom's nucleus occupy a single spherical orbital. *B.* Additional orbitals are dumbbell-shaped, with axes that are perpendicular to one another. The atom's nucleus is at the intersection of the axes.

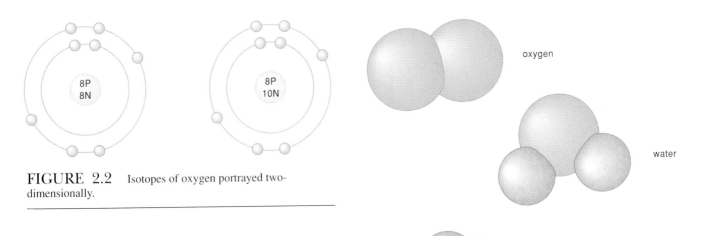

FIGURE 2.2   Isotopes of oxygen portrayed two-dimensionally.

FIGURE 2.3   Models of oxygen, water, and hydrogen molecules. A water molecule is 0.6 nanometer in diameter.

called a **compound**. A **molecule** consists of two or more atoms bound together and is the smallest independently existing particle of a compound or element. The molecules of the gases oxygen and hydrogen, for example, exist in nature as combinations of two atoms of oxygen or two atoms of hydrogen, respectively. Water molecules consist of two atoms of hydrogen and one atom of oxygen (Fig. 2.3).

Molecules are in constant motion, the motion speeding up or slowing down with an increase or decrease in temperature. The more molecular movement there is, the greater the chances are that some molecules will collide with each other. Also, the chances of random collisions increase in proportion to the density of the molecules (i.e., the number of molecules present in a given space).

Random collisions between molecules capable of sharing electrons are the basis for all chemical reactions. The reactions often result in new molecules being formed. Each chemical reaction in a cell usually takes place in a watery fluid and is controlled by a specific *enzyme*. Enzymes are organic *catalysts*. (A catalyst speeds up a chemical reaction without being used up in the reaction. See the discussion of enzymes on page 23.)

When a water molecule is formed, two hydrogen atoms become attached to an oxygen atom at an angle averaging 105° in liquid water (for ice, the angle is precisely 105°). The electrons of the three atoms are shared and form an electron cloud around the core, giving the molecule an asymmetrical shape. Although the electron and proton charges balance each other, the asymmetrical shape and unequal sharing of the electrons in the bond between oxygen and hydrogen cause one side of the water molecule to have a slight positive charge and the other a slight negative charge. Such molecules are said to be *polar*. Since negative charges attract positive charges, polarity affects the way in which molecules become aligned toward each other; polarity also causes molecules other than water to be water soluble.

Water molecules form a cohesive network as their slightly positive hydrogen atoms are attracted to the slightly negative oxygen atoms of other water molecules. The cohesion between water molecules is partly responsible for their movement through fine (capillary) tubes, such as those present in the wood and other parts of plants. The attraction between the hydrogen atoms of water and other negatively charged molecules, such as those of fibers, also causes *adhesion* (attraction of dissimilar molecules to each other) and is the basis for substances that can be wet by water. When there is no attraction between water and other substances (e.g., between water and the waxy surface of a cabbage leaf), the cohesion between the water molecules results in droplets beading in the same way that raindrops bead on a freshly waxed automobile.

Valence

The combining capacity of an atom or ion is called valence. Atoms of the element calcium, for example, have a valence of two, while those of the element chlorine have a valence of one. In order for the atoms of these two elements to combine, there must be a balance between electrons lost or gained (i.e., the valences must balance); for example, it takes two chlorine atoms to combine with *one* calcium atom. The compound formed by the union of calcium and chlorine is called *calcium chloride*. It is customary to use standard abbreviations taken from the Latin names of the elements when giving chemical formulas or equations. Calcium chloride, for example, is $CaCl_2$, indicating that one atom of calcium (Ca^{++}) requires two atoms of chorine (Cl^-) to form a calcium chloride molecule.

Bonds and Ions

Bonds that hold atoms together in molecules form in several different ways. There are three types of chemical bonds that are of major importance:

1. *Hydrogen bonds* form as a result of attraction between positively charged hydrogen atoms and other negatively charged atoms. Negatively charged oxygen and/or nitrogen atoms of one molecule may attract positively but weakly charged hydrogen atoms of other molecules, forming a weak bond. Hydrogen bonds are very important in nature because of their abundance in many biologically significant molecules. They have, however, only about 5% of the strength of covalent bonds.

2. *Covalent bonds* involve the sharing of a pair of electrons occurring in the outermost orbital; they hold together two or more atomic nuclei and travel between them, keeping them at a stable distance from each other. For example, the single orbital of a hydrogen atom, which has just one electron, is usually filled by attracting an electron from another hydrogen atom. As a result, two hydrogen atoms share their single electrons, making a combined orbital with two electrons. The combined orbital, with its two hydrogen atoms, makes a molecule of hydrogen gas, shown as H_2.

   Covalent bonds are the strongest of the three types of bonds discussed here and the principal force binding together atoms that make up some important biological molecules discussed later in this chapter (Fig. 2.4).

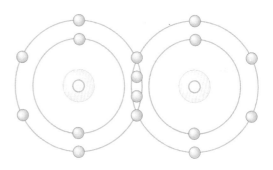

FIGURE 2.4 A covalent bond between two oxygen atoms. In a covalent bond, electrons are shared as outer shells of atoms overlap. In this instance, two pairs of electrons are shared between the two atoms, and the shared electrons are counted as belonging to each atom.

3. *Ionic bonds* form whenever one or more electrons are donated to another atom and result whenever two oppositely charged ions come in contact. In nature, some electrons in the outermost orbital are not really shared but instead are completely removed from one atom and transferred to another, particularly between elements that can strongly attract or easily give up an electron. Molecules that lose or gain electrons become positively or negatively charged particles called ions. Ions are shown with their charges as superscripts. For example, table salt (sodium chloride) is formed by ionic bonding between an ion of sodium (Na^+) and an ion of chlorine (Cl^-). The sodium becomes a positively charged ion when it loses one of its electrons, which is gained by an atom of chlorine. This extra electron makes the chlorine ion negatively charged, and the sodium ion and chlorine ion become bonded together by the force of the opposite charge.

Some ions such as those of magnesium (Mg^{++}), for example, give up two electrons and therefore have two positive charges. Such ions can form ionic bonds with two single negatively charged ions such as those of chlorine (Cl^-), forming magnesium chloride ($MgCl_2$). Most biologically important molecules exist as ions in living matter. Types of bonds other than hydrogen, covalent, and ionic need not be considered here.

Mixtures

A **mixture** differs from a compound in that not all of its molecules or atoms are united in definite ratios. For example, granite is composed of several different materials that vary in proportion to one another throughout the rock; likewise, a cake consists of ingredients that can vary in proportion to one another. Accordingly, granite, cakes, and a myriad of other substances with variable proportions of molecules are mixtures.

Acids, Bases, and Salts

As previously indicated, water molecules are held together by weak hydrogen bonds (see also the cohesion-tension theory in Chapter 9). In pure water, however, a few molecules sometimes separate into hydrogen (H^+) and hydroxyl (OH^-) ions, with the number of H^+ ions precisely equaling the number of OH^- ions. **Acids,** which taste sour like cranberry or lemon juice, are defined as substances that release hydrogen (H^+) ions when dissolved in water, with the result that there are proportionately more hydrogen than hydroxyl ions present. Conversely, **bases** (also referred to as *alkaline compounds*), which usually feel slippery or soapy, are defined as compounds that release negatively charged hydroxyl (OH^-) ions when dissolved in water. Bases may also be defined as compounds that accept H^+ ions.

The pH Scale

The concentration of H^+ ions present is used to define degrees of acidity or alkalinity on a specific scale, called the **pH scale.** The scale ranges from 0 to 14, with each unit representing a tenfold change in H^+ concentration. Pure water has a pH of 7—the point on the scale where the number of H^+ and OH^- ions is exactly the same, or the neutral point.[1] The lower a number is below 7, the higher the degree of acidity, and conversely, the higher a number is above 7, the higher the degree of alkalinity. Vinegar, for example, has a pH of 3, tomato juice has a pH of 4.3, and egg white has a pH of 8.

When an acid and a base are mixed, the H^+ ions of the acid bond with the OH^- ions of the base—forming water (H_2O). The remaining ions bond together, forming a **salt.** If hydrochloric acid (HCl) is mixed with a base—for example, sodium hydroxide (NaOH)—water (H_2O) and sodium chloride (NaCl), a salt, are formed. The reaction is represented by symbols in an equation that shows what occurs:

$$HCl + NaOH \longrightarrow H_2O + NaCl$$

Energy

Energy, which can be defined as "the ability to do work" or "the ability to produce a change in motion or matter," is required for all the activities of living things to take place, whether at the level of whole organisms, cells, or molecules. The ultimate source on earth of that energy is the sun.

Scientists have characterized energy with laws of thermodynamics. The *first law of thermodynamics* states that energy is constant—it cannot be increased or diminished—but it can be converted from one form to another. Among its forms are *chemical, electrical, heat,* and *light* energy. The *second law of thermodynamics* states that when energy is converted from one form to another in a given system in which no energy enters or leaves, it flows in one direction and the amount of useful energy remaining after the conversion will always be less than before the conversion. For example, heat will always flow from a hot iron to cold clothing but never from the cold clothing to the hot iron. Such energy-yielding reactions are vital to the normal functions of cells and provide the energy needed for other cell reactions. Both types of reactions are discussed in Chapter 10. Forms of energy include *kinetic* (motion) and *potential* energy. Potential energy is defined as the "capacity to do work owing to the position or state of a particle." For example, a ball resting at the top of a hill possesses potential energy that is converted to kinetic energy if the ball rolls down the hill. Some chemical reactions release energy and others require an input of energy.

Although all electrons have the same weight and electrical charge, they vary in the amount of potential energy they possess. Electrons with the least potential energy are located within a single spherical orbital closest to the atom's nucleus. At the next highest energy level, there are up to four orbitals, each with two electrons. Depending on the kind of atom, there can be higher numbers of orbitals at still higher energy levels, but the outermost level is limited to four orbitals and a maximum of four electrons. Each energy level is usually referred to as an *electron shell.* The outermost electron shell determines how or if an atom reacts with another atom. Atoms with eight electrons in the outer shell have no reaction with other atoms.

The various energy levels are due to the attraction between the positive charges of the protons and negative charges of the electrons. The farther away from the nucleus an electron is, the greater the amount of energy required to keep it there (Fig. 2.5). Some of the numerous energy exchanges and carriers occurring in living cells are discussed in later chapters.

Chemical Components of Protoplasm

The living substance of all cells is called **protoplasm.** Protoplasm is organized into numerous bodies of various sizes, most of which are discussed in Chapter 3. About 96% of protoplasm is composed of the elements carbon, hydrogen, oxygen, and nitrogen; 3% consists of phosphorus, potassium, and sulphur. The remaining 1% includes calcium, iron, magnesium, sodium, chlorine, copper, manganese, cobalt, zinc, and minute quantities of other elements. When a plant first absorbs these elements from the soil or atmosphere (see the section on essential elements in Chapter 9) or when it utilizes breakdown products within the cell, the elements are in the form of simple molecules or ions. These simple forms may be converted to very large, complex molecules through the metabolism of the cells.

The large molecules invariably have "backbones" of carbon atoms within them and are said to be **organic.** Other molecules that contain no carbon atoms are called **inorganic.**

---

1. Note that although distilled water is theoretically "pure," its pH is always less than 7 because carbon dioxide from the air with which it is in contact dissolves in it, forming carbonic acid (H_2CO_3); the actual pH of distilled water is usually approximately 5.7.

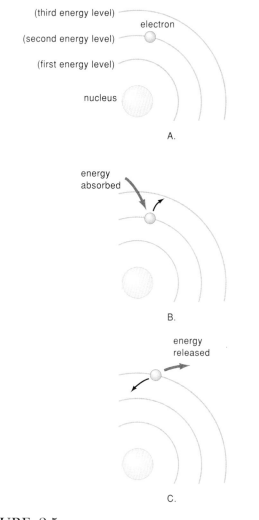

(third energy level)

electron

(second energy level)

(first energy level)

nucleus

A.

energy
absorbed

B.

energy
released

C.

FIGURE 2.5 Energy levels of electrons. The closer electrons are to the nucleus, the less energy they possess and vice versa. The energy levels are referred to as electron shells. *A.* An electron at a second energy level. *B.* An electron can absorb energy from sunlight or some other source and be boosted to a higher energy level. *C.* The absorbed energy can be released, with the electron dropping back to its original level (see Fig. 10.8).

Exceptions include carbon dioxide (CO_2) and sodium bicarbonate ($NaHCO_3$).

The name **organic** was given to most of the chemicals of living things when it was believed that only living organisms could produce molecules containing carbon. Today, many organic compounds can be produced artificially in the laboratory, and scientists sometimes hesitate to classify as either organic or inorganic some of the 4 million carbon-containing compounds thus far identified. Most scientists, nevertheless, agree that inorganic compounds usually do not contain carbon.

Macromolecules

The large molecules making up the majority of cell components are called *macromolecules* or **polymers.** Polymers are formed when two or more small units called **monomers** bond together. The bonding between monomers occurs when a hydrogen (H^+) is removed from one and a hydroxyl (OH^-) is removed from another, creating an electrical attraction between them. Since the components of water (H^+ and OH^-) are removed (*dehydration*) in the formation (*synthesis*) of a bond, the process is referred to as *dehydration synthesis.* Dehydration synthesis is controlled by an enzyme (see page 23).

Hydrolysis, which is essentially the opposite of dehydration synthesis, occurs when a hydrogen from water becomes attached to one monomer and a hydroxyl group to the other. Energy is released when a bond is broken by hydrolysis. This energy may be stored temporarily or used in the manufacture or renewal of cell components.

Four of the most important classes of polymers found in protoplasm are *carbohydrates, lipids, proteins,* and *nucleic acids.*

Carbohydrates *Carbohydrates* are the most abundant organic compounds in nature. They contain C, H, and O in or close to a ratio of 1C:2H:1O. Carbohydrates include *monosaccharides,* which are simple sugars with backbones consisting of three to seven carbon atoms. Among the most common monosaccharides are glucose and fructose, each of which has six carbon atoms. Glucose is a primary source of energy in cells. *Fructose,* which is found in fruits, is an *isomer* of glucose. Isomers are molecules with identical numbers and kinds of atoms but with different structures and shapes (Fig. 2.6).

Disaccharides are formed when two monosaccharides become bonded together by dehydration synthesis. The common table sugar **sucrose** is a disaccharide formed from a molecule of glucose and a molecule of fructose. Sucrose is the form in which sugar is usually transported throughout plants.

Polysaccharides are formed when several monosaccharides are bonded together. Polysaccharide polymers sometimes consist of thousands of simple sugars attached to one another in long chains or coils. *Starches,* for example, are polysaccharides that usually consist of several hundred to several thousand coiled glucose units. When numerous glucose molecules become a starch molecule, each glucose gives up a molecule of water. The formula for starch is $(C_6H_{10}O_5)n,$ the n representing many units. *Cellulose,* a principal substance in plant cell walls, is a polysaccharide consisting of 3,000 to 10,000 unbranched chains of glucose molecules. In order for a starch molecule to become available as an energy source in cells, it has to be hydrolyzed; that is, it has to be broken up into individual glucose molecules through the restoration of a water molecule for each unit.

Lipids *Lipids* are fatty or oily substances that are mostly insoluble in water because they have no polarized components. They store more energy than carbohydrates and play an important role in the longer term energy reserves and structural components of cells. Like carbohydrates, their molecules contain principally carbon, hydrogen, and oxygen, but there is proportionately much less oxygen present.

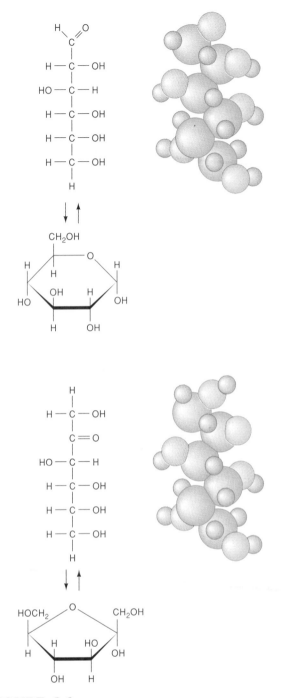

FIGURE 2.6 Structural formulas and models of glucose and fructose molecules. The molecules can exist either as straight chains or in the form of rings. Each chain is 3 nanometers long. The ring forms are more common in cells.

Examples of lipids include **fats** (Fig. 2.7) and **oils,** whose molecules are manufactured from sugars and are composed of a unit of glycerol (an alcohol) with three fatty acids attached. The fatty acids have carbon atoms to which hydrogen atoms can become attached.

Most fatty acid molecules consist of a chain with 16 to 18 carbon atoms. If hydrogen atoms are attached to every available attachment point of these fatty acid carbon atoms, the fat is said to be *saturated*. However, if fewer hydrogen atoms are attached, as is the case with some vegetable oils that are currently receiving publicity, the fat is said to be *polyunsaturated*. Like polysaccharides and proteins (discussed in the next section), lipids are broken down by hydrolysis.

Waxes are lipids consisting of very long-chain fatty acids bonded to a very long-chain alcohol other than glycerol. Waxes, which are solid at room temperature, are found on the surfaces of aboveground plant organs. They are usually embedded in a matrix of *cutin* or *suberin,* which are also insoluble lipid polymers. The combinations of wax and cutin or wax and suberin function in waterproofing, reduction of water loss, and protection against microorganisms and small insects.

Phospholipids are constructed like fats, but one of the three fatty acids is usually replaced by a phosphate group; this can cause the molecule to become a polarized ion. When phospholipids are placed in water, they form a double-layered sheet resembling a membrane. Indeed, phospholipids are important components of all membranes found in living organisms.

Amino Acids and Proteins   The cells of living organisms contain from several hundred to many thousand different kinds of **proteins**. These important molecules are usually very large and are composed of monomers called **amino acids** (Fig. 2.8). Proteins consist of carbon, hydrogen, oxygen, and nitrogen atoms and sometimes also sulphur atoms. Aside from water, proteins form the bulk of protoplasm.

There are 20 different kinds of amino acids, and from 50 to 50,000 or more of them are present in various combinations in each protein molecule. Every amino acid has two special functional groups of atoms plus a remainder called the *R group*. One functional group is called the *amino group* (NH$_2$); the other, which is acid, is called the *carboxyl group* (COOH). The makeup of the R group is distinctive for each of the 20 amino acids. Some R groups are polar while others are not.

Peptides   A *peptide* consists of two or more amino acids bonded together. Bonds between amino acids are called *peptide bonds;* many amino acids may form long *polypeptide* chains. Each polypeptide usually coils, bends, and folds in a specific fashion within a protein, which characteristically has three levels of structure and sometimes four:

1.   A sequence of amino acids fastened together by peptide bonds forms the *primary structure* of a protein.

2.   As hydrogen bonds form between oxygen and nitrogen atoms of different amino acids, the polypeptide chain coils like a spiral staircase, and the *secondary structure* develops. Some secondary structures include polypeptide chains that double back and form hydrogen bonds between the two lengths in what is referred to as a *beta sheet*, or *pleated sheet.*

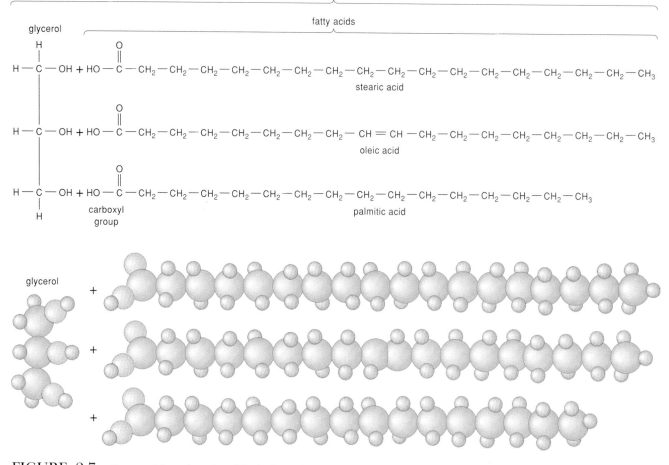

FIGURE 2.7   Structural formula and model of a fat molecule. H = hydrogen, C = carbon, O = oxygen. A typical fatty acid is 4 nanometers long.

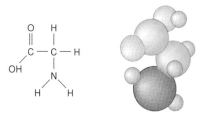

FIGURE 2.8   Structural formula and model of the amino acid glycine.

3. *Tertiary structure* develops as the polypeptide further coils and folds. The tertiary structure is maintained by bonds between R groups.

4. If a protein happens to have more than one kind of polypeptide, a fourth or *quaternary structure* may form (Fig. 2.9).

Anything that disturbs the normal pattern of bonds between parts of the protein molecule and thereby alters the characteristic coiling and folding, will *denature* the protein (adversely affect its function or properties). Denaturing, which is often brought about by high temperatures or chemicals, may kill the cell of which the protein is a part.

Enzymes   *Enzymes* are large, complex proteins (or in a few instances ribonucleic acid molecules—see the section on nucleic acids that follows). Enzymes function as organic catalysts under specific conditions of pH and temperature. They facilitate cellular chemical reactions, even at very low concentrations, and they can increase the rate of reaction as much as a billion times. Enzymes are essential, for none of the 2,000 or more chemical reactions in cells can take place unless the enzyme specific for each one is present and functional in the cell in which it is produced. In addition, enzymes do not usually break down in the reactions they accelerate and are often used repeatedly.

The names of enzymes normally end in *-ase*. One of the most common is maltase, which catalyzes the hydrolysis of maltose to glucose; maltose is a disaccharide composed of two glucose monomers. Enzymes lower the *energy of activation,* which is the energy needed to cause molecules to react with one another. An enzyme brings about its effect by bonding with potentially reactive molecules at a surface site. The reactive molecules temporarily fit into the active site, where a short-lived complex is formed. The reaction then occurs

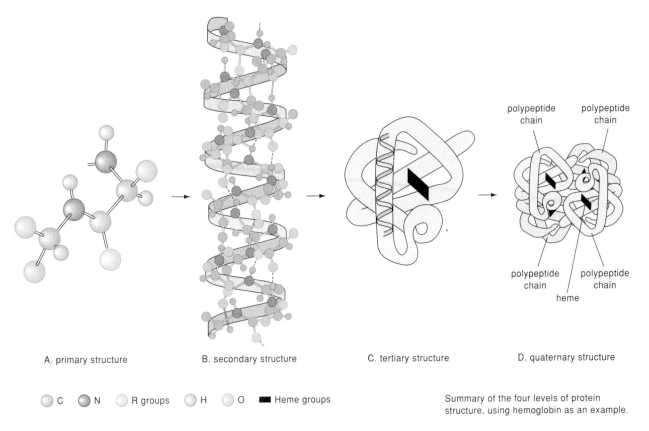

A. primary structure   B. secondary structure   C. tertiary structure   D. quaternary structure

○ C   ● N   ○ R groups   ○ H   ○ O   ■ Heme groups

Summary of the four levels of protein structure, using hemoglobin as an example.

FIGURE 2.9   The four levels of protein structure. The example shown is for hemoglobin. *A.* The primary structure consists of a chain of amino acids bonded together. *B.* As the amino acid chain grows it coils and often doubles back, with hydrogen bonds forming between the two lengths and creating a *beta-* or *pleated sheet. C.* The coil or helix folds further, forming a somewhat globular structure. *D.* Several chains combine into a single functional protein molecule.

After Caret, R. L., K. J. Denniston, and J. J. Topping. 1993. *Inorganic, Organic & Biological Chemistry.* Copyright 1993 by Wm C. Brown Publishers, Dubuque, IA.

rapidly, often at rates exceeding 500,000 times per second. The complex then breaks down as the products of the reaction are released, with the enzyme remaining unchanged and capable of once more facilitating the reaction (Fig. 2.10).

Many enzymes, derived mostly from bacteria and fungi, have very important industrial uses. For example, waste treatment plants, the dairy industry, and manufacturers of detergents all use enzymes that have been mass-produced by microorganisms in large vats. One such commercially marketed enzyme produced by the activities of *Aspergillus,* a mold, breaks down complex sugars found in beans, broccoli, and many other vegetables consumed by humans. A few drops of the enzyme placed on these foods while they are being consumed effectively reduces the gas produced when enzymes in human digestive tracts are otherwise unable to accomplish the breakdown.

Nucleic Acids   *Nucleic acids* are exceptionally large complex polymers originally thought to be confined to the nuclei of cells but now known also to be associated with other cell parts. They are vital to the normal internal communication and functioning of all living cells. The two types of nucleic acids—deoxyribonucleic acid (DNA) and

ribonucleic acid (RNA)—are briefly introduced here and discussed in more detail in Chapter 13.

Deoxyribonucleic acid (DNA) molecules consist of double helical (spiral) coils of repeating subunits called **nucleotides,** each composed of a base, a sugar, and a phosphate. Four kinds of nucleotides, each with a unique nitrogenous base, occur in DNA. DNA molecules contain, in units known as **genes,** the coded information that precisely determines the nature and proportions of the myriad substances found in cells and also the ultimate form and structure of the organism itself. If this coded information were written out, it would fill over 1,000 books of 300 pages each—at least for the more complex organisms. DNA molecules can replicate (duplicate themselves) in precise fashion. When a cell divides, the hereditary information contained in the DNA of the new cells is an exact copy of the original and can be passed on from generation to generation without change, except in the event of a *mutation* (discussed in Chapter 13).

Ribonucleic acid (RNA) is similar to DNA but differs in its sugar and one of its nucleotide components. It usually occurs as a single strand. Forms of RNA facilitate protein synthesis.

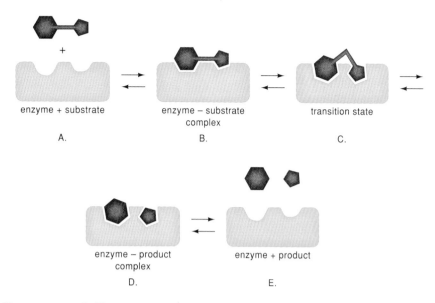

FIGURE 2.10 How an enzyme facilitates a reaction. *A.* An enzyme and the raw material (substrate) for which it is specific. *B.* The substrate fits into the active site on the enzyme. *C.* The enzyme then changes shape, putting stress on the linkage between parts of the substrate. *D.* The bonds (linkage) are broken. *E.* The enzyme returns to its original shape and the products are released. When an enzyme is combining substrates the events shown proceed in reverse.

Summary

1. Activities of living organisms stem from protoplasm, the physical basis or the "stuff of life." Structure and growth are among the attributes of living organisms. Growth has been described as an increase in volume; it results primarily from the production of new protoplasm. Variations in form may be inherited or result from response to the environment.

2. Reproduction involves offspring that are always similar in form to their parents; if reproduction ceases, the organism becomes extinct. Plants generally respond to stimuli more slowly than animals and in a different fashion.

3. All living organisms exhibit metabolic activities, including production of new protoplasm, respiration, digestion, and assimilation (green organisms can, in addition, carry on photosynthesis); they also all exhibit movement to varying degrees. Cyclosis is the streaming motion of protoplasm within living cells. Living organisms have a much more complex structure than nonliving objects and are adapted to their individual environments.

4. The basic "stuff of the universe" is called *matter;* it occurs in three states: solid, liquid, and gas. It is composed of elements, the smallest stable subdivision of which is an atom. Atoms contain, in a tiny nucleus, positively charged protons and uncharged neutrons, surrounded by much larger orbitals or regions of whirling, negatively charged electrons. Isotopes are forms of elements that have slight variations in the number of neutrons in their atoms.

5. The combining capacities of atoms or ions are called *valence.* Atoms of elements can bond to other atoms, and those of most elements do not exist independently; compounds are substances composed of two or more elements combined in a definite ratio by chemical bonds. Molecules are the smallest independently existing particles. If a molecule loses or gains electrons, it becomes an ion, which may form an ionic bond with another ion. In a covalent bond, pairs of electrons link two or more atomic nuclei; nitrogen and/or oxygen atoms of one molecule may form weak hydrogen bonds with hydrogen atoms of other molecules.

6. Water molecules are asymmetrical in shape, causing them to have slight electrical charges on each side (i.e., they are polar). Water molecules cohere to each other and adhere to other molecules.

7. The atoms of mixtures are not all chemically united in definite ratios. Acids release positively charged hydrogen ions when dissolved in water. Bases release negatively charged hydroxyl ions when dissolved in water. The pH scale is used to measure degrees of acidity or alkalinity. Salts and water are formed when acids and bases are mixed.

8. Energy can be defined as "ability to produce a change in motion or matter" or simply as "ability to do work." Its forms include chemical, electrical, heat, light, kinetic, and potential. The farther away from the nucleus an electron is, the greater the amount of energy required to keep it there.

9. Protoplasm is composed primarily of carbon, hydrogen, oxygen, and nitrogen, with a little phosphorus and potassium, plus small amounts of other elements. A plant may convert the simple molecules or ions it recycles or absorbs from the soil to very large, complex molecules. Organic molecules are usually large polymers that have a "backbone" of carbon atoms.

10. Carbohydrates contain carbon, hydrogen, and oxygen in a ratio of 1C:2H:1O. Carbohydrates occur as monosaccharides (simple sugars such as glucose) and disaccharides (two simple sugars joined together such as sucrose). Some polysaccharides, (e.g., starch, cellulose) consist of many simple sugars condensed together, and others (e.g., pectin) are a little more complex. Lignins are often associated with structural polysaccharides. Simple sugars, when they are attached to one another, each give up a molecule of water, forming starch. Hydrolysis is the process of restoring a water molecule to each simple sugar when starch is broken down during digestion.

11. Lipids (e.g., fats, oils, and waxes) consist of a unit of glycerol or other alcohol with three fatty acids attached. They are insoluble in water and contain carbon, hydrogen, and oxygen, with proportionately much less oxygen than found in carbohydrates. Saturated fats are those that have hydrogen atoms attached to every available attachment point of their carbon atoms; if there are very few places for hydrogen atoms to attach, the fat is said to be polyunsaturated. Phospholipids have a phosphate group replacing one fatty acid.

12. Proteins are usually large molecules composed of subunits called *amino acids*. Each amino acid has two special groups of atoms: an amino group (NH_2) and a carboxyl group (COOH). These groups bond amino acids together, forming polypeptide chains; the bonds are called peptide bonds. Enzymes are large protein molecules that function as organic catalysts. Their names end in -ase. Some have important industrial uses.

13. There are two nucleic acids (DNA and RNA) associated primarily with cell nuclei. DNA and RNA molecules consist of chains of building blocks called *nucleotides,* each of which has a nitrogenous base, a 5-carbon sugar, and a phosphate group. Four kinds of nucleotides, each with a unique nitrogenous base, occur in DNA. Helical coils of DNA contain coded information determining the nature and proportions of substances in cells and the ultimate form and structure of the organism. RNA has a different sugar and nucleotide.

Review Questions

1. What distinguishes a living organism from a nonliving object, such as a rock or a tin can?

2. What is meant by the term *organic?*

3. How are acids, bases, and salts distinguished from one another?

4. Differentiate among carbohydrates, lipids, and proteins.

5. What is energy and what forms does it take?

6. How are macromolecules formed?

7. How is a protein molecule different from a nucleic acid molecule?

Discussion Questions

1. Can part of an organism be alive while another part is dead? Explain.

2. What is the difference between inherited form and form resulting from response to the environment?

3. What might happen if all enzymes were to work at half their usual speed?

Additional Reading

Day, W. 1984. *Genesis on planet earth.* New Haven, CT: Yale University Press.

Dickerson, R. E. 1981. Chemical evolution and the origin of life. *Scientific American* 239(3): 70.

Lehninger, A. L., D. L. Nelson, and M. M. Cox. 1992. *Principles of biochemistry,* 2d ed. New York: Worth Publishers.

Margulis, L. 1992. *Diversity of life: The five kingdoms.* Hillside, NJ: Enslow Publications.

Oparin, A. I. 1953. *Origin of life,* 2d ed. Mineola, NY: Dover Publications.

Raven, P. H., R. F. Evert, and S. E. Eichhorn. 1992. *Biology of plants,* 5th ed. New York: Worth Publishers.

Sackheim, G. 1991. *Introduction to chemistry for biology students,* 4th ed. Redwood City, CA: Benjamin/Cummings.

Smith, C. A., and E. J. Wood (Eds.). 1991. *Biological molecules.* New York: Chapman and Hall.

Chapter Outline

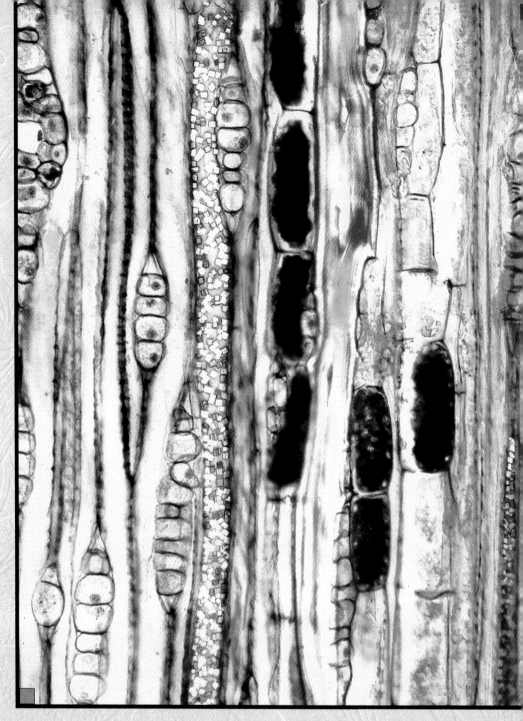

Phloem of Douglas fir (Pseudotsuga menziesii), ×500. *(Polarized light photomicrograph by G. S. Ellmore.)*

Cells

Overview

This chapter gives a brief review of the history of the discovery of cells and the development of cell theory. Differences between prokaryotic and eukaryotic cells are mentioned, and observations on cell size and structure follow. Each of a cell's particulates are discussed, beginning with the cell wall. Included are the plasma membrane, endoplasmic reticulum, ribosomes, Golgi bodies, mitochondria, plastids, microtubules, microfilaments, the nucleus, other organelles, vacuoles, and vacuolar membranes. Distinctions between plant and animal cells are then given. The chapter next discusses mitosis and cytokinesis and concludes with a brief review of intercellular communication.

Some Learning Goals

1. Trace the development of modern cell theory and show how the advances of early researchers have led us to our current understanding.
2. Know the following cell structures and organelles and indicate the function of each: plasma membrane, mitochondria, plastids, ribosomes, endoplasmic reticulum, Golgi bodies, vacuoles.
3. Describe the components of a nucleus and understand the function of each component.
4. Contrast plant cells with animal cells.
5. Understand the cell cycle and the events that take place in each phase of mitosis.

FIGURE 3.1   Robert Hooke's microscope, as illustrated in one of his works.
(Courtesy National Library of Medicine)

A ll living organisms, from aardvarks and almond trees to zebras and zinnias, are composed of cells, and all living organisms, including each of us, also generally begin life as a single cell. This single cell divides repeatedly until it develops into an organism consisting of perhaps billions of cells. During the first few hours of any organism's development, the cells all resemble each other, but changes soon occur, not only in the appearance of the cells but also in their function. The modifications of some, for example, permit them to serve as conduits for food and water, while others come to function in secretion or support. Some cells live and function for many years; others mature and degenerate in just a few days. Even as you read this, millions of new cells are being produced in your body. Some cells add to your total body mass (if you have not yet stopped growing), but most replace the millions of older cells that are destroyed every second you remain alive. The variety and form of cells seem almost infinite, but certain features are shared by most of them. A discussion of these features forms the body of this chapter.

CELLS

History

The discovery of cells is associated with the development of the microscope in the 17th century (see Chapter 1). In 1665,

the English physicist Robert Hooke, using a primitive microscope (Fig. 3.1), examined thin slices of cork he had cut with a sharp penknife. Hooke compared the boxlike compartments he saw to the surface of a honeycomb and is credited with applying the term *cell* to those compartments. He also estimated that a cubic inch of cork would contain approximately 1,259 million such cells. What Hooke saw in the cork were really only the walls of dead cells, but he also observed "juices" in living cells of elderberry plants and thought he had found something similar to the veins and arteries of animals.

Two physicians, Marcello Malpighi in Italy and Hooke's compatriot Nehemiah Grew in England, along with Anton van Leeuwenhoek reported for 50 years on the organization of cells in a variety of plant tissues. In the 1670s, they also reported on the form and structure of single-celled organisms, which they referred to as "animalcules."

After this period, little more was reported on cells until the early 1800s. This lack of progress was due in large part to the imperfections of the primitive microscopes and also to the crude methods of tissue preparation used. But both microscopes and tissue preparations slowly improved, and by 1809, the famous French biologist Jean Baptiste de Lamarck had seen a wide enough variety of cells and tissues to conclude that "no body can have life if its constituent parts are not cellular tissue or are not formed by cellular tissue." In 1824, René J. H. Dutrochet, also of France, reinforced Lamarck's conclusions that all animal and plant tissues are composed of cells of various kinds. Neither of them, however, realized that each cell could, in most cases, reproduce itself and exist independently.

In 1831, the English botanist Robert Brown discovered that all cells contain a relatively large body that he called the *nucleus.* Shortly thereafter, the German botanist Matthias Schleiden observed a smaller body within the nucleus that he called the *nucleolus.* Schleiden and a German zoologist, Theodor Schwann, were not the first to grasp the significance of cells, but they explained them with greater clarity and perception than others before them had done. They are generally credited with developing the *cell theory,* beginning with their publications of 1838 to 1839. In essence, this theory holds that all living organisms are composed of cells and that cells form a unifying structural basis of organization.

In 1858, an important augmentation of the cell theory appeared in a classic textbook by another German scientist, Rudolf Virchow. He argued cogently that every cell comes from a preexisting cell ("*omnis cellula e cellula*") and that there is no spontaneous generation of cells. Virchow's publication stirred up a great controversy, because prior to this time there was a widespread belief among scientists and nonscientists alike that animals could originate spontaneously from dust. Many who had microscopes were thoroughly convinced they could see "animalcules" appearing in decomposing substances.

The controversy became so heated that in 1860, the Paris Academy of Sciences offered a prize to anyone who could, through experiments, shed light on the matter. Just two years later, the brilliant French scientist Louis Pasteur was awarded the prize. Pasteur, using swan-necked flasks, demonstrated conclusively that boiled media remained sterile indefinitely if microorganisms from the air were excluded from the media.

In 1871, Pasteur proved that natural alcoholic fermentation always involves the activity of yeast cells. In 1897, the German scientist Eduard Buchner accidentally discovered that the yeast cells did not need to be alive for fermentation to occur. He found that extracts from the yeast cells would convert sugar to alcohol. This discovery was a major surprise to the biologists of the time and quickly led to the identification and description of enzymes (discussed in Chapter 2), the organic catalysts (substances that aid chemical reactions without themselves being changed) found in all living cells; it also led to the belief that cells were little more than miniature packets of enzymes. During the first half of the 20th century, however, great advances were made in the refinement of microscopes and in tissue preparation techniques. Numerous structures and bodies, in addition to the nucleus, were observed in cells, and the relationship between structure and function came to be realized and understood on a much broader scale than previously had been possible.

Modern Microscopes

Without microscopes, very little would be known about cells. Our present vast knowledge of cells and all aspects of biological investigations associated with them is directly related to the development of these instruments.

Light microscopes increase magnification as light passes through a series of transparent lenses, currently made of various types of glass or calcium fluoride crystals. The curvatures of the lens materials and their composition are designed to minimize distortion of image shapes and colors.

Light microscopes are of two basic types: *compound microscopes,* which require the material being examined to be sliced thinly enough for light to pass through, and *dissecting microscopes (stereomicroscopes),* which permit three-dimensional viewing of opaque objects. The best compound microscopes in use today can produce useful magnifications of up to 1,500 times under ideal conditions. Most dissecting microscopes used in teaching laboratories magnify up to 30 times, but higher magnifications are possible with both types of microscopes. Magnifications of more than 1,500 times, however, are considered "empty" because *resolution* (the capacity of lenses to aid in separating closely adjacent tiny objects) does not improve with magnification beyond a certain point. Light microscopes will continue to be useful, particularly for observing living cells, into the foreseeable future (Fig. 3.2).

Since the 1950s, the production and development of high resolution modern electron microscopes has resulted in observation of much greater detail than is possible with light microscopes. Instead of using light, electron microscopes use a beam of electrons produced when electricity of high voltage is passed through a wire. This electron beam is directed through a vacuum in a large tube or column. When the beam passes through a specimen, an image is formed on a plate. Magnification is controlled by powerful electromagnetic lenses located on the column.

Like light microscopes, electron microscopes are of two basic types. *Transmission electron microscopes* (Fig. 3.3 *A*) permit magnifications of 200,000 or more times, but the material to be viewed must be sliced extremely thinly and introduced into the column's vacuum, so living objects cannot be observed. *Scanning electron microscopes* (Fig. 3.3 *B*) usually do not achieve such high magnifications (3,000 to 10,000 times is the usual range), but opaque objects can be observed as a scanner renders the object visible on a cathode tube like a television screen. The techniques for such observation have become so refined that even preserved material can appear exceptionally lifelike, and high resolution three-dimensional images can be obtained.

In 1986, the Nobel Prize in physics was awarded to two International Business Machine scientists, Gerd Binnig and Heinrich Rohrer, for their invention in 1982 of a *scanning tunneling microscope.* This microscope uses a minute probe rather than electrons or light to scan across a surface and then reproduces an image of it down to the atomic level, doing so without damaging the probed area. The probe can scan areas barely twice the width of an atom and theoretically could be used to print on the head of an ordinary pin the words contained in more than 50,000 single-spaced pages of books.

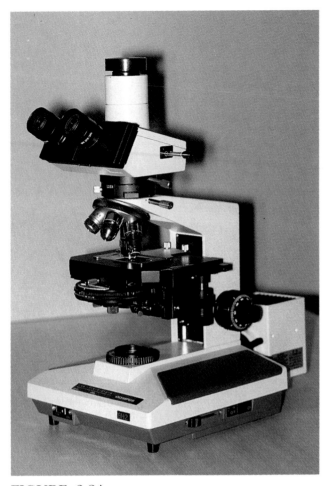

FIGURE 3.2A   A compound light microscope.

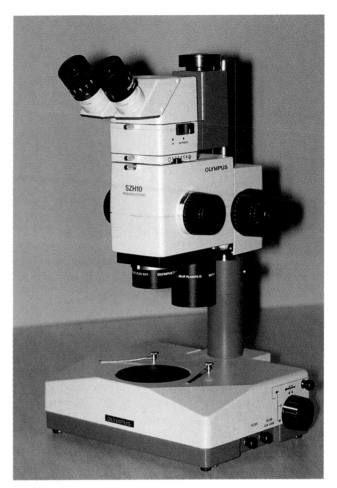

FIGURE 3.2B   A stereomicroscope (dissecting microscope).

Early in 1989, the first picture of a segment of DNA showing its helical structure was taken with a scanning tunneling microscope by an undergraduate student associated with the Lawrence Laboratories in northern California. Several variations of this microscope, each using a slightly different type of probe, have now been produced. Significant new discoveries by cell biologists using one or more of all three types of microscopes in their research have now become frequent events.

EUKARYOTIC VERSUS PROKARYOTIC CELLS

Nearly all higher plant and animal cells exhibit most of the various features that are discussed in the sections that follow. There are some very primitive organisms, however, whose cells lack a number of these features (e.g., true nuclei and other bodies bound by membranes). Such cells, called **prokaryotic** to distinguish them from the typical **eukaryotic** cells discussed here, may have been the origin of several cell components almost universally found in cells of less

primitive organisms. These and other aspects of prokaryotic cells are covered in Chapter 17.

CELL SIZE AND STRUCTURE

Cell Size

Most plant cells, and the vast majority of animal cells, are so tiny they are invisible to the unaided eye. Cells of higher plants generally vary in length between 10 and 100 micrometers.[1] Since there are roughly 25,000 micrometers to the inch, it would take about 500 average-sized cells to extend across 2.54 centimeters (1 inch) of space; 30 of them could easily be placed across the head of a pin. Some bacterial cells are less than 0.5 micrometer in diameter, while cells of the green alga mermaid's wineglass (*Acetabularia*)

---

1. See Appendix 5 for metric conversion tables.

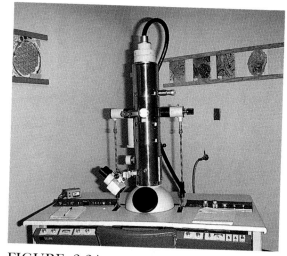

FIGURE 3.3A   A modern transmission electron microscope.

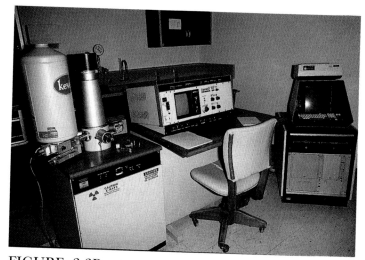

FIGURE 3.3B   A scanning electron microscope.

are mostly between 2 and 5 centimeters in length, and fiber cells of some nettles are about 20 centimeters long.

Because cells are so minute, the numbers occurring in full-grown organisms are astronomical. For example, it has been calculated that a single mature leaf of a pear tree contains 50 million cells and that the total number of cells in the roots, stem, branches, leaves, and fruit of a full-grown pear tree exceeds 15 trillion. Can you imagine how many cells there are in a 3,000-year-old redwood tree of California, which may reach heights of 90 meters (300 feet) and measure up to 4.5 meters (15 feet) in diameter near the base?

Some cells are boxlike with six walls, but others assume a wide variety of shapes, depending on their location and function. The most abundant cells in the younger parts of plants and fruits may be more or less spherical when they are first formed, but they are packed together in such a way that they commonly have 14 sides by the time they are mature. These are discussed in the next chapter.

As indicated at the beginning of Chapter 2, the living part of the cell within the wall is called *protoplasm*. Two main components of protoplasm are readily discernible: the *nucleus,* which controls the cell's activities, and the **cytoplasm,** which is a souplike fluid containing water, dissolved substances, and many small **organelles** (persistent structures of various shapes and sizes with specialized functions in the cell; most, but not all, are bound by membranes). The organelles are the sites of many different activities that take place within the cell. A brief examination of each of the various cell components follows (Figs. 3.4 and 3.5 show these various components).

The Cell Wall

A popular novelty song of 50 or more years ago had several verses listing food items the author purportedly disliked,

with each verse ending, "But I like bananas because they have no bones!" Indeed, bananas and all parts of plants differ from animals in that they have no bones or similar internal skeletal structures. Yet large trees support branches and leaves weighing many tons. They are able to do this because most plant cells have semirigid or rigid walls that perform the functions of bones; that is, they provide strength and support for the plants (and also protect the delicate cell contents within). When millions of these cells function together as a tissue, their collective strength is enormous. The largest trees alive today, the redwoods, exceed the mass, or volume, of the largest land animals, the elephants, by more than a hundred times. The wood of one tree could support the combined weight of a thousand elephants.

Cell walls are composed of *cellulose,* a substance having long molecules made up of as many as 10,000 simple glucose molecules attached end to end, the polysaccharides **pectin** (the complex organic material that gives stiffness to fruit jellies) and *hemicellulose* (a gluelike substance unrelated to cellulose that holds cellulose fibrils together), and *glycoproteins* (proteins that have sugars associated with their molecules).

When new cell walls are first formed, a **middle lamella,** consisting primarily of pectin, appears. This middle lamella is normally shared by two adjacent cells and is so thin that it may not be visible with an ordinary light microscope unless it is specially stained. A fine network of cellulose is laid down on either side of the middle lamella (Fig. 3.6). The long cellulose molecules are grouped together in bundles known as *microfibrils,* which, in turn, are twisted together in ropelike fashion, forming larger bundles. The larger bundles are held together by pectin and related substances and make up the bulk of the cell wall.

Sometimes, the cellulose is deposited in two stages, forming a *primary cell wall* and then a *secondary cell wall* inside the primary wall. When this happens, the secondary

Aerial roots of a tropical fig tree (Ficus sp.).

5 Roots and Soils

Overview────────────────

This chapter discusses roots, beginning with the functions and continuing with the development of roots from a seed. It covers the function and structure of the root cap, region of cell division, region of elongation, and region of maturation (with its tissues). The endodermis and pericycle are also discussed.

Specialized roots (food-storage roots, water-storage roots, propagative roots, pneumatophores, aerial roots, contractile roots, buttress roots, parasitic roots, mycorrhizae) are given brief treatment. This is followed by some observations on the economic importance of roots.

After a brief examination of soil horizons, the chapter concludes with a discussion of the development of soil, its texture, its composition, its structure, and its water.

Some Learning Goals

1. Know the primary functions and forms of roots.
2. Learn the root regions, including the root cap, region of cell division, region of cell elongation, and region of maturation (including root hairs and all tissues), and know the function of each.
3. Discuss the specific functions of the endodermis and the pericycle.
4. Understand the differences among the various types of specialized roots.
5. Know at least 10 practical uses of roots.
6. Understand how a good agricultural soil is developed from raw materials.
7. Contrast the various forms of soil particles and soil water with regard to specific location and availability to plants.

Y ou have probably seen pictures of the destruction caused by a tornado as it cut a swath through a village or a city, but have you ever seen what a twister can do to a forest? Large trees may be snapped off above the ground or knocked down, and branches may be stripped bare of leaves. Depending on the type of forest and the composition of the soil, however, you will probably see few trees completely torn up by the roots and blown elsewhere. And in the tropics it is indeed rare to find healthy palm trees uprooted after a hurricane.

Roots anchor trees firmly in the soil, usually by forming an extensive branching network that constitutes about one-third of the total dry weight of the plant. The roots of most plants do not usually extend down into the earth beyond a depth of 3 to 5 meters (10 to 16 feet); those of many herbaceous species are confined to the upper 0.6 to 0.9 meter (2 to 3 feet). The roots of a few plants, such as alfalfa, however, frequently extend more than 6 meters (20 feet) into the earth. When the Suez Canal was being built, workers encountered roots of tamarisk at depths of nearly 30 meters (100 feet), and mesquite roots have been seen 53.4 meters (175 feet) deep in a pit mine in the southwestern United States. Some plants, such as cacti, form very shallow root systems but still effectively achieve *anchorage* by means of

a densely branching mass of roots radiating out in all directions as far as 15 meters (50 feet) from the stem.

In addition to serving as anchors, roots function extensively in the *absorption of water and minerals in solution,* with the bulk of such "feeder" roots being confined to the upper meter (3.3 feet) of soil. The roots of some plants have specialized functions, such as food or water storage.

Although some aquatic plants (e.g., duckweeds) normally produce roots in water, and others (e.g., many orchids) produce aerial roots, the great majority of vascular plants develop their root systems in soils. Soils vary considerably in composition, texture, and other characteristics; they are discussed toward the end of this chapter.

Root Origins

When a seed germinates, a part of the **embryo** within it, the **radicle,** grows out and develops into the first root. This may develop either into a thickened *taproot*, from which thinner branch roots arise, or into numerous *adventitious roots*. A *fibrous* root system with numerous fine roots of similar diameter then develops from the adventitious roots (Fig. 5.1). Many mature plants have a combination of taproot and fibrous root systems. The number of roots produced by a single plant may be prodigious. For example, a single mature ryegrass plant may have as many as 15 million individual roots and branch roots, with a combined length of 644 kilometers (400 miles) and a total surface area larger than a volleyball court, all contained within 0.57 cubic meter (2 cubic feet) of soil. Significant additional surface area is provided by *root hairs* discussed in the section on the region of maturation.

Most *dicot* plants (plants such as peas and carrots, whose seeds have two "seed leaves") have taproot systems with one, or occasionally more, primary roots from which secondary roots develop. *Monocots* (plants such as corn and rice, whose seeds have one "seed leaf"), on the other hand, have fibrous root systems. Other types of roots, such as adventitious roots (discussed in Chapter 6), may develop in both dicots and monocots. In English and other ivies, lateral adventitious roots that aid in climbing appear along the aerial stems, and in certain plants with specialized stems (e.g., rhizomes, corms, and bulbs; see Fig. 6.14), adventitious roots are the only kind produced.

ROOT STRUCTURE

Botanists have traditionally recognized four regions or zones in developing young roots. Three of the regions are not sharply defined at their boundaries. The cells of each region gradually develop the form of those of the next region, and the extent of each region varies considerably, depending on the species involved. These regions are called (1) the *root cap*, (2) the *region of cell division*, (3) the *region of elongation*, and (4) the *region of maturation* (Fig. 5.2).

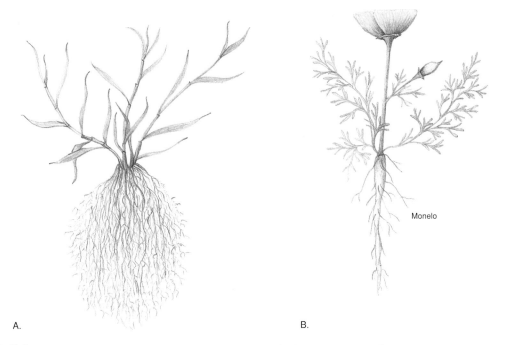

A.

B.

Monelo

FIGURE 5.1   Root systems. *A.* A fibrous root system of a grass. *B.* A taproot system of a poppy.

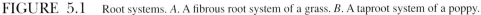

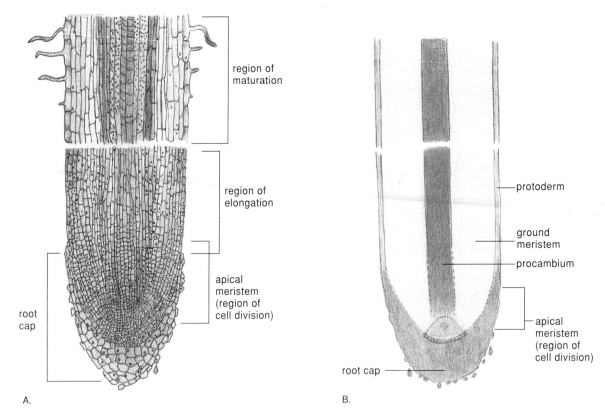

region of
maturation

region of
elongation

apical
meristem
(region of
cell division)

root
cap

A.

protoderm

ground
meristem

procambium

apical
meristem
(region of
cell division)

root cap

B.

FIGURE 5.2   A longitudinal section through a dicot root tip. *A.* Regions of the root. *B.* Locations of the primary meristems of the root.

The Root Cap

The **root cap** is composed of a thimble-shaped mass of parenchyma cells covering the tip of each root. It is quite large and obvious in some plants, while in others it is nearly invisible. One of its functions is to *protect* from damage the delicate tissues behind it as the young root tip pushes through often angular and abrasive soil particles.

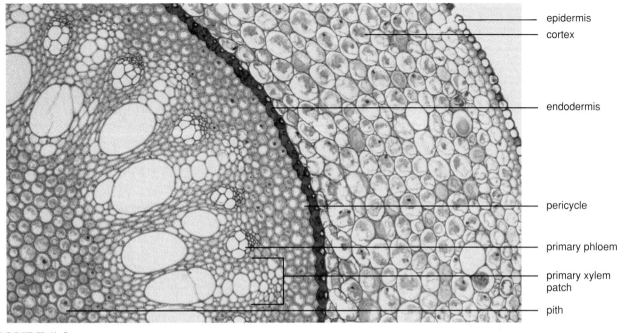

FIGURE 5.3   A cross section of a portion of a root of greenbrier (*Smilax*), a monocot, ×500. (Photomicrograph by G.S. Ellmore)

The root cap has no equivalent in stems. The Golgi bodies of the root cap's outer cells secrete and release a slimy substance that lodges in the walls and eventually passes to the outside. The cells, which are replaced from the inside, constantly slough off, forming a slimy lubricant that facilitates movement through the soil. This mucilaginous lubricant also provides a medium favorable to the growth of beneficial bacteria.

   The root cap, whose cells have an average life of less than a week, can be slipped off or cut from a living root, and when this is done, a new root cap is produced. Until the root cap has been renewed, however, the root seems to grow randomly instead of downward, suggesting that the root cap has another function, namely, the *perception of gravity* (see *gravitropism* in Chapter 11). It is known that *amyloplasts* (plastids containing starch grains) collect on the sides of root cap cells facing the direction of gravitational force. When a root that has been growing vertically is artificially tipped horizontally, the amyloplasts tumble or float down to the "bottom" of the cells in which they occur. Within 30 minutes to a few hours, the root begins growing downward again. The exact nature of this gravitational response is not known, but there is some evidence that calcium ions known to be present in the amyloplasts influence the distribution of growth hormones in the cells.

The Region of Cell Division

The root cap is produced by cells in the region of *cell division,* which is in the center of the root tip and surrounded by the root cap. It is composed of an **apical meristem.** Most of the cell divisions take place at the edges of an inverted, cup-shaped zone a short distance behind the actual base of the meristem just adjacent to the root cap. Here the cells divide every 12 to 36 hours, while at the base of the meristem, they may divide only once in every 200 to 500 hours. The divisions are often rhythmic, reaching a peak once or twice each day, usually toward noon and midnight, with relatively quiescent intermediate periods. Cells in this region are mostly cube shaped, with relatively large centrally located nuclei and few, if any, small vacuoles. In both roots and stems, the apical meristem soon subdivides into three meristematic areas: (1) the **protoderm** gives rise to an outer layer of cells, the *epidermis;* (2) the **ground meristem,** to the inside of the protoderm, produces parenchyma cells of the *cortex;* and (3) the **procambium,** which appears as a solid cylinder in the center of the root, produces *primary xylem* and *phloem* (Fig. 5.3). *Pith* tissue, which is seen in most stems, is absent in most dicot roots but is found in those of many monocots, such as grasses.

The Region of Elongation

The *region of elongation,* which merges with the apical meristem, usually extends about 1 centimeter (0.4 inch) or less from the tip of the root. Here the cells become several times their original length and also somewhat wider. At the same time, the tiny vacuoles merge and grow until one or two large vacuoles, occupying up to 90% or more of the volume of each cell, have been formed. Only the root cap and apical meristem are actually pushing through the soil, since no further increase in cell size takes place above the region of elongation. The usually extensive remainder of each root remains stationary for the life of the plant. If a cambium is present, however, there normally is a gradual increase in girth through the addition of **secondary tissues.**

The Region of Maturation

Most of the cells mature into the various distinctive cell types of the primary tissues in this region, which is sometimes called the *region of differentiation*, or *root-hair zone*. It is given this last name because of the numerous, hairlike, delicate protuberances that develop from many of the epidermal cells. The **root hairs,** as they are called, adhere tightly to soil particles (Fig. 5.4) with the aid of microscopic fibers they produce.

The root hairs are not separate cells; in fact, the nucleus of the epidermal cell to which each is attached often moves out into the protuberance. They are so numerous that they appear as fine down to the naked eye, typically numbering more than 38,000 per square centimeter (250,000 per square inch) of surface area in roots of plants such as corn, and they seldom exceed 1 centimeter (0.4 inch) in length. A single ryegrass plant occupying less than 0.6 cubic meter (2 cubic feet) of soil has been found to have more than 14 billion root hairs, with a total surface area almost the size of a football field. Obviously, the root hairs greatly increase the absorptive surface of the root.

When a seedling or plant is moved, many of the delicate root hairs are torn off or die within seconds if exposed to the sun, thus greatly reducing the plant's capacity to absorb water and minerals in solution. This is why plants should be watered, shaded, and pruned after transplanting until new root hairs have formed. In any growing root, the extent of the root-hair zone remains fairly constant, with new root hairs being formed toward the root cap and older root hairs dying back in the more mature regions. The life of the average root hair is usually not more than a few days, although a few live for a maximum of perhaps three weeks.

An examination of the region of maturation (Fig. 5.5) reveals that the cuticle (shown in Fig. 4.10), which may be relatively thick on the epidermal cells of stems and leaves, is very thin on the root hairs and epidermal cells of roots. Presumably, any significant amount of fatty substance present in or on the walls of water-absorbing cells would interfere with their function.

The **cortex,** a tissue composed of parenchyma cells adjacent to the epidermis, functions primarily in food storage. It may be many cells thick and is similar to the cortex of

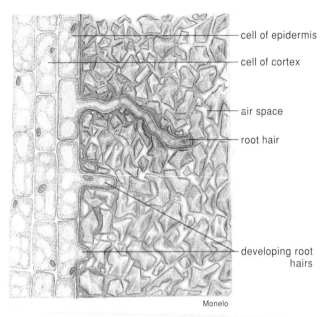

cell of epidermis

cell of cortex

air space

root hair

developing root hairs

Monelo

FIGURE 5.4   Root hairs in contact with soil particles.

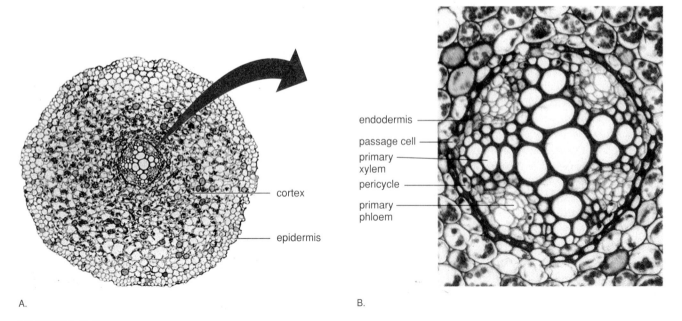

A.

endodermis

passage cell

primary xylem

pericycle

primary phloem

cortex

epidermis

B.

FIGURE 5.5   A cross section through the region of maturation of a root of buttercup (*Ranunculus*), a dicot, ×500.

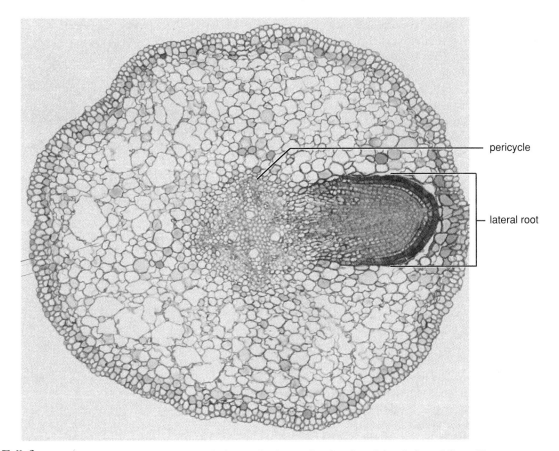

FIGURE 5.6   A cross section through a willow (*Salix*—a dicot) root showing the origin of a lateral (branch) root. (Photomicrograph by G.S. Ellmore)

stems except for the presence of an **endodermis** at its inner boundary. An endodermis is rare in stems but so universal in roots that only three species of plants are known to lack a root endodermis. It is a cylinder formed by a single layer of compactly arranged cells whose primary walls are impregnated with suberin.

The suberin forms bands around the walls perpendicular to the root's surface (i.e., the radial and transverse walls). These suberin bands are called **Casparian strips.** The plasma membranes of the endodermal cells are fused to the Casparian strips, which prevent water from passing through the otherwise permeable (porous) cell walls. The Casparian strip barrier forces all water and dissolved substances entering and leaving the central core of tissues to pass through the plasma membranes of the endodermal cells or their plasmodesmata. This has a regulatory effect on the types of minerals absorbed and transported by the root to the stems, leaves, and seeds. Harmful minerals may be excluded by the plasma membranes while useful ones generally are retained.

In some roots, the epidermis, cortex, and endodermis are sloughed off as their girth increases, but in those roots where the endodermis is retained, the inner walls of the endodermal cells eventually become thickened by the addition of alternating layers of suberin and wax. Later, cellulose and sometimes lignin are also deposited. Some endodermal

cells, called **passage cells,** may remain thin-walled and retain their Casparian strips for a while, but they, too, eventually tend to become suberized.

On the inside of the endodermis and immediately adjacent to it is a cylinder of parenchyma cells called the **pericycle.** This is usually one cell wide, although in some plants it may extend for several cells. It is a very important tissue, for the cells retain their capacity to divide even after they have matured, and it is within the pericycle that the *lateral (branch) roots* and part of the *vascular cambium* of dicots arise (Fig. 5.6).

The *primary xylem*, with its water-conducting cells such as tracheids, forms a solid core in the center of most dicot and conifer roots. Unlike stem xylem, however, root xylem of most dicots first forms with short spirelike ridges or arms, which point toward the pericycle. There are often four of these arms, but there can be from two to several. Branch roots usually arise in the pericycle opposite the xylem arms. In many monocot and a few dicot roots, the primary xylem forms a cylinder of tissue in which the arms may not be well-defined; this tissue surrounds the pith parenchyma cells. *Primary phloem*, which contains food-conducting cells, forms in discrete patches between the xylem arms of both dicot and monocot roots (as shown in Fig. 5.5).

In woody dicots, some herbaceous dicots, and conifers, parts of the pericycle and parenchyma cells between the xylem arms and the phloem patches become a vascular cambium. This cambium starts producing secondary phloem to the outside and secondary xylem to the inside. Soon, the cambium follows the outline of the primary xylem, and eventually, instead of appearing as patches and arms, the secondary conducting tissues appear as concentric cylinders. The primary tissues are crushed out of existence or sloughed off and lost as secondary tissues are added.

In woody plants, a cork cambium normally arises in the pericycle and gives rise to *cork* tissue (*periderm*) similar to that of stems. In fact, the cross section of a woody root several years old, with its annual rings of xylem along with its sapwood and heartwood, may be distinguished from that of a stem of the same species only by the relatively larger proportion of parenchyma cells present in the root. Although there are exceptions, monocot roots generally have no secondary meristems and therefore no secondary growth.

Natural grafting may occur between roots of different trees of the same species, particularly in the tropics. When the roots come in contact with one another, they unite through secondary growth, but the details of the uniting process are not yet known. One unfortunate aspect of this grafting is that if one tree becomes diseased, the disease can be transmitted through these grafts to all the other trees connected to it.

Branch or lateral roots arise internally in the pericycle, pushing their way out to the surface through the endodermis, cortex, and epidermis (Fig. 5.6). This process is in contrast with that in stems, where branches develop at the surface from axillary buds. Some specialized roots, which at first glance may appear similar to specialized stems, such as rhizomes, bulbs, and corms (discussed in Chapter 6), can be distinguished by their complete lack of leaves.

SPECIALIZED ROOTS

As indicated earlier, most plants produce either a fibrous root system, a taproot system, or, more commonly, a combination of the two types. Some plants, however, have roots with modifications that permit specific functions in addition to the absorption of water and minerals in solution.

Food-Storage Roots

Most roots and stems store some food, but in certain plants the roots are enlarged and store large quantities of starch and other carbohydrates (Fig. 5.7). In sweet potatoes and yams, for example, extra cambial cells develop in parts of the xylem of branch roots and produce numerous parenchyma cells. These cause the organs to swell, providing storage areas for considerable quantities of carbohydrates. Similar food-storage roots are found in the deadly poisonous water hemlocks, in dandelions, and in salsify. Some food-storage

FIGURE 5.7   A sweet potato plant. Note the food-storage roots.

organs, such as those of carrots, beets, turnips, and radishes, are actually a combination of root and stem. In carrots, for example, approximately 2 centimeters (0.8 inch) at the top are derived from stem tissue, yet a casual inspection will not reveal any differences of origin.

Water-Storage Roots

Some members of the Pumpkin Family produce huge water-storage roots. This is particularly characteristic of those that grow in arid regions or in those areas where there may be no precipitation for several months of the year. In certain man-roots, for example, roots weighing 30 kilograms (66 pounds) or more are frequently produced (Fig. 5.8), and a major root of one calabazilla plant was found to weigh 72.12 kilograms (159 pounds). The water in the roots is apparently used by the plants when the supply in the soil is inadequate.

Propagative Roots

Many plants produce **adventitious buds** (buds appearing in unusual places) along the roots that grow near the surface of the ground. The buds develop into aerial stems called *suckers,* which have additional rootlets at their bases. The rooted suckers can be separated from the original root and grown individually. A number of fruit trees, such as cherry and pear trees, frequently produce such roots. The adventitious roots of horseradish and rice-paper plants can become a nuisance in gardens, the latter often producing propagative roots within a radius of 10 meters (33 feet) or more from the

FIGURE 5.8   A manroot water-storage root weighing over 25.3 kilograms (60 pounds).

(Courtesy Robert A. Schlising)

FIGURE 5.9   Pneumatophores (foreground) of tropical mangroves rising above the sand at low tide. The pneumatophores are spongy outgrowths from the roots beneath the surface. Pneumatophores facilitate the exchange of oxygen and carbon dioxide for the roots, which grow in areas where little oxygen is otherwise available to them.

(Courtesy Lani Stemmerman)

parent plant. Tree of Heaven, Canada thistle, and some other weeds have a remarkable facility to reproduce in this fashion as well as by means of seeds. This ability made it difficult to control them in the past, but some **biological controls** now being investigated (see Appendix 2) may be an answer to the problem in the future.

Pneumatophores

Water, even after air has been bubbled through it, contains less than one 30th of the amount of free oxygen found in the air. Plants growing with their roots in water may therefore encounter less than the amount of oxygen necessary for normal respiration. Some swamp plants, such as the black mangrove and the yellow water weed, develop special "spongy" roots, called *pneumatophores,* which rise above the surface and facilitate gas exchange between the atmosphere and the subsurface roots to which they are connected (Fig. 5.9). The bald cypress, which occur in southern swamps produces "knees" (shown in Fig. 22.19) that were once thought to function as pneumatophores, but there is no conclusive evidence for this theory.

Aerial Roots

The various kinds of aerial roots produced by plants include the *velamen* roots of orchids, *prop* roots of corn and banyan trees (Fig. 5.10), *adventitious* roots of ivies, and *photosynthetic* roots of certain orchids. Velamen roots have an epidermis several cells thick. It was assumed that this thick

epidermis aids in the absorption of rainwater, but it now appears it may function more in preventing loss of moisture from the root. Corn prop roots, produced toward the base of the stems, support the plants in a high wind. A number of tropical plants, including the screw pines and various mangroves, produce sizable prop roots extending for several feet above the surface of the ground or water. Debris collects between them and helps to create additional soil.

Many of the tropical figs or banyan trees produce roots that grow down from the branches until they contact the soil. Once they are established, they continue secondary growth and look just like additional trunks. Banyan trees may live for hundreds of years and can become very large. In India and southeast Asia, several such trees display almost 1,000 root-trunks and have circumferences approaching 450 meters (1,476 feet). The oldest is estimated to be about 2,000 years old.

FIGURE 5.10   A banyan tree, with numerous large prop roots that have developed from the branches.

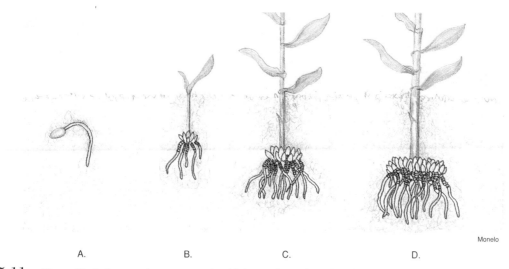

Monelo

| A. | B. | C. | D. |

FIGURE 5.11   How a lily bulb, over three seasons, is withdrawn deeper into the soil through the action of contractile roots. *A.* A seed germinates. *B.* Contractile roots pull the newly formed bulb down several millimeters during the first season. *C.* The bulb is pulled down farther the second season. *D.* The bulb is pulled down even farther the third season. The bulb will continue to be pulled down in succeeding seasons until it reaches an area of relatively stable soil temperatures.

The vanilla orchid, from which we obtain vanilla flavoring, produces chlorophyll in its aerial roots and thus, through photosynthesis, can manufacture food with them. The adventitious roots of English ivy, Boston ivy, and Virginia creeper appear along the stem and aid the plants in climbing.

Contractile Roots

A number of herbaceous dicots and monocots have *contractile* roots that pull the plant deeper into the soil. Many lily bulbs are pulled a little deeper into the soil each year as

FIGURE 5.12   Buttress roots of a tropical fig tree.

additional sets of contractile roots are developed (Fig. 5.11). The bulbs continue to be pulled down until an area of relatively stable temperatures is reached. The leaves of plants such as dandelions always seem to be coming out of the ground as the top of the stem is pulled down a small amount each year when the root contracts. The contractile part of the root may lose as much as two-thirds of its length within a few weeks. Mechanisms of contraction include the thickening and constriction of parenchyma cells, causing the xylem elements to spiral somewhat like a corkscrew.

Buttress Roots

Some tropical trees produce huge buttress roots toward the base of the trunk, giving them great stability (Fig. 5.12). Except for their angular appearance, these roots look like a part of the trunk.

Parasitic Roots

A number of plants, such as the dodders, broomrapes, and pinedrops, have no chlorophyll (necessary for photosynthesis) and have become dependent on plants with chlorophyll for their nutrition. They parasitize their host plants via somewhat rootlike projections called **haustoria** (singular: **haustorium**), which develop along the stem in contact with the host. The haustoria penetrate the outer tissues and establish connections with the water-conducting and food-conducting tissues (Fig. 5.13). Some green plants, such as

Indian warrior and the mistletoes, also form haustoria. These haustoria, however, apparently function primarily in obtaining water and dissolved minerals from the host plants, since the partially parasitic plants are capable of manufacturing at least some of their own food through photosynthesis.

Mycorrhizae

More than three quarters of all seed plant species have various fungi associated with their roots. The association is *mutualistic;* that is, both the fungus and the root benefit from it and are dependent upon the association for normal development. (Mutualism is a form of *symbiosis*; see page 267.)

The fungus is able to absorb and concentrate phosphorus much better than it can be absorbed by the root hairs. In fact, if mycorrhizal fungi have been killed by fumigation or are otherwise absent, many plants appear to have considerable difficulty absorbing phosphorus, even when the element is abundant in the soil. The phosphorus is stored in granular form until it is utilized by the plant. The fungus also often forms a mantle of millions of threadlike strands that facilitate the absorption of water and nutrients. The plant furnishes sugars and amino acids without which the fungus cannot survive.

These "fungus-roots," or **mycorrhizae** (Fig. 5.14), are essential to the normal growth and development of forest trees and many herbaceous plants. Orchid seeds, for example, will not germinate until mycorrhizal fungi invade their cells.

A.

B.

FIGURE 5.13 *A.* Pale stems of a dodder plant twining around other vegetation. *B.* A close-up view of dodder, showing the peglike *haustoria* that penetrate the tissues of the host plant, ×40.

In virtually all of the woody trees and shrubs found in forests, the fungal threads grow between the walls of the outer cells of the cortex but rarely penetrate into the cells themselves. If they should happen to penetrate, they are apparently broken down and digested by the host plants. In herbaceous plants,

the fungi do penetrate the cortex cells as far as the endodermis, but they cannot grow beyond the Casparian strips. Once inside the cells, the fungi branch repeatedly but do not break down the plasma or vacuolar membranes.

Some plants do not seem to need mycorrhizae unless essential elements in the soil are present in amounts barely sufficient for healthy growth. Plants with mycorrhizae develop few root hairs as compared with those growing without an associated fungus. Mycorrhizae have proved to be particularly susceptible to acid rain (discussed in Chapter 25); this may portend major problems for our coniferous forests in the future if the problem of acid rain is not solved.

Root Nodules

Although almost 80% of our atmosphere consists of nitrogen gas, plants are not capable of converting the nitrogen gas to forms they can use. Bacteria, however, produce enzymes that facilitate the transformation of nitrogen into nitrates and other substances that can readily be absorbed by plant roots. Members of the Legume Family, including peas and beans, and a few other plants such as alders form associations with soil bacteria that result in the production of numerous small swellings called *root nodules* that are clearly visible when such plants are uprooted (Fig. 5.15).

A substance exuded into the soil by plant roots stimulates *Rhizobium* bacteria which, in turn, respond with another substance that prompts root hairs to bend sharply. A bacterium may attach to the concave side of a bend and then invade the cell with a tubular *infection thread* that does not actually break the host cell wall and plasma membrane. The infection thread grows through to the cortex, which is stimulated to produce new cells that become a part of the root nodule; here the bacteria multiply and engage in nitrogen conversion. (See also the discussion of the nitrogen cycle in Chapter 25.)

Root nodules should not be confused with *root knots*, which are also swellings and may be seen in the roots of tomatoes and many other plants. Root knots develop in response to the invasion of tissue by small, parasitic roundworms (nematodes). Unlike bacterial nodules, root knots are not beneficial and the activities of the parasites within them can eventually lead to the premature death of the plant.

HUMAN RELEVANCE OF ROOTS

Roots have been important sources of food for humans since prehistoric times, and some, such as the carrot, have been in cultivation in Europe for at least 2,000 years. A number of cultivated root crops involve biennials (i.e., plants that complete their life cycles from seed to flowering and back to seed in two seasons). Such plants store food in a swollen taproot during the first year of growth and then utilize the stored food to produce flowers in the second season. Among the best-known biennial root crops are sugar beets, beets, turnips, rutabagas, parsnips, horseradishes, and carrots.

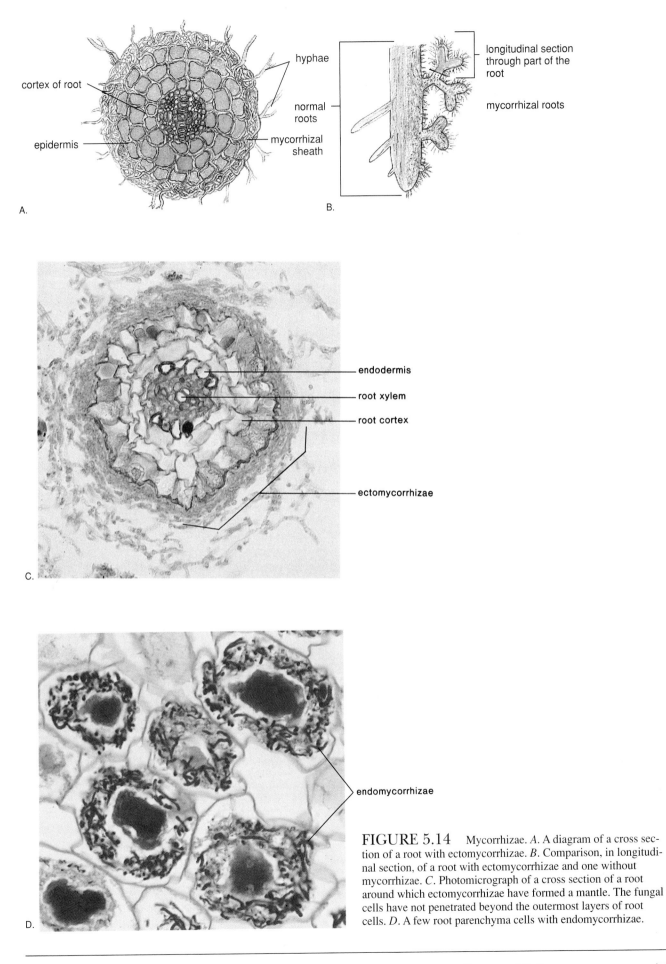

A.

B.

C.

D.

hyphae

cortex of root

epidermis

mycorrhizal
sheath

normal
roots

longitudinal section
through part of the
root

mycorrhizal roots

endodermis

root xylem

root cortex

ectomycorrhizae

endomycorrhizae

FIGURE 5.14 Mycorrhizae. *A.* A diagram of a cross section of a root with ectomycorrhizae. *B.* Comparison, in longitudinal section, of a root with ectomycorrhizae and one without mycorrhizae. *C.* Photomicrograph of a cross section of a root around which ectomycorrhizae have formed a mantle. The fungal cells have not penetrated beyond the outermost layers of root cells. *D.* A few root parenchyma cells with endomycorrhizae.

FIGURE 5.15   Root nodules on a bur clover plant. The nodules contain bacteria that convert nitrogen from the air into forms that can be used by plants, ×5.

Other important root crops include sweet potatoes, yams, and cassava. Cassava (Fig. 5.16), from which tapioca is made, forms a major part of the basic diet for millions of inhabitants of the tropics. It yields more starch per hectare (about 45 metric tons, the equivalent of 20 tons per acre) than any other cultivated crop, with a minimum of human labor. A number of minor root crops are cultivated in South America and other parts of the world. Among such plants are relatives of wild mustards, nasturtiums, and sorrel.

A number of spices, such as sassafras, sarsaparilla, licorice, and angelica, are obtained from roots. Sweet potatoes are used in the production of alcohol in Japan. Several important red to brownish dyes are obtained from roots of members of the Madder Family, to which coffee plants belong. Drugs obtained from roots include aconite, ipecac, gentian, and reserpine, a tranquilizer. A valuable insecticide, rotenone, is obtained from the barbasco plant, which has been cultivated for centuries as a fish poison by primitive South American tribes. When thrown into a dammed stream, the roots containing rotenone cause the fish to float but in no way poison them for human consumption. Other uses of roots are discussed in Chapter 24.

FIGURE 5.16   Cassava plants. Note the food-storage roots on the plant that has been dug up.
(Courtesy Monica E. Emerson)

SOILS

The soil is a dynamic, complex, constantly changing part of the earth's crust, which extends from a few centimeters deep in some places to hundreds of meters deep in others. It is essential not only to our existence but also to the existence of most living organisms. It has a pronounced effect on the plants that grow in it, and they have an effect on it.

If you dig up a shovelful of soil from your yard and examine it, you will probably find a mixture of ingredients, including several grades of sand, rocks and pebbles, powdery silt, clay, humus, dead leaves and twigs, clods consisting of soil particles held together by clay and organic matter, plant roots, and small animals, such as ants, pill bugs, millipedes, and earthworms. Also present, but not visible, would be millions of microorganisms, particularly bacteria, and, of course, air and water.

The soil became what it is today through the interaction of a number of factors: climate, parent material, topography of the area, vegetation, living organisms, and time. Because there are thousands of ways in which these factors may interact, there are many thousands of different soils. The solid portion of a soil consists of mineral matter and organic matter. Pore spaces, shared by variable amounts of water and air, occur between the solid particles. The smaller pores often contain water, and the larger ones usually contain air. The sizes of the pores and the connections between them largely determine how well the soil is aerated.

If we were to dig down 1 or 2 meters (3 to 6 feet) in an undisturbed area, we would probably expose a soil profile of three intergrading regions called *horizons* (Fig. 5.17). The horizons show the soil in different stages of development, and the composition varies accordingly. The upper layer, usually extending down 10 to 20 centimeters (4 to 8 inches), is called the *A horizon*, or *topsoil*. It is usually subdivided into a darker upper portion called the *A₁ horizon* and a lighter lower portion called the *A₂ horizon*. The *A₁* portion contains more organic matter than do the layers below.

The next 0.3 to 0.6 meter (1 to 2 feet) is called the *B horizon*, or *subsoil*. It usually contains more clay and is lighter in color than the topsoil. The *C horizon* at the bottom may vary from about 10 centimeters (4 inches) to several meters (6 to 10 feet or more) in depth; it may even be absent. It is commonly referred to as the *soil parent material* and extends down to *bedrock*.

Parent Material

The first step in the development of soil is the formation of *parent material*. This material accumulates through the weathering of three types of rock that originate from various sources. These sources and types include volcanic activity (*igneous rocks*); fragments deposited by glaciers, water, or wind (*sedimentary rocks*); and changes in igneous or sedimentary rocks brought about by great pressures, heat, or both (*metamorphic rocks*).

Climate

Climate varies greatly throughout the globe, and its role in the weathering of rocks varies correspondingly. In desert

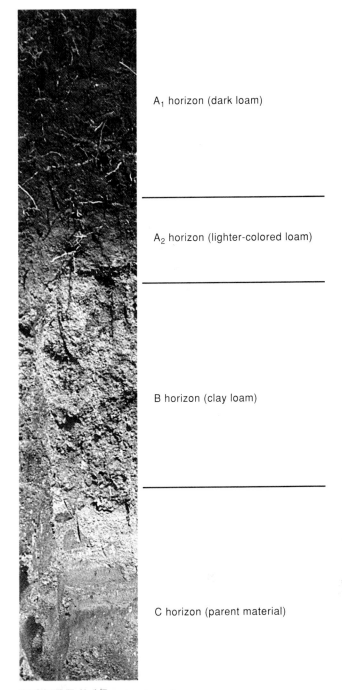

A₁ horizon (dark loam)

A₂ horizon (lighter-colored loam)

B horizon (clay loam)

C horizon (parent material)

FIGURE 5.17 A soil profile.

Reproduced from Marbut Memorial Slide Set, 1968, SSSA, by permission of Soil Sciences of America, Inc.

areas, for example, there is little weathering by rain, and soils are poorly developed. In areas of high rainfall, however, well-developed soils are common. In areas where there are great temperature ranges, rocks may split or crack as their outer surfaces expand or contract at rates different from those of the material beneath the surface. Low temperatures cause water in rock crevices to freeze, which also causes cracks and splits.

their parenchyma cells continuing to divide and enlarge without a true cambium developing. Several popular houseplants (e.g., ti plants, *Dracaena, Sansevieria*) are monocots in which a secondary meristem develops as a cylinder that extends throughout the stem. Unlike the vascular cambium of dicots and conifers, this secondary meristem produces only parenchyma cells to the outside and secondary vascular bundles to the inside.

Several commercially important cordage fibers (e.g., broomcorn, Mauritius and Manila hemps, sisal) come from the stems and leaves of monocots, but the individual cells are not separated from one another by *retting* (a process that utilizes the rotting power of microorganisms thriving under moist conditions to break down thin-walled cells) as they are when fibers from dicots are obtained. Instead, during commercial preparation, entire vascular bundles are scraped free of the surrounding parenchyma cells by hand; the individual bundles then serve as unit "fibers." If such fibers are treated with chemicals or bleached, the cementing middle lamella between the cells breaks down. Monocot fibers are not as strong or as durable as most dicot fibers.

SPECIALIZED STEMS

While an erect shoot system is characteristic of most higher plants, many species have stems that perform specialized functions; these stems are modified accordingly (Fig. 6.14). Although the overall appearance of specialized stems may differ markedly from that of the stems discussed so far, all stems have *nodes, internodes,* and *axillary buds;* these features distinguish them from roots and leaves, which do not have them. The leaves at the nodes of these specialized stems are often small and scalelike. They are seldom green, but full-sized functioning leaves are also produced. Descriptions of some of the specialized stems follow.

Rhizomes

Rhizomes (Fig. 6.14) are horizontal stems that grow below ground, often near the surface of the soil. Superficially, they resemble roots, but a close examination will reveal scalelike leaves and axillary buds at each node, at least during some stage of development, with short to long internodes in between. **Adventitious** roots are produced all along the rhizome, mainly on the lower surface. As indicated in Chapter 5, the word *adventitious* refers to structures arising at unusual places, such as roots growing from stems or leaves, or buds appearing at places other than leaf axils and tips of stems. A rhizome may be a relatively thick, fleshy, food-storage organ, as in irises, or it may be quite slender, as in many perennial grasses and some ferns.

Runners and Stolons

Runners are horizontal stems that differ from rhizomes in that they grow above ground, generally along the surface; they also have long internodes (Fig. 6.14). In strawberries,

This tree is 62 years old. It's been through fire and drought, plague and plenty. And all of this is recorded in its rings.

Each spring and summer a tree adds new layers of wood to its trunk. The wood formed in spring grows fast, and is lighter because it consists of large cells. In summer, growth is slower; the wood has smaller cells and is darker. So when the tree is cut, the layers appear as alternating rings of light and dark wood.

Count the dark rings, and you know the tree's age. Study the rings, and you can learn much more. Many things affect the way the tree grows, and thus alter the shape, thickness, color and evenness of the rings.

For St. Regis these rings have a special significance. They record the steady accumulation of those fibers we use to create noteworthy printing papers, kraft paper and boards, fine papers, packaging products, building materials, and products for consumers.

Essentially, then, the life of the forest is St. Regis' life. That is why we—together with the other members of the forest products industry—are vitally concerned with maintaining the beauty and utility of America's forests for the generations to come.

1904
The tree—a loblolly pine—is born.

1909
The tree grows rapidly, with no disturbance. There is abundant rainfall and sunshine in spring and summer. The rings are relatively broad, and are evenly spaced.

FIGURE 6.7 Climatic history illustrated by a cross section of a 62-year-old tree.

(Courtesy St. Regis Paper Company)

1914
When the tree was 6 years old, something pushed against it, making it lean. The rings are now wider on the lower side, as the tree builds "reaction wood" to help support it.

1924
The tree is growing straight again. But its neighbors are growing too, and their crowns and root systems take much of the water and sunshine the tree needs.

1927
The surrounding trees are harvested. The larger trees are removed and there is once again ample nourishment and sunlight. The tree can now grow rapidly again.

1930
A fire sweeps through the forest. Fortunately, the tree is only scarred, and year by year more and more of the scar is covered over by newly formed wood.

1942
These narrow rings may have been caused by a prolonged dry spell. One or two dry summers would not have dried the ground enough to slow the tree's growth this much.

1957
Another series of narrow rings may have been caused by an insect like the larva of the sawfly. It eats the leaves and leafbuds of many kinds of coniferous trees.

...ical size.

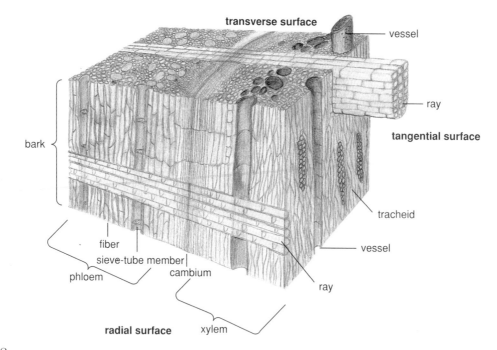

FIGURE 6.8 A three-dimensional, magnified view of a block of a woody dicot.

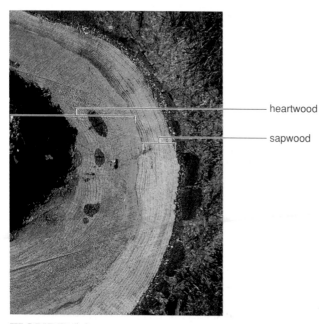

FIGURE 6.9 This tree was 100 years old when it was cut down. Note the proportion of *sapwood*, which consists of functional cells, to *heartwood*, in which the cells are no longer capable of conduction.

FIGURE 6.10 An African baobab tree.

runners are usually produced after the first flowering of the season has occurred. Several may radiate out from the parent plant, attaining lengths of up to 1 meter (3 feet) or more within a few weeks. Adventitious buds appear at alternate nodes along the runners and develop into new strawberry plants, which can be separated and grown independently. In some house plants, such as the saxifrages, runners may grow out and hang over the edge of the pot, producing new plants at intervals.

Stolons are similar to runners but usually grow more or less vertically beneath the surface of the ground. In Irish potato plants, *tubers* are produced at the tips of stolons.

Tubers

As food accumulates at the tips of stolons, such as those produced by Irish or white potato plants, several internodes swell, becoming **tubers** (Fig. 6.14). After the tuber is mature, the stolon dies, isolating it. The "eyes" of the potato are actually nodes formed in a spiral around the modified stem.

FIGURE 6.11   Resin canals in a portion of a pine (*Pinus*) stem, ca. ×400.

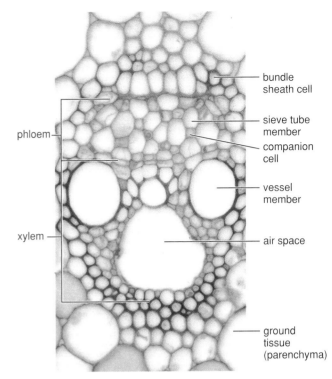

FIGURE 6.13   A single vascular bundle of corn (*Zea mays*) enlarged, ca. ×800.

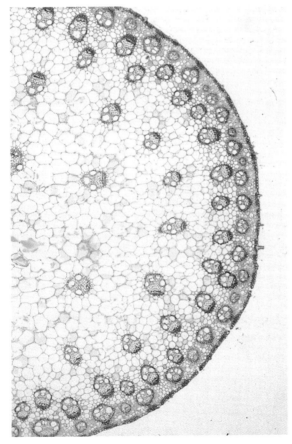

FIGURE 6.12   A portion of a cross section of a monocot (corn—*Zea mays*) stem, ca. ×100.

(Photomicrograph by G.S. Ellmore)

Each eye consists of an axillary bud in the axil of a scalelike leaf, although the latter is visible only in very young tubers, the small ridges seen on mature tubers being leaf scars.

Bulbs

Bulbs (Fig. 6.14) are actually large buds with a small stem at the lower end surrounded by numerous fleshy leaves. Adventitious roots grow from the bottom of the stem, but the fleshy leaves comprise the bulk of the bulb tissue, which functions in food storage. In onions, the fleshy leaves usually are surrounded by the scalelike leaf bases of long, green, aboveground leaves. Other plants producing bulbs include lilies, hyacinths, and tulips.

Corms

Corms, which superficially resemble bulbs, differ from them in being composed almost entirely of stem tissue except for the few papery scalelike leaves sparsely covering the outside (Fig. 6.14). Adventitious roots are produced at the base, and corms, like bulbs, function in food storage. Well-known plants producing corms include crocuses and gladioli.

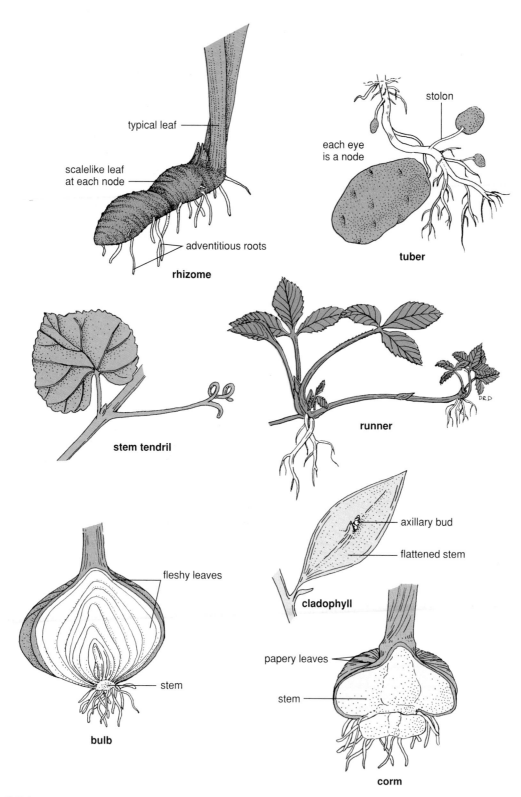

rhizome

typical leaf

scalelike leaf at each node

adventitious roots

stolon

each eye is a node

tuber

stem tendril

runner

axillary bud

flattened stem

cladophyll

fleshy leaves

stem

bulb

papery leaves

stem

corm

FIGURE 6.14   Types of specialized stems.

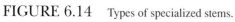

FIGURE 6.15 The flattened stems of prickly pear cacti (*Opuntia*) are cladophylls on which the leaves have been reduced to spines.

Cladophylls

In butcher's-broom plants, stems are flattened and leaflike in appearance. Such stems are called **cladophylls** (or *cladodes* or *phylloclades*) (Fig. 6.14). In the center of each butcher's-broom cladophyll is a node bearing very small scalelike leaves with axillary buds. The feathery appearance of asparagus is due to the presence of numerous small cladophylls. Cladophylls also occur in greenbriers, certain orchids, prickly pear cacti (Fig. 6.15), and several lesser-known plants.

Other Specialized Stems

In many cacti and in some of the spurges, the stems are stout and fleshy. Such stems are modified for water and food storage. Other stems may be modified in the form of *thorns*, as in the honey locust whose branched thorns may exceed 3 decimeters (1 foot) in length, but all thornlike objects are not necessarily modified stems. For example, the black locust has a pair of *spines* at the base of each petiole of most leaves (these are parts of the leaf called *stipules*,

mentioned in the discussion of twigs and discussed further in Chapter 7). The prickles of raspberries and roses, both of which originate from the epidermis, are neither thorns nor spines. Tiger lilies produce small aerial bulblets in the axils of their leaves.

Climbing plants have stems modified in various ways that adapt them for their growth habit. Some stems, called *ramblers*, simply rest on the tops of other plants, but many produce **tendrils** (Fig. 6.14). These are specialized stems in the grape and Boston ivy but are modified leaves or leaf parts in plants such as peas and cucumbers. In Boston ivy, the tendrils have adhesive disks. In English ivy, the stems climb with the aid of adventitious roots that arise along the sides of the stem and become embedded in the bark or other support material over which the plant is growing.

WOOD AND ITS USES

The use of wood by humans dates back into antiquity, and present uses are so numerous that it would be impossible to list in a work of this type more than the most important ones. Before looking at the economic importance of wood, a brief discussion of its properties is in order.

In a living tree, up to 50% of the weight of the wood comes from the water content. Before the wood can be used, the moisture content is reduced to 10% or less through *seasoning*, either by air-drying the wood in ventilated piles or stacks or by drying it in special ovens known as *kilns*. The seasoning has to be done gradually and under carefully controlled conditions or the timber may warp and split along the rays, making it unfit for most uses. The dry part of wood is composed of 60% to 75% *cellulose* and about 15% to 25% *lignin,* an organic substance that is deposited in the walls of xylem cells and makes the walls tough and hard. Other substances present in smaller amounts include resins, gums, oils, dyes, tannins, and starch. The proportions and amounts of these and other substances determine how various woods will be used (Fig. 6.16).

Density

Among the most important physical properties of wood is its *density.* Technically, the density is the weight per unit volume. The weight is compared with that of an equal volume of water and is stated as a fraction of 1.0. Because of the considerable air space within the cells, most woods have a *specific gravity,* as the comparative density is called, of less than 1.0. The range of specific gravities of known woods varies from 0.04 to 1.40, the lightest commercially used wood being balsa with a specific gravity of about 0.12. Woods with specific gravities of less than 0.50 are considered light; those with specific gravities of above 0.70 are considered heavy. Among the heaviest woods are the South

White Pine. A softwood widely used in home construction and for virtually everything from masts and matches to boxes and crates. Its soft, uniform texture and straight grain cuts easily in every direction, polishes well, and warps or swells little.

White Oak. Makes good barrels because the wood is resilient, durable, and impermeable to liquids. This hardwood, which is about twice as dense as white pine, has many other uses ranging from flooring to fine cabinet work.

Hard Maple. In bowling pins and flooring for bowling alleys its uniform texture and hardness result in resistance to abrasion. The Romans used it for spears and lances. We turn it on lathes to make spools, bobbins, cue sticks, and croquet balls.

Baldcypress. Because it is weather-resistant without treatment, this wood was widely used for cross ties in the early days of railroading. Today it is used for water tanks and other applications requiring prolonged contact with water.

Black Walnut. A choice hardwood for fine furniture and interior paneling, because of the beauty of the heartwood grain, its ability to stay in place after seasoning, and its good machining properties. It is harder than oak, and shock-resistant.

White Ash. Perfect for baseball bats, tennis racquets, oars and long tool handles. This hardwood's major virtues are straight grain, stiffness, strength, moderate weight, good bending qualities, and capacity for wearing smooth.

FIGURE 6.16 Uses of some common North American woods.
(Courtesy St. Regis Paper Company)

American ironwood and lignum vitae, with specific gravities of over 1.25. Lignum vitae, obtained from West Indian trees, is extremely hard wood and is used instead of metal in the manufacture of main bearings for drive shafts of submarines, because it is self-lubricating and less noisy.

Durability

A wood's ability to withstand decay organisms and insects is referred to as its *durability*. Moisture is needed for the enzymatic breakdown of cellulose and other wood

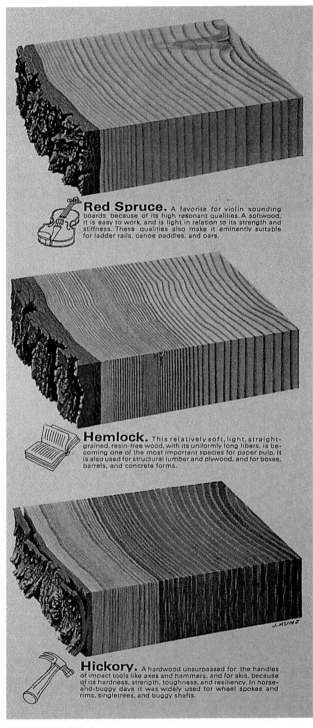

Red Spruce. A favorite for violin sounding boards because of its high resonant qualities. A softwood, it is easy to work, and is light in relation to its strength and stiffness. These qualities also make it eminently suitable for ladder rails, canoe paddles, and oars.

Hemlock. This relatively soft, light, straight-grained, resin-free wood, with its uniformly long fibers, is becoming one of the most important species for paper pulp. It is also used for structural lumber and plywood, and for boxes, barrels, and concrete forms.

Hickory. A hardwood unsurpassed for the handles of impact tools like axes and hammers, and for skis, because of its hardness, strength, toughness, and resiliency. In horse-and-buggy days it was widely used for wheel spokes and rims, singletrees, and buggy shafts.

FIGURE 6.16   continued

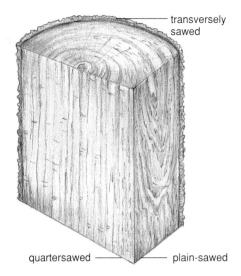

FIGURE 6.17   How the surfaces of plain-sawed, quartersawed, and transversely sawed wood appear (See also Fig. 6.8).

substances by decay organisms, but the seasoning process usually reduces the moisture to a level below that necessary for the fungi and other decay organisms to survive. Other natural constituents of wood that repel decay organisms include tannins and oils. Wood with a tannin content of 15%

or more may survive on a forest floor for many years after the tree has been felled by diseases of the phloem or other causes. Among the most durable of American woods are cedar, catalpa, black locust, red mulberry, and Osage orange. The least durable woods include cottonwood, willow, fir, and basswood.

Types of Sawing

Logs are usually cut longitudinally in one of two ways: along the radius or perpendicular to the rays (Fig. 6.17). Radially cut, or *quartersawed*, boards show the annual rings in side view; they appear as longitudinal streaks and are the most conspicuous feature of the wood. Only a few perfect quartersawed boards can be obtained from a log, making them quite expensive. Boards cut perpendicular to the rays (tangentially cut boards) are more common. In these, the annual rings appear as irregular bands of light and dark alternating streaks or patches, with the ends of the rays visible as narrower and less conspicuous vertical streaks. Lumber cut tangentially is referred to as being *plain-sawed* or *slab cut*. Slabs are the boards with rounded sides at the outside of the log; they are usually made into chips for pulping.

Knots

Knots are the bases of lost branches that have become covered, over a period of time, by new annual rings of wood produced by the cambium of the trunk. They are found in greater concentration in the older parts of the log toward the center, because in the forest, the lowermost branches of a tree (produced while the trunk was small in girth) often die from lack of sufficient light. When a branch dies and falls

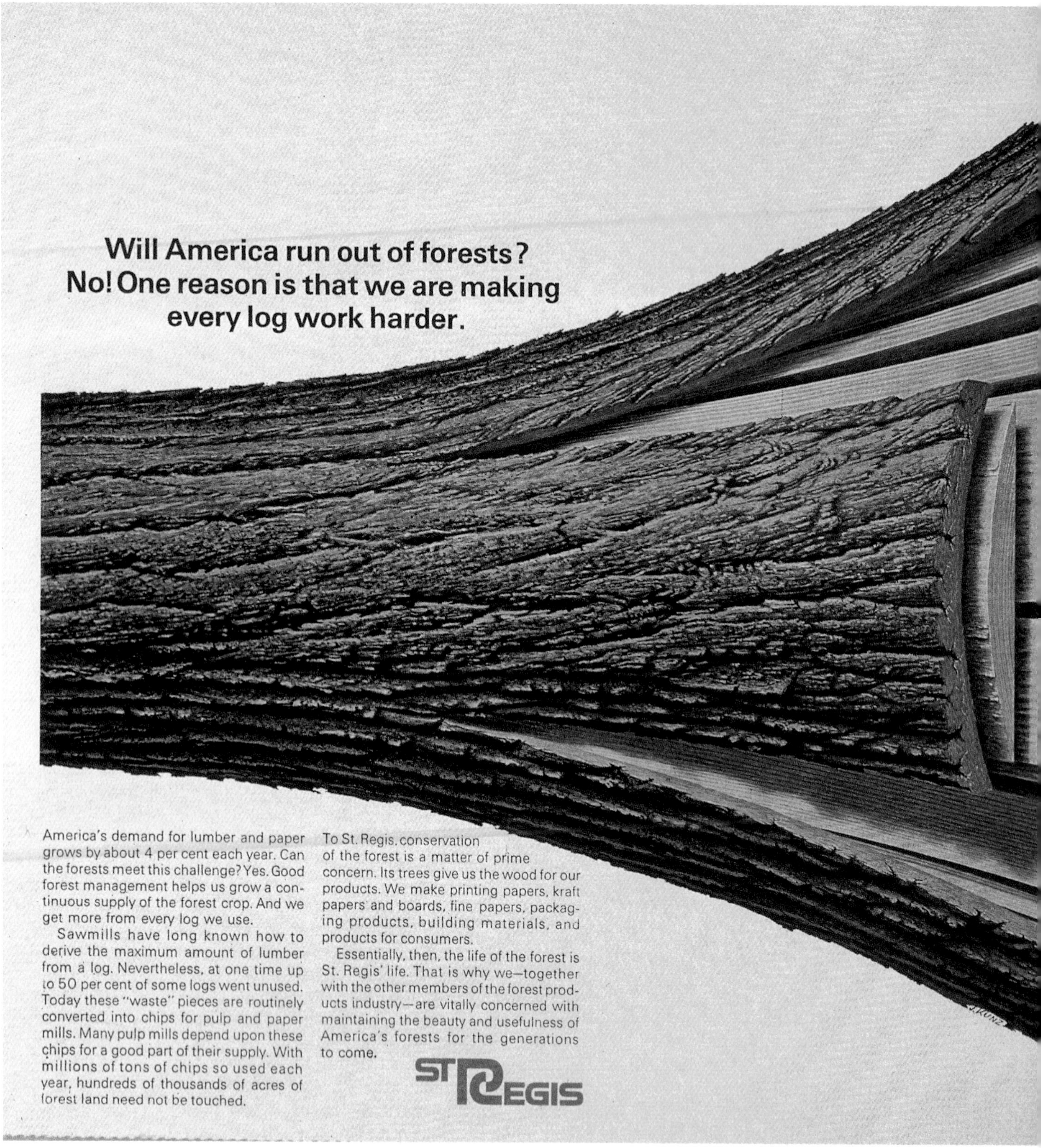

**Will America run out of forests?
No! One reason is that we are making
every log work harder.**

America's demand for lumber and paper grows by about 4 per cent each year. Can the forests meet this challenge? Yes. Good forest management helps us grow a continuous supply of the forest crop. And we get more from every log we use.

Sawmills have long known how to derive the maximum amount of lumber from a log. Nevertheless, at one time up to 50 per cent of some logs went unused. Today these "waste" pieces are routinely converted into chips for pulp and paper mills. Many pulp mills depend upon these chips for a good part of their supply. With millions of tons of chips so used each year, hundreds of thousands of acres of forest land need not be touched.

To St. Regis, conservation of the forest is a matter of prime concern. Its trees give us the wood for our products. We make printing papers, kraft papers and boards, fine papers, packaging products, building materials, and products for consumers.

Essentially, then, the life of the forest is St. Regis' life. That is why we—together with the other members of the forest products industry—are vitally concerned with maintaining the beauty and usefulness of America's forests for the generations to come.

ST **R**EGIS

FIGURE 6.18   How a log is used. Note that some logging practices have become controversial due to their effects on the habitats of threatened species of living organisms, as discussed in Chapter 25.

(Courtesy St. Regis Paper Company)

How the log is used

Debarking the log is essential to its full utilization because bark cannot be used for paper-making, and therefore any piece dropped in the chipper has to be free of bark. But the bark can be used for fuel and soil mulch.

The rounded sides of the log, called "slabs," are the first pieces sent to the chipper as the log goes through the saw-mill. This idealized picture shows the entire log being used for lumber, except for the slabs. Actually, as cutting continues, other pieces go to the chipper, including edgings, trim ends, and other parts of the log not usable as lumber. Each log presents different problems and can be handled differently.

The outer portions of the log have the fewest knots. This "clear" lumber is usually made into boards or planks varying in thickness from one to three inches.

Toward the center of the log, knots increase and the wood is less suitable for boards. Heavier planks, and square or rectangular beams are normally sawed from this section. The center of the log is used primarily for structural beams strong enough so that they are not weakened by knots. Knots are most frequent here because this is the oldest section of the tree. Branches that were removed during the early years of the tree's life left knots that were covered over as the tree grew outward.

Plywood is, in effect, a sandwich of thin wooden veneers. Veneer is made by "peeling," that is, holding a long blade against a rotating log. The wood is continuously peeled off, down to an eight-inch core. The core is then treated as though it were a small log. It can be made into lumber and, of course, the rounded portions go to the chipper.

off, the cambium at its base also dies, but the cambium of the trunk remains alive and increases the girth of the tree, slowly enveloping the dead tissue of the branch base until it may be completely buried and not visible from the surface. Knots usually weaken the boards in which they occur.

Wood Products

In the United States, about half of the wood produced is used as lumber, primarily for construction; the sawdust and other waste formed in processing the boards is converted to particle board and pulp. A considerable amount of lumber goes into the making of furniture, which may be constructed of solid wood or covered with a *veneer*. A veneer is a very thin sheet of desirable wood that is glued to cheaper lumber; it is carefully cut so as to produce the best possible view of the grain (Fig. 6.18).

The next most extensive use of wood is for *pulp*, which, among other things, is converted by various processes to paper, synthetic fibers, plastics, and linoleum. In recent years, it has been added as a filler to commercial ice cream and bread. Some hardwoods are treated chemically or heated under controlled conditions to yield a number of chemicals, such as wood alcohol and acetic acid, but other sources of these products are now usually considered more economical. Charcoal, excelsior, cooperage (kegs, casks, and barrels), railroad ties, boxes and crates, musical instruments, bowling pins, tool handles, pilings, cellophane, photographic film, and Christmas trees are but a few of the additional wood products worth billions of dollars annually on the world market (see Fig. 6.16).

In developing countries, approximately half of the timber cut is used for fuel, but in the United States, a little less than 10% is currently used for that purpose. In colonial times, wood was the almost exclusive source of heating energy. In the 1980s, Brazil's major cities were still using scrub timber from the surrounding forests to energize their utilities, but rapidly depleting supplies and problems related to the greenhouse effect (discussed in Chapter 25) have pointed to the need for alternate sources of energy. Many types of coal are wood that has been compressed for millions of years until nearly pure carbon remains. The formation of coal and other fossils is discussed in Chapter 21. Although the world's supply of coal is still plentiful, the rate at which this fossil fuel is being consumed makes it obvious that resources will eventually be exhausted unless our energy demands find renewable or less destructive alternatives.

Some of the vast array of secondary products from stems, including dyes, medicines, spices, and foods, are discussed in later chapters and in the appendices.

14. Laticifers are latex-secreting cells or ducts found in various flowering plants. The latex of some plants has considerable commercial value.

15. Monocot stems have scattered vascular bundles and no cambia. The parenchyma tissue is not divided into pith and cortex. Each vascular bundle is surrounded by a sheath of sclerenchyma cells. Numerous bundles and a band of sclerenchyma cells and thicker-walled parenchyma cells just beneath the surface of monocot stems aid in withstanding stresses.

16. Palm trees are monocots that become large because their parenchyma cells continue to divide. Other monocots develop a secondary meristem that produces parenchyma cells and secondary vascular bundles. Grasses have at the base of each internode intercalary meristems that contribute to rapid increases in length. Several commercially important cordage fibers are obtained from monocots.

17. Specialized stems include rhizomes, stolons, tubers, bulbs, corms, cladophylls, and tendrils. Such stems may have adventitious roots.

18. The dry part of wood consists primarily of cellulose and lignin. Resins, gums, oils, dyes, tannins, and starch are also present. Properties of wood that play a role in its use include density, specific gravity, and durability.

19. Logs are usually cut longitudinally along the radius (quartersawed) or perpendicular to the rays (tangentially, plain-sawed, or slab cut). Knots are bases of lost branches that have become covered over by new wood; they usually weaken the boards in which they occur.

20. About half the timber produced in the United States is used as lumber. Sawdust and waste are converted to particle board and pulp for paper, synthetics, and linoleum. Other timber is used for cooperage, charcoal, railroad ties, boxes, tool handles, and so on. Developing countries use a greater proportion of their timber for fuel than do other countries.

Review Questions

1. What is the function of bud scales?

2. How can you tell the age of a twig?

3. Distinguish among procambium, vascular cambium, and cork cambium.

4. How can you tell, when you look at a cross section of a young stem, whether it is a dicot or a monocot?

5. What are laticifers?

6. An Irish or white potato is a stem, but a sweet potato is a root. How can you tell the difference?

7. Distinguish among corms, bulbs, and tubers.

8. If you were examining the top of a wooden desk, how could you tell if the wood had been radially or tangentially cut (quartersawed or plain-sawed)?

9. What differences are there between heartwood and sapwood?

10. What is meant by the specific gravity of wood?

Discussion Questions

1. If the cambium of a tropical tree were active all year long, how would its wood differ from that of a typical temperate climate tree?

2. It was mentioned that a nail driven into the side of a tree will remain at exactly the same distance from the ground for the life of the tree. Why?

3. Do climbing plants have any advantages over erect plants? Any disadvantages?

4. If two leaves are removed from a plant and one is coated with petroleum jelly while the other is not, the uncoated leaf will shrivel considerably sooner than will the coated one. Would it be helpful to coat the stems of young trees with petroleum jelly? Explain.

5. Suggest some reasons for heartwood being preferred to sapwood for making furniture.

Additional Reading

Core, H.A., W.A. Cote, and A.C. Day. 1979. *Wood structure and identification*, 2d ed. Syracuse, NY: Syracuse University Press.

Cutter, E.G. 1978. *Plant anatomy: Experiment and interpretation. Part I: Cells and tissues*, 2d ed. Reading, MA: Addison-Wesley Publishing Co.

Fahn, A. 1990. *Plant anatomy*, 4th ed. Elmsford, NY: Pergamon Press.

Flynn, J. H., Jr. (Ed.). 1994. *A guide to useful woods of the world*. Portland, ME: King Philip Publishing Co.

Hughs, M.K., et al. (Eds.). 1982. *Climate from tree rings*. Fair Lawn, NY: Cambridge University Press.

Lewin, M. 1991. *Wood structure and composition*. New York: Dekker, Marcel, Inc.

Mauseth, J.D. 1988. *Plant anatomy*. Menlo Park, CA: Benjamin/Cummings Publishing Co., Inc.

Metcalfe, C.R., and L. Chalk (Eds.). 1988–1989. *Anatomy of the dicotyledons*, 2 vols. Fair Lawn, NY: Oxford University Press.

Meylan, B.A., and B.G. Butterfield. 1972. *Three-dimensional structure of wood: A scanning electron microscope study*. Syracuse, NY: Syracuse University Press.

Schweingruber, F. H. 1993. *Trees and wood in dendrochronology*. New York: Springer-Verlag.

Steeves, T.A., and I.M. Sussex. 1989. *Patterns in plant development*, 2d ed. Englewood Cliffs, NJ: Prentice-Hall.

Chapter Outline

Leaves of Croton, *a colorful member of the Spurge Family. The intense red anthocyanin pigments mask the visibility of the green chlorophyll pigments, which are also present.*

Leaves

7

FIGURE 7.6 Waxes, in addition to those of the cuticle, are sometimes produced on the surfaces of leaves, stems, and fruits, giving them a whitish appearance, as seen on this black raspberry cane.

although it may not be visible with ordinary light microscopes without being specially stained. Many plants produce, in addition to the cuticle, other waxy substances on their surfaces (Fig. 7.6).

In studies of the effects on plants of smog and auto exhaust fumes, it was found that these waxes may be produced in abnormal fashion on beet leaves within as little as 24 hours of exposure to the pollutants. Beet leaves also respond to aphid damage by producing wax around each tiny puncture. Occasionally, epidermal cells contain crystals of waste materials. Different types of **glands** may also be present in the epidermis. Glands occur in the form of depressions, protuberances, or appendages either directly on the leaf surface or on the ends of hairs (see Fig. 4.11). Glands often secrete sticky substances.

STOMATA

The lower epidermis of most plants generally resembles the upper epidermis, but it is perforated by numerous tiny pores, the **stomata** (Fig. 7.7). They also occur in both leaf surfaces of some plants (e.g., alfalfa, corn), exclusively on the upper epidermis of other plants such as water lilies where the lower epidermis is in contact with water, and are absent altogether from the submerged leaves of aquatic plants. Stomata are very numerous, ranging from about 1,000 to more than 1.2 million per square centimeter (6,300 to 8 million per square inch) of surface. An average-sized sunflower leaf has about two million of these pores throughout its lower epidermis. Each stoma is bordered by two sausage- or dumbbell-shaped cells that usually are smaller than most of the neighboring epidermal cells. These are called **guard cells,** and unlike other cells of either epidermis, they contain chloroplasts.

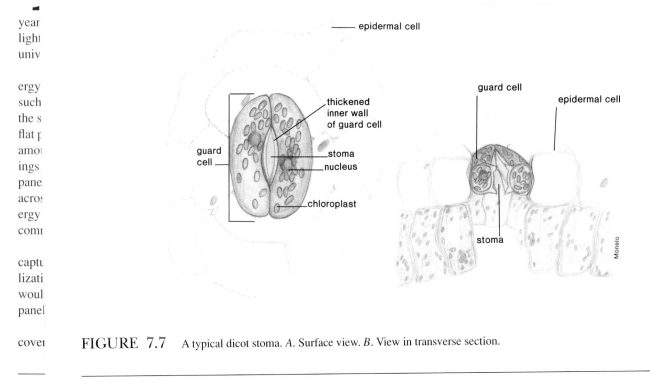

FIGURE 7.7 A typical dicot stoma. *A*. Surface view. *B*. View in transverse section.

The photosynthesis that takes place in the guard cells aids in the functioning of the cells. The primary functions include permitting gas exchange between the interior of the leaf and the atmosphere and regulation of evaporation of most of the water entering the plant at the roots.

The guard cell walls are distinctly thickened but quite flexible on the side adjacent to the pore. As the guard cells expand or contract with changes in the amount of water within the cells, their unique construction causes the stomata to open or close. When the guard cells are expanded, the stomata are open; when the water content of the guard cells decreases, the cells contract and the stomata close. (For more detailed discussions of this stomatal mechanism see the sections on regulation of transpiration in Chapter 9 and turgor movements in Chapter 11.)

MESOPHYLL AND VEINS

Most photosynthesis takes place in the **mesophyll** between the two epidermal layers. When two regions of the mesophyll are distinguishable, as is often the case, the uppermost region consists of compactly stacked parenchyma cells; the cells are mostly shaped like short posts and are commonly in two rows. This region is called the **palisade mesophyll** and may contain more than 80% of the leaf's chloroplasts. The lower region, consisting of loosely arranged parenchyma cells with abundant air spaces between them, is called the **spongy mesophyll.** Its cells also contain numerous chloroplasts.

Parenchyma tissue containing numerous chloroplasts is referred to as *chlorenchyma* tissue. Besides being present in leaves, chlorenchyma tissue is found in the outer parts of the cortex in the stems of herbaceous plants. Inside the leaf, the surfaces of mesophyll cells in contact with the air are moist. If moisture decreases below a certain level, the stomata close, thereby significantly reducing further drying.

Veins (*vascular bundles*) of various sizes are scattered throughout the mesophyll (Fig. 7.8). They consist of xylem and phloem tissues surrounded by a jacket of fibers called the **bundle sheath.** The veins give the leaf its "skeleton." The carbohydrates produced in the mesophyll cells are transported in solution throughout the plant by the phloem. Water, sometimes located more than 100 meters (330 feet) away in the ground below, is brought up to the leaf by the xylem, which, like the phloem, is part of a vast network of "plumbing" throughout the plant. Since veins run in all directions in the network, particularly in dicots, it is common, when examining a cross section of a leaf under the microscope, to see veins cut transversely, lengthwise, and at a tangent, all in the same section.

Monocot leaves, in addition to having *parallel veins,* usually do not have the mesophyll differentiated into palisade and spongy layers. Some monocot leaves (e.g., those of grasses) have large, thin-walled *bulliform cells* on either side of the main central vein (midrib) toward the upper surface (Fig. 7.9). The bulliform cells partly collapse under dry

FIGURE 7.8   Net venation in a leaf, ×50.

conditions, causing the leaf blade to fold or roll, thus reducing transpiration (see Fig. 11.16).

SPECIALIZED LEAVES

If the leaves of all plants could function normally under any environmental condition, various leaf modifications would provide no special benefits to a plant. But the form and structure of plants of tropical rain forests do not permit them to thrive in a desert and cacti soon die if planted in a creek because their structure, form, and life cycles are attuned to specific combinations of environmental factors, such as temperature, humidity, light, water, and soil conditions. In addition, the modifications of leaves occupying any single ecological niche may be very diverse, resulting in such a rich variety of leaf forms and specializations throughout the Plant Kingdom that only a few may be mentioned here.

Shade Leaves

Some leaves with inconspicuous modifications may occur along with unmodified leaves on the same plant. For example, leaves in the shade tend to be thinner and have fewer

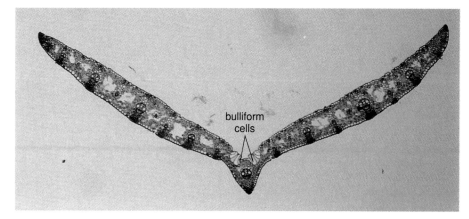

FIGURE 7.9   A cross section of a grass leaf.

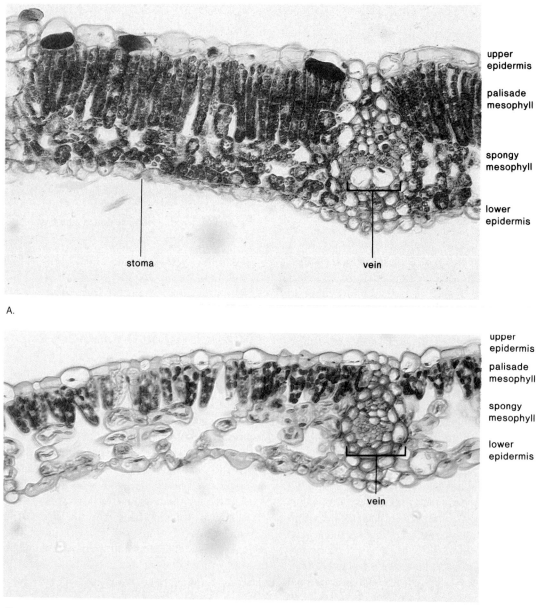

A.

B.

FIGURE 7.10   Portions of cross sections of maple (*Acer*) leaves. The chloroplasts are stained red. *A*. A leaf exposed to full sun. *B*. A leaf exposed to shade. Note the reduction in mesophyll cells and chloroplasts in the shade leaf.

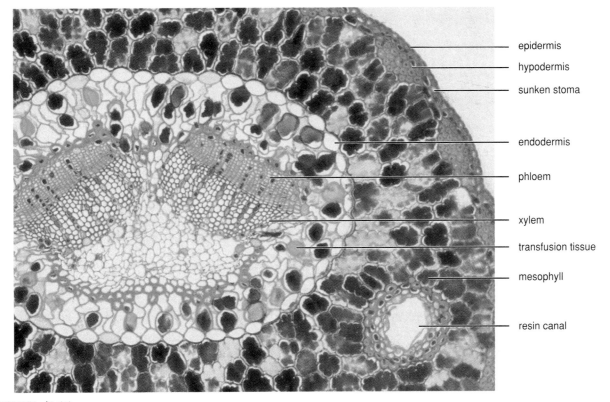

epidermis
hypodermis
sunken stoma
endodermis
phloem
xylem
transfusion tissue
mesophyll
resin canal

FIGURE 7.11   A pine needle in cross section.
(Photomicrograph by G.S. Ellmore)

hairs than leaves on the same tree that are exposed to direct light. They also tend to be larger and to have fewer well-defined mesophyll layers and fewer chloroplasts than their counterparts in the sun (Fig. 7.10).

Leaves of Arid Regions

In different climatic zones or habitats, leaf modifications are generally more pronounced. Plants growing in arid regions may have thick and leathery leaves and have stomata that are fewer in number or sunken below the surface in special depressions. They also may have succulent water-retaining leaves or no leaves at all (with the stems taking over the function of photosynthesis), or they may have dense, hairy coverings. Pine trees, whose water supply may be severely restricted in the winter when the soil is frozen, have some leaf modifications similar to those of desert plants. The modifications include sunken stomata, a thick cuticle, and a layer of thick-walled cells (the **hypodermis**) beneath the epidermis (Fig. 7.11).

The leaves of the compass plant face the East and West, with the blades perpendicular to the ground, so when the sun is overhead it strikes only the thin edge of the leaf, minimizing moisture loss (Fig. 7.12). In plants that grow in water, the submerged leaves usually have considerably less xylem than phloem. Large spaces are found in the mesophyll, which is not differentiated into palisade and spongy layers. Other modifications are described in the following sections.

Tendrils

There are many plants whose leaves are partly or completely modified as **tendrils.** These leaves, when curled tightly around more rigid objects, help the plant in climbing or in supporting weak stems. In garden peas, the leaves are compound (Fig. 7.13) and the upper leaflets are reduced to whip-like strands that, like all tendrils, are very sensitive to contact. In fact, if you lightly stroke a healthy tendril, there is a sudden rapid growth of cells on the opposite side, and it starts curling in the direction of the contact within a minute or two. If the contact is very brief, the tendril undergoes a reverse movement and straightens out again. If, however, the tendril encounters a suitable solid support (e.g., a twig), the stimulation is continuous, and the tendril coils tightly around the support as it grows.

In the yellow vetchling, the whole leaf is modified as a tendril, and photosynthesis is carried on by the leaflike stipules at the base. In the potato vine and the garden nasturtium, the petioles serve as tendrils, while in some greenbriers, stipules are modified as tendrils. In *Clematis,* the rachises of some of the compound leaves serve very effectively as tendrils. Members of the Pumpkin Family, which include

FIGURE 7.13   Tendrils at the tip of a garden pea plant leaf.

FIGURE 7.12   A compass plant. The leaves, which have their blades parallel to the sun, are oriented toward the East and West.

squashes, melons, and cucumbers, produce tendrils that may be up to 3 decimeters (1 foot) in length.

As the tendrils develop, they become coiled like a spring. When contact with a support is made, the tip not only curls around it, but the direction of the coil reverses (see Fig. 11.8); sclerenchyma cells then develop in the vicinity of contact. This makes a very strong but flexible attachment that protects the plant from damage during high winds. It should be noted that the tendrils of many other plants (e.g., grapes) are not modified leaves but have developed from stems.

Spines, Thorns, and Prickles

Many desert plants have their leaves modified as **spines.** This reduction in leaf surface correspondingly reduces water loss from the plants, and the spines also apparently protect the plants from browsing animals. Photosynthesis in such desert plants, which would otherwise take place in leaves, occurs in the green stems. In a number of woody plants (e.g.,

mesquite, black locust), the stipules at the bases of the leaves are modified as short, paired spines. Most *spines* are modifications of the whole leaf, in which much of the normal leaf tissue is replaced with sclerenchyma. As is the case with tendrils, many spinelike objects arising in the axils of leaves are modified stems rather than modified leaves. Such modifications should be referred to as *thorns* to distinguish them from true spines. The *prickles* of roses and raspberries, however, are neither leaves nor stems but outgrowths from the epidermis or cortex just beneath it (Fig. 7.14).

Storage Leaves

As previously mentioned, desert plants may have *succulent leaves* (i.e., leaves that are modified so that they retain water). The modifications that permit water storage involve large, thin-walled parenchyma cells without chloroplasts to the interior of chlorenchyma tissue just beneath the epidermis. These nonphotosynthetic cells contain large vacuoles that can store proportionately substantial quantities of water. If removed from the plant and set aside, the leaves will often retain much of the water for several weeks. Many plants with succulent leaves carry on a special form of photosynthesis discussed in Chapter 10.

Other fleshy leaves, such as those that comprise the bulk of onion and lily bulbs, store large amounts of carbohydrates, which are utilized by the plant in the subsequent growing season.

Flower Pot Leaves

Some of the leaves of *Dischidia* (Fig. 7.15), a greenhouse curiosity from tropical Australasia, develop into urnlike pouches that become the homes of ant colonies. The ants carry in soil and add nitrogenous wastes, while moisture collects in the leaves through condensation of the water vapor

FIGURE 7.14 A. The *spines* of this barberry are modified leaves.

FIGURE 7.14 C. *Prickles* of raspberry stems. The prickles are neither leaves nor stems; they are outgrowths from the epidermis or the cortex just beneath it.

FIGURE 7.14 B. *Thorns* (modified stems) produced in the axils of leaves.

coming through stomata from the mesophyll. This situation creates a good growing medium for roots, which develop adventitiously from the same node as the leaf and grow down into the soil contained in the urnlike pouch. In other words, this extraordinary plant not only reproduces itself by conventional means but also, with the aid of ants, provides its own fertilized growing medium and flower pots and then produces special roots, which "take advantage" of the situation.

Window Leaves

In the Kalahari desert of South Africa, there are at least three plants belonging to the Carpetweed Family that have unique adaptations to living in dry, sandy areas. Their leaves, which are shaped like ice cream cones, are about 3.75 centimeters (1.5 inches) long (Fig. 7.16) and are buried in the sand; only the dime-sized wide end of a leaf is exposed at the surface. This exposed end is covered with a thick epidermis and cuticle. It has a few stomata and is relatively transparent. Below the exposed end is a mass of tightly packed water-storage cells; these allow light coming through the "windows" to penetrate to the chloroplasts in the mesophyll, located all around the inside of the shell of the leaf. This arrangement,

FIGURE 7.20 Sundew leaves.

by bacteria and by enzymes secreted by digestive glands near the bottom of the plant.

Malaysian tree frogs lay their eggs in the pitchers of pitcher plants. The eggs, presumably, contain something that neutralizes digestive enzymes. In North America, the pitcher plants produce their pitchers in erect clusters on the ground, but their mechanisms for trapping insects are similar to those of Asian species.

Merchandisers have been shameless in their wholesale collection of these plants for sale, and pitcher plants may become extinct in the wild. The cobra plant, a pitcher plant restricted to a few swampy areas in California and Oregon, has been placed on official threatened species lists.

Sundews

The tiny plants called *sundews* (Fig. 7.20) often do not measure more than 2.5 to 5.0 centimeters (1 to 2 inches) in diameter. The roundish to oval leaves are covered with about 150 to 200 upright glandular hairs that look like miniature clubs. There is a clear glistening drop of sticky fluid containing digestive enzymes at the tip of each hair. As the droplets sparkle in the sun, they may attract insects, which find themselves stuck if they alight. The hairs are exceptionally sensitive to contact, responding to weights of less than one 1,000th of a milligram, and bend inward, surrounding any trapped insect

within the space of a few minutes. The digestive enzymes break down the soft parts of the insects, and after digestion has been completed (within a few days), the glandular hairs return to their original positions. If bits of nonliving debris happen to catch in the sticky fluid, the hairs barely respond, showing they can distinguish between protein and something "inedible." Some sundew owners regularly feed their plants tiny bits of hamburger and boiled egg white.

Portuguese peasants use relatives of sundews with less specialized leaves in their homes as an effective substitute for flypaper. In response to contact by living insects, the edges of specialized leaves of similar plants called *butterworts* rapidly curl over and trap unwary victims.

Venus Flytraps

The *Venus flytrap* (Fig. 7.21), which has leaves constructed along the lines of an old-fashioned steel trap, is found in nature only in wet areas of North Carolina and South Carolina. The two halves of the blade have the appearance of being hinged along the midrib, with stiff hairlike projections located along their margins. There are three tiny trigger hairs on the inner surface of each half. If two trigger hairs are touched simultaneously or if any one of them is touched twice within a few seconds, the blade halves suddenly snap together, trapping the insect or other small animal. As the

FIGURE 7.21   A Venus flytrap plant.

FIGURE 7.22   A bladderwort.

organism struggles, the trap closes even more tightly. Digestive enzymes secreted by the leaf break down the soft parts of the insect, which are then absorbed. After digestion has been completed, the trap reopens, ready to repeat the process. As is the case with sundews, the traps do not normally close for bits of debris that might accidentally fall on the leaf, presumably because nonliving material does not move about and stimulate the trigger hairs.

Bladderworts

Bladderworts (Fig. 7.22) are plants with finely dissected leaves that are found submerged and floating in the shallow water along the margins of lakes and streams. Toward the bases of many of the leaves are tiny stomach-shaped bladders, each with a trapdoor over the opening at one end. The bladders, which are between 0.3 and 0.6 centimeter (0.125 to 0.250 inch) in diameter, trap aquatic insects and other small animals through a complex mechanism. Four curled but stiff hairs located toward one end of the trapdoor act as triggers when an insect touches one of them. The trapdoor springs open, and water rushes into the bladder. The stream of water propels the victim into the trap, and the door snaps shut behind it. The action takes place in less than one 100th of a second and makes a distinct popping sound, which can be heard with the aid of a sensitive underwater microphone. The trapped insect eventually dies, it is broken down by bacteria, and the breakdown products are absorbed by cells in the walls of the bladder.

Science fiction writers have contributed to superstitions and beliefs that deep in the tropical jungles there are plants capable of trapping humans and other large animals. No such plants have been proved to exist, however. The largest pitcher plants known hold possibly 1 liter (roughly 1 quart) of fluid in their pitchers, and small frogs have been known to decompose in them, but the trapping of anything larger than a mouse or possibly a small rabbit seems very unlikely.

AUTUMNAL CHANGES IN LEAF COLOR

When leaf cells break down and die, the leaves tend eventually to turn some shade of brown or tan due to a reaction between leaf proteins and tannins stored in the cell vacuoles. This is akin to the formation of leather when tannins react with animal hides. Before turning brown, however, the leaves may exhibit a variety of other colors.

The chloroplasts of mature leaves contain several groups of pigments, such as green *chlorophylls,* and *carotenoids*—which include yellow *carotenes* and pale yellow *xanthophylls*. Each of these groups plays a role in photosynthesis. The chlorophylls are usually present in much larger quantities than the other pigments, and their intense green color masks or hides the presence of the carotenes and xanthophylls. In the fall, however, the chlorophylls break down, and other colors are revealed. The exact cause of the chlorophyll breakdown is not known, but it does appear to involve, among other factors, a gradual reduction in day length. Water-soluble *anthocyanin* and *betacyanin* pigments may also accumulate in the vacuoles of the leaf cells in the fall.

Anthocyanins, the more common of the two groups, are red if the cell sap is slightly acid, blue if it is slightly alkaline, and of intermediate shades if it is neutral. Betacyanins are usually red; they apparently are restricted to several plant families, such as the cacti, the Goosefoot Family (to which beets belong), the Four-o'clock Family, and the Portulaca Family.

Some plants (e.g., birch trees) consistently exhibit a single shade of color in their fall leaves, but many (e.g., maple, ash, sumac) vary considerably from one locality to another or even from one leaf to another on the same tree, depending on the combinations of carotenes, xanthophylls, and other pigments present. Some of the most spectacular fall colors in North America occur in the Eastern Deciduous Forest, particularly in New England and the upper reaches of the Mississippi Valley. In parts of Wisconsin and Minnesota, one can observe the brilliant reds, oranges, and golds of maples; the deep maroons (and also yellows) of ashes; the bright yellows of aspen; and the seemingly glowing reds of sumacs and wahoos—all in a single locality. Some fall coloration is found almost anywhere in temperate zones where deciduous trees and shrubs exist.

ABSCISSION

Plants whose leaves drop seasonally are said to be **deciduous.** In temperate climates, new leaves are produced in the spring and are shed in the fall, but in the tropics, the cycles coincide with wet and dry seasons rather than with temperature changes. Even so-called evergreen trees shed their leaves; they do so a few at a time, however, so that they never have the bare look of deciduous trees in their winter condition. The process by which the leaves are shed is called **abscission.**

Abscission occurs as a result of changes that take place in an *abscission zone* near the base of the petiole of each leaf (Fig. 7.23). Sometimes, the abscission zone can be seen externally as a thin band of slightly different color on the petiole. Hormones that apparently inhibit the formation of the specialized layers of cells that facilitate abscission are produced in young leaves. As the leaf ages, hormonal changes take place, and at least two layers of cells become differentiated. Closest to the stem, the cells of the *protective layer,* which may be several cells deep, become coated and impregnated with fatty *suberin.* On the leaf side, a *separation layer* develops in which the cells swell, sometimes divide, and become gelatinous. In response to any of several environmental changes (such as lowering temperatures, decreasing day lengths or light intensities, lack of adequate water, or damage to the leaf), the pectins in the middle lamella of the cells of the separation layer are broken down by enzymes. All that holds

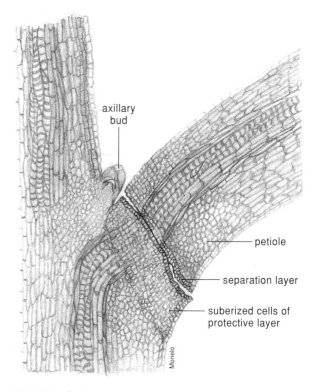

FIGURE 7.23   The abscission zone of a leaf.

the leaf on to the stem at this point are some strands of xylem. Wind and rain then easily break the connecting strands, leaving tiny bundle scars within a leaf scar (see Fig. 6.1), and the leaf falls to the ground.

HUMAN AND ECOLOGICAL RELEVANCE OF LEAVES

Humans use shade trees and shrubs in landscaping for cooling as well as for aesthetic effects. The leaves of shade plants planted next to a dwelling can make a significant difference in energy costs to the home owner. Humans also use for food the leaves of cabbage, parsley, lettuce, spinach, and chard and the petioles of celery and rhubarb, to mention a few. A large number of spices and flavorings are derived from leaves, including thyme, marjoram, oregano, tarragon, peppermint, spearmint, wintergreen, basil, dill, sage, cilantro, and savory.

Many dyes (e.g., a yellow dye from bearberry, a reddish dye from henna, and a pale blue dye from blue ash) are extracted from leaves (see Appendix 3), although nearly all commercial dyes are now derived from coal tar. Many cordage fibers for ropes and twines come from leaves, with various species of *Agave* (century plants) accounting for about 80% of the world's production. Bowstring fibers are obtained from a relative of the common houseplant *Sansevieria,* and Manila hemp fibers, which are used both in fine-quality cordage and in textiles, are obtained from the leaves of a close relative of the banana. Panama hats are made from the leaves of the panama hat palm, and palms and grasses are used in the tropics as thatching material for huts and other buildings.

In the high mountains of Chile and Peru, the leaves of the yareta plant are used for fuel. They produce a resin that causes the leaves to burn with an unusually hot flame. Leaves of many plants produce oils. Petitgrain oil (from the leaves of a variety of orange tree) and lavender oil, for example, are used for scenting soaps and colognes. Patchouli and lemongrass oils are used in perfumes, as is citronella oil, which was once the leading mosquito repellent before synthetic repellents gained favor. Eucalyptus oil, camphor, cajeput, and pennyroyal (Fig. 7.24) are all used medicinally.

Leaves are an important source of drugs used in medicine and also of narcotics and poisons. Cocaine, obtained from plants native to South America, has been used medicinally and as a local anesthetic, but its use as a narcotic has, in recent years, become a major problem in western cultures. Andean natives chew coca leaves while working and are reported capable of performing exceptional feats of labor with little or no food while under the influence of cocaine. Apparently, the drug, which is highly addictive in forms such as "crack," anesthetizes the nerves that convey hunger pangs to the brain.

Belladonna is a drug complex obtained from leaves of the deadly nightshade, a native of Europe. The plant has been used in medicine for centuries, and several drugs are now isolated from belladonna. Included among the isolates are atropine, which is used in shock treatments, to dilate eyes, to relieve pain locally, and to check secretions. Scopolamine, also a belladonna derivative, is used in tranquilizers and sleeping aids. Another European plant, the foxglove, is the source of digitalis. This drug has been used for centuries in regulating blood circulation and heartbeat.

Tobacco is another widely used leaf. At present, more than 940 million kilograms (2 billion pounds) of tobacco for smoking, chewing, and use as snuff are produced annually around the world. Also at present, about 125,000 Americans die annually from lung cancer, and almost all of them have a history of cigarette smoking. Cigarette smoking is also evidently a principal contributor to cardiovascular diseases. During recent years, the increase in the use of chewing tobacco has seen a corresponding increase in the development of mouth and throat cancers. Federal law forbids concentrations of more than five parts per billion of nitrosamines (cancer-causing chemicals) in cured meats, but the levels of nitrosamines in the five most popular brands of chewing tobacco range from 9,600 to 289,000 parts per billion. Though humans have long used tobacco, it has only been in the last two decades that its health threat has become appreciated.

Similarly, marijuana, the controversial plant widely used as an intoxicant, has been utilized in various ways for thousands of years. The active principle, tetrahydrocannabinol (THC), although found in the leaves, is concentrated in hair secretions among the female flowers. In recent decades, marijuana has found increasing acceptance in the western

FIGURE 7.24   Flowers of a pennyroyal plant.

hemisphere; it appears, however, not to be the harmless intoxicant many thought it to be. Regular use of marijuana for a year has been shown to have the same effect on human lungs as smoking one and a half packs of cigarettes a day for 13 years. A decision to legalize its prescription in pill form for the alleviation of nausea caused by chemotherapy treatments has been criticized by some medical authorities.

The drug lobeline sulphate is obtained from the leaves of a close relative of garden lobelias. It is used in compounds taken by people who are trying to stop smoking. The leaves of several species of *Aloe,* especially *Aloe vera,* yield a juice that is used to treat various types of skin burns, including those accidentally received from X-ray equipment.

Several beverages are extracted by the brewing of leaves. Numerous teas have been obtained from a wide variety of plants, but most now in use come from a close relative of the garden camellia. Maté, the popular South American tea, is brewed from the leaves and twigs of a relative of holly. The alcoholic beverages pulque and tequila find their origin in the mashed leaves of *Agave* plants, and absinthe liqueur receives its unique flavor from the leaves of wormwood, a relative of western sagebrush, and other flavorings, such as anise.

Insecticides of various types are also derived from leaves. A type of rotenone and a substance related to nicotine are obtained from tropical plants; both are effective against a variety of insects. Mexico's cockroach plant has leaves that,

A water lily flower (Nymphaea sp.).

Flowers, Fruits, and Seeds

FIGURE 8.3   Male catkins of an alder. Each catkin consists of numerous tiny, inconspicuous, wind-pollinated flowers that have no petals.

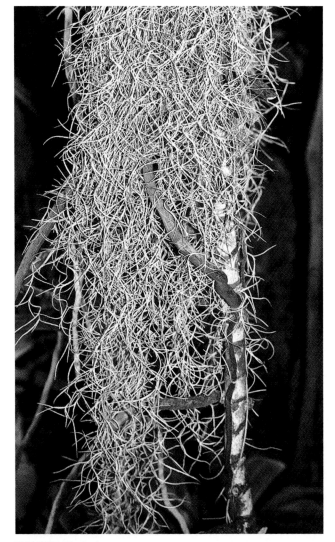

FIGURE 8.4   Spanish moss. Spanish moss is a nonparasitic flowering plant that goes through its life cycle suspended from other plants or objects such as wires. This plant should not be confused with *lichens* (see Chapter 18), which may also hang suspended from other objects.

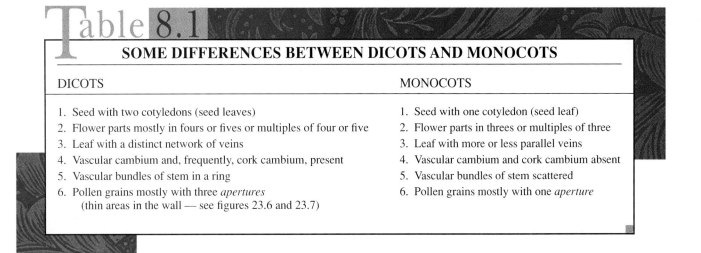

Table 8.1

SOME DIFFERENCES BETWEEN DICOTS AND MONOCOTS

DICOTS	MONOCOTS
1. Seed with two cotyledons (seed leaves)	1. Seed with one cotyledon (seed leaf)
2. Flower parts mostly in fours or fives or multiples of four or five	2. Flower parts in threes or multiples of three
3. Leaf with a distinct network of veins	3. Leaf with more or less parallel veins
4. Vascular cambium and, frequently, cork cambium, present	4. Vascular cambium and cork cambium absent
5. Vascular bundles of stem in a ring	5. Vascular bundles of stem scattered
6. Pollen grains mostly with three *apertures* (thin areas in the wall — see figures 23.6 and 23.7)	6. Pollen grains mostly with one *aperture*

DIFFERENCES BETWEEN DICOTS AND MONOCOTS

Slightly less than three-fourths of all flowering plant species are dicots. Dicots include many annual plants and virtually all flowering trees and shrubs. Monocots, which are primarily herbaceous, include plants that produce bulbs (e.g., lilies), grasses and related plants, orchids, irises, and palms; they are believed to have developed from ancestors derived from primitive dicots (Table 8.1).

STRUCTURE OF FLOWERS

Regardless of form, all flowers share certain basic features. A typical flower develops several different parts, each with its own function (Fig. 8.5). Each flower, which begins as an embryonic *primordium* that develops into a *bud* (discussed in Chapter 6), occurs as a specialized branch at the tip of a stalk called a **peduncle.** The peduncle swells at its tip into a small pad known as the **receptacle.** The other parts of the flower, some of which are in *whorls,* are attached to this receptacle (a whorl consists of three or more plant parts, such as leaves, encircling another plant part, such as a twig, at the same point on an axis).

The outermost whorl consists, as a rule, of three to five small, usually green, somewhat leaflike **sepals.** The sepals of a flower are collectively referred to as the **calyx.** In many flowers, the calyx functions in protection, particularly while the flower is in the bud. The next whorl of flower parts consists of petals, which are known collectively as the **corolla.** When corollas are showy, they attract pollinators, such as bees, moths, or birds. Some may have special markings that are invisible to humans but can be seen by bees whose vision functions in the ultraviolet range of the light spectrum. The corolla is often missing in wind-pollinated plants. In peach flowers, the petals are distinct separate units, but in other flowers, such as petunias or jimson weeds, the petals are fused together so that the corolla consists of a single, flared, trumpetlike sheet of tissue (Fig. 8.6). Sepals also may be fused together.

Several to many **stamens** are attached to the receptacle around the base of the often greenish *pistil* located in the center of the flower. Each stamen consists of a semirigid but otherwise slender **filament** with a sac called an **anther** at the top. The development of **pollen grains** in anthers is described in Chapter 23. In most flowers, the release of pollen is facilitated by lengthwise slits that develop on the anthers, but pores that function in the same way develop on the anthers of members of the Heath Family and those of a few other groups.

The **pistil,** which often is shaped like a tiny vase that is closed at the top, consists of three regions that merge with one another. At the top is a slight swelling called the **stigma,** which is connected by a slender stalklike **style** to the swollen base called the **ovary**; the ovary later develops into a *fruit.*

If the calyx and corolla are attached to the receptacle at the base of the ovary, the ovary is said to be **superior,** but in other flowers, the receptacle grows up around the ovary so that the calyx and corolla appear to be attached at the top; such flowers are said to have **inferior** ovaries (ovary positions are shown in Fig. 23.11). Within the ovary is a cavity containing an egg-shaped **ovule,** which is attached to the wall of the cavity by means of a short stalk. The ovule (the development of which is described in Chapter 23) eventually becomes a **seed.**

While peach flowers are produced singly on their own peduncles, those of lilacs, grapes, bridal wreaths, and many others are produced in clusters called **inflorescences** (Fig. 8.7). In an inflorescence, the peduncle has numerous little stalks serving individual flowers. These additional little stalks, one to each flower, are called **pedicels.**

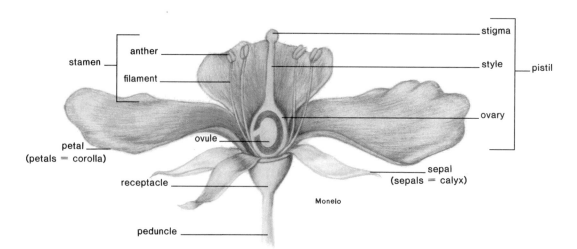

FIGURE 8.5   Parts of a typical flower. The interior structure of the ovule and the sexual processes involved are discussed in Chapter 23.

FIGURE 8.6   A jimson weed flower.

FRUITS

Introduction

In 1893, the U.S. Supreme Court, in the case of *Nix v. Hedden,* ruled that a tomato was legally a vegetable rather than a fruit. This was in keeping with the general conception in the public's mind that fruits tend to be relatively sweet and in the nature of a dessert food, while vegetables tend to be more savory as salad or main course foods. Be that as it may, botanically speaking, a **fruit** is any ovary that has developed and matured. It also usually contains seeds. By this definition, many so-called vegetables, including tomatoes, string beans, cucumbers, and squashes, are really fruits. Vegetables can consist of leaves (e.g., lettuce, cabbage), leaf petioles (e.g., celery), specialized leaves (e.g., onion), stems (e.g., white potato), roots (e.g., sweet potato), stems and roots (e.g., beets), flowers and their peduncles (e.g., broccoli), flower buds (e.g., globe artichoke), or other parts of the plant.

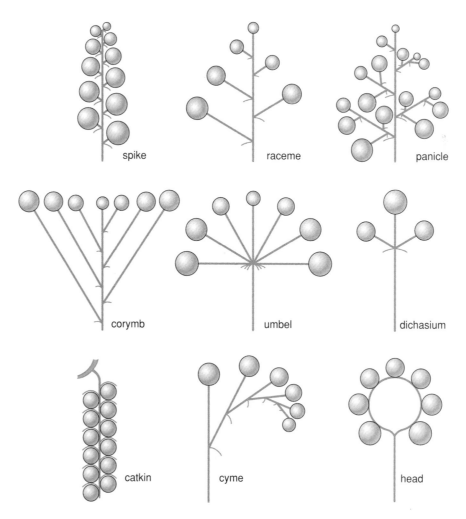

FIGURE 8.7   Inflorescence types. Each ball represents a flower.

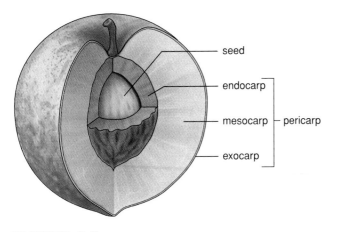

FIGURE 8.8   Regions of a mature fruit.

Fruit Regions

All fruits arise from flowers and thus are found exclusively in the flowering plants. *Fertilization* (see Chapter 23) usually indirectly determines whether the ovary or ovaries (and sometimes additional tissues such as the receptacle) of a flower will develop into a fruit. If at least a few of the ovules are not fertilized, the flower normally withers and drops without developing further. Pollen grains contain specific stimulants called *hormones* (discussed in detail in Chapter 11) that may initiate fruit development, and sometimes a quantity of dead pollen is all that is needed to stimulate an ovary into becoming a fruit. It is the hormones produced by the developing seeds, however, that promote the greatest fruit growth. These hormones, stimulate the production of more fruit growth hormones by the ovary wall.

Fruit Regions

By the time an ovary has matured into a fruit, the bulk of it is divided into three regions, which are sometimes difficult to distinguish from one another (Fig. 8.8). The skin forms the **exocarp,** while the inner boundary around the seed(s) forms the **endocarp.** The endocarp may be hard and stony (as in a peach pit around the seed). It also may be papery (as in apples), or it may not be distinct from the **mesocarp,** which is the name given to the often fleshy tissue between the exocarp and the endocarp. The three regions are collectively called the **pericarp.** In dry fruits, the pericarp is usually quite thin.

Some fruits consist of only the ovary and its seeds. Others have adjacent flower parts, such as the receptacle or calyx, fused to the ovary, or different parts may be modified in various ways. Fruits may be either fleshy or dry at maturity, and they may split, exposing the seeds, or no split may occur. They may be derived from a single ovary or from more than one. Traditionally, all these features have been used in the classification of fruits, but not all fruits lend themselves to neat pigeonholing by such characteristics. Some of these problems are pointed out in the classification that follows.

Kinds of Fruits

Fleshy Fruits

Fruits whose mesocarp is at least partly fleshy at maturity are classified as fleshy fruits.

Simple Fleshy Fruits   Simple fleshy fruits are fruits that develop from a flower with a single pistil. The ovary may be superior or inferior, and it may be *simple* (derived from a single modified leaf called a **carpel**) or it may consist of one or more carpels and thus be *compound* (for a discussion of the derivation of carpels and compound ovaries, see Chapter 23). The ovary alone may develop into the fruit, or other parts of the flower may develop with it.

Drupe. A **drupe** is a simple fleshy fruit with a single seed enclosed by a hard, stony endocarp, or pit (Fig. 8.9). It usually develops from flowers with a superior ovary containing a single ovule. The mesocarp is not always obviously fleshy, however. In coconuts, for example, the *husk* (consisting of the mesocarp and the exocarp), which is usually removed before the rest of the fruit is sold in markets, is very fibrous (the fibers are used in making mats and brushes). The seed ("meat") of the coconut is hollow and contains a watery *endosperm* (see Chapter 23) commonly but incorrectly referred to as "milk." It is surrounded by the thick, hard endocarp typical of drupes. Other examples of drupes are the stone fruits (e.g., apricots, cherries, peaches, plums, olives, and almonds). In almonds, the husk, which dries somewhat and splits at maturity, is removed before marketing, and it is the endocarp that we crack to obtain the seed.

Berry. **Berries** usually develop from a compound ovary and commonly contain more than one seed. The entire pericarp is fleshy, and it is difficult to distinguish between the mesocarp and the endocarp (Fig. 8.10). Three types of berries may be recognized.

A *true berry* is a fruit with a thin skin and a pericarp that is relatively soft at maturity. Although most contain more than one seed, notable exceptions are dates and avocados, which have only one seed. Typical examples of true berries are tomatoes, grapes, persimmons, peppers, and eggplants. Some fruits that popularly include the word *berry* in their common name (e.g., strawberry, raspberry, blackberry) botanically are not berries at all.

Some berries are derived from flowers with inferior ovaries so other parts of the flower also contribute to the flesh. They can usually be distinguished by the remnants of flower parts or their scars that persist at the tip. Examples of such berries are gooseberries, blueberries, cranberries, pomegranates, and bananas. In the cultivated banana, fruit development is parthenocarpic (that is, the fruits develop without fertilization—see Chapter 23), so there are no seeds. Several other species of banana produce an abundance of seeds.

A.

B.

C.

FIGURE 8.9 Representative drupes. *A.* Peach. *B.* Almond. *C.* Olive.

The *pepo* has a relatively thick rind. Fruits of members of the Pumpkin Family, including pumpkins, cucumbers, watermelons, squashes, and cantaloupes, are pepos.

The *hesperidium* is a berry with a leathery skin containing oils. Numerous outgrowths from the inner lining of the ovary wall become saclike and swollen with juice as the fruit develops. All members of the Citrus Family produce this type of fruit. Examples are oranges, lemons, limes, grapefruit, tangerines, and kumquats.

Pome. **Pomes** are simple fleshy fruits, the bulk of whose flesh comes from the enlarged receptacle that grows up around the ovary. The endocarp around the seeds is papery or leathery. Examples are apples, pears, and quinces. In an apple, the ovary consists of the core and a little adjacent tissue. The remainder of the fruit has developed primarily from the receptacle (Fig. 8.11). Botany texts often refer to pomes, pepos, some berries, and other fruits derived from more than an ovary alone as *accessory fruits* or as fruits having *accessory tissue.*

Aggregate Fruits An **aggregate fruit** is one that is derived from a single flower with several to many pistils. The individual pistils develop into tiny drupes or other fruitlets, but they mature as a clustered unit on a single receptacle (Fig. 8.12). Examples are raspberries, blackberries, and strawberries. In a strawberry, the cone-shaped receptacle becomes fleshy and red, while each pistil becomes a little whitish *achene* (achenes are described in the section on dry fruits) on the strawberry surface. In other words, the strawberry, while being an aggregate fruit, is also partly composed of accessory tissue.

A.

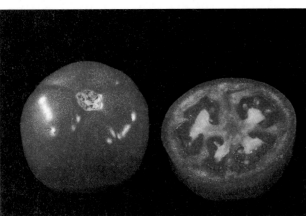

B.

FIGURE 8.10   Representative berries. *A.* Grapes. *B.* Tomatoes.

Multiple Fruits   **Multiple fruits** are derived from several to many individual flowers in a single inflorescence. Each flower has its own receptacle, but as the flowers mature separately into fruitlets, they develop together into a single larger fruit, as in aggregate fruits. Examples of multiple fruits are mulberries, osage oranges (Fig. 8.13), pineapples, and figs. Pineapples, like bananas, usually develop parthenocarpically (see Chapter 23), and there are no seeds. The individual flowers are fused together on a fleshy axis, and the fruitlets coalesce into a single fruit.

Figs mature from a unique "outside in" inflorescence. The individual flowers of the inflorescence are enclosed by the common receptacle, which has an opening to the outside at the tip (Fig. 8.14). Some varieties develop parthenocarpically, but others are pollinated by tiny wasps that crawl in and out through the opening. Some multiple fruits, such as those of the sweet gum, are dry at maturity.

Dry Fruits

Fruits whose mesocarp is definitely dry at maturity are classified as *dry fruits*.

Dry Fruits That Split at Maturity   The fruits in this group are distinguished from one another by the manner in which they split.

Follicle. The **follicle** splits along one side or seam only, exposing the seeds within (Fig. 8.15). Examples are larkspur, columbine, milkweed, and peony.

Legume. The **legume** splits along two sides or seams (Fig. 8.16). Literally thousands of members of the Legume Family produce this type of fruit. Examples are peas, beans, garbanzo beans, lentils, carob, kudzu, and mesquite. Peanuts are legumes, but they are atypical in that the fruits develop and mature underground. The seeds are usually released in nature by bacterial breakdown of the pericarp instead of through an active splitting action.

Silique. **Siliques** also split along two sides or seams, but the seeds are borne on a central partition, which is exposed when the two halves of the fruit separate (Fig. 8.17 A). Such fruits, when they are less than three times as long as they are wide, are called *silicles* (Fig. 8.17 B). Siliques and silicles are typically produced by members of the Mustard Family, which includes broccoli, cabbage, radish, shepherd's purse, and watercress.

Capsule. **Capsules** are the most common of the dry fruits that split (Fig. 8.18). They consist of at least two carpels and split in a variety of ways. Some split along the partitions between the carpels, while others split through the cavities (*locules*) in the carpels. Still others form a cap toward one end that pops off and permits release of the seeds, or they form a row of pores through which the seeds are shaken out as the capsule rattles in the wind. Examples are irises, orchids, lilies, poppies, violets, and snapdragons.

endocarp

seed

vascular bundle
in outer part of
the ovary

FIGURE 8.11 Apples (representative of pomes). The bulk of the flesh is derived from the receptacle and the bases of the calyx, corolla, and stamen.

A.

B.

FIGURE 8.12 Blackberry. *A*. A flower. Note the numerous green pistils. *B*. Aggregate fruits.
(*B*. courtesy Robert A. Schlising)

FIGURE 8.13   An osage orange.

A.

individual
flowers

FIGURE 8.14   Section through a developing fig.

B.

FIGURE 8.15   Follicles. *A.* Milkweed. *B.* Magnolia. The fruit of the magnolia is actually an aggregate fruit consisting of approximately 40 to 80 individual one-seeded follicles on a common axis. The follicles and axis fall from the tree as a unit.

Dry Fruits That Do Not Split at Maturity   In this type of dry fruit, the single seed is, to varying degrees, united with the pericarp.

Achene. The single seed of the **achene** is attached to its surrounding pericarp only at its base. The husk (pericarp) is relatively easily separated from the seed. Examples are sunflower "seeds" (the edible kernel plus the husk constitute the achene) (Fig. 8.19), buttercup, and buckwheat.

FIGURE 8.16 Legumes of a coral tree.

A.

B.

FIGURE 8.17 *A.* A silique after it has split open. The seeds are borne on a central, membranous partition. *B.* Silicles of *Lunaria* (dollar plant).

Nut. **Nuts** are one-seeded fruits similar to achenes, but they are generally larger, and their pericarp is much harder and thicker. They develop with a cup, or cluster, of bracts at their base. Examples are acorns (Fig. 8.19), hazelnuts, hickory nuts, and chestnuts. Many nuts in the popular sense are not nuts, botanically speaking. We have already seen that peanuts are atypical legumes, and that coconuts and almonds are drupes. Walnuts and pecans are also drupes, whose "flesh" withers and dries after the seed matures. Brazil nuts are the seeds of a large capsule, and a cashew nut is the single seed of a peculiar drupe. It appears as a curved appendage at the end of a swollen pedicel, which is eaten raw in the tropics or made into preserves or wine. Pistachio nuts are also the seeds of drupes.

Grain (Caryopsis). The pericarp of the **grain** is tightly united with the seed and cannot be separated from it (Fig. 8.19). All members of the Grass Family, including corn, wheat, rice, oats, and barley, produce grains (also called **caryopses**).

A.

B.

C.

D.

FIGURE 8.18   Capsules. *A.* Butterfly iris. *B. Bletilla* orchid. *C.* Autograph tree. *D.* Unicorn plant.

Samara. In **samaras,** the pericarp surrounding the seed extends out in the form of a wing or membrane, which aids in dispersal (Fig. 8.20). Samaras are produced in pairs in maples. In ashes, elms, and the tree of heaven, they are produced singly.

Schizocarp. The twin fruit called a *schizocarp* (Fig. 8.21) is unique to the Parsley Family. Members of this family include parsley, carrots, anise, caraway, and dill. Upon drying, the twin fruits break into two one-seeded segments.

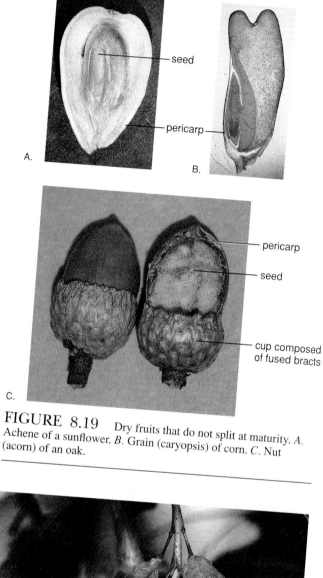

seed

pericarp

A.

B.

pericarp

seed

cup composed
of fused bracts

C.

FIGURE 8.19 Dry fruits that do not split at maturity. *A.* Achene of a sunflower. *B.* Grain (caryopsis) of corn. *C.* Nut (acorn) of an oak.

FIGURE 8.20 Samaras of a big-leaf maple.

FIGURE 8.21 Schizocarps of carrots. A schizocarp separates at maturity into two one-seeded fruitlets.

FRUIT AND SEED DISPERSAL

Why are so many species of orchids rare, while dandelions, shepherd's purse, and other weeds occur all over the world? Why are some plants confined to single continents, mountain ranges, or small niches occupying less than a hectare (2.47 acres) of land? The answers to these questions involve many different factors, including climate, soil, the adaptability of the plant, and its means of seed dispersal. How fruits and seeds are transported from one place to another is the subject of the following sections. Other factors are discussed in chapters 13 and 25.

Dispersal by Wind

Fruits and seeds have a variety of adaptations for wind dispersal (Fig. 8.22). The samara of a maple has a curved wing that causes the fruit to spin as it is released from the tree. In a brisk wind, samaras may be carried up to 10 kilometers (6 miles) away from their source. In hop hornbeams, the seed is enclosed in an inflated sac that gives it some buoyancy in the wind. In some members of the Buttercup and Sunflower Families, the fruits have plumes, and in the Willow Family, the fruits are surrounded by cottony or woolly hairs that aid in wind dispersal. In button snakeroots and Jerusalem sage, the fruits are too large to be airborne, but they are spherical enough to be rolled along the ground by the wind.

Seeds themselves may be so tiny and light that they can be blown great distances by the wind. Orchids and heaths, for example, produce seeds that are as fine as dust and equally light in weight. In trees such as catalpas and jacarandas, the seeds themselves are winged rather than the fruits, which remain on the branches and split, releasing their contents. Dandelion fruitlets have plumes that radiate out at the ends like tiny parachutes; these catch even a slight breeze. In tumble mustard and other tumbleweeds, the whole aboveground portion of the plant may break off and be blown away by the wind, releasing seeds as it bumps along.

Dispersal by Animals

The adaptations of fruits and seeds for animal dispersal are legion. Birds, mammals, and ants all act as disseminating

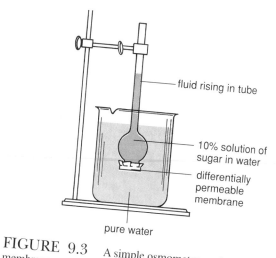

FIGURE 9.3 A simple osmometer made by tying a membrane over the mouth of a thistle tube.

fluid rising in tube

10% solution of sugar in water

differentially permeable membrane

pure water

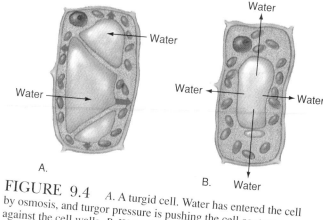

FIGURE 9.4 A. A turgid cell. Water has entered the cell by osmosis, and turgor pressure is pushing the cell contents against the cell walls. B. Water has left the cell and turgor pressure has dropped, leaving the cell limp.

Water

Water

A.

B.

Water

Water

Water

Water

potential is the pressure required to prevent osmosis from taking place.

Water enters a cell by osmosis until the osmotic potential is balanced by the resistance of the cell wall to expansion. Water gained by osmosis may keep a cell firm, or **turgid,** and the **turgor pressure** that develops against the walls as a result of water entering the vacuole of the cell is called *pressure potential.*

The release of turgor pressure can be heard each time you bite into a crisp celery stick or the leaf of a young head of lettuce. When we soak carrot sticks, celery, or lettuce in pure water to make them crisp, we are merely assisting the plant in bringing about an increase in the *turgor* of the cells (Fig. 9.4).

The *water potential* of a plant cell is essentially its osmotic potential and pressure potential combined. If we have two adjacent cells of different water potentials, water will move from the cell having the higher water potential to the cell having the lower water potential.

Osmosis is the primary means by which water enters plants from their surrounding environment. In land plants, water from the soil enters the cell walls and intercellular spaces of the epidermis and the root hairs and travels along the walls until it reaches the endodermis. Here it crosses the differentially permeable membranes and protoplasts of the endodermal cells on its way to the xylem, where it flows to the leaves, evaporates within the leaf air spaces, and diffuses out (*transpires*) through the stomata into the atmosphere. The movement of water occurs because there is a water potential gradient between relatively high soil water potential to successively lower water potentials in roots, stems, leaves, and the atmosphere.

Plasmolysis

If you place turgid carrot and celery sticks in a 10% solution of salt in water, they soon lose their rigidity and become limp enough to curl around your finger. The water potential inside the carrot cells is greater than the water potential out-

side, so diffusion of water out of the cells into the salt solution takes place. If you were to examine such cells with the aid of a microscope, you would see that the vacuoles, which are largely water, had disappeared and that the protoplasm had shrunk away from the walls and was clumped in the middle of the cell. Such cells are said to be *plasmolyzed.* This loss of water through osmosis, which is accompanied by the shrinkage of protoplasm away from the cell wall, is called **plasmolysis** (Fig. 9.5). If plasmolyzed cells are placed in fresh water before permanent damage is done, water reenters the cell by osmosis, and the cells become turgid once more.

Imbibition

Osmosis is not the only force involved in the absorption of water by plants. *Colloidal* materials (i.e., materials that contain a permanent suspension of fine particles) and large molecules, such as cellulose and starch, usually develop electrical charges when they are wet, and they attract water molecules, which adhere to the internal surfaces of the materials. Because water molecules are *polar,* they can become both highly adhesive to large organic molecules such as cellulose and cohesive with one another. (Polar molecules have slightly different electrical charges at each end due to their asymmetry—see the discussion in Chapter 2.) This process, known as **imbibition,** results in the swelling of tissues, whether they are alive or dead, often to several times their original volume. Imbibition is the initial step in the germination of seeds (Fig. 9.6).

The physical forces developed during germination can be tremendous, even to the point of causing a seed to split a rock weighing several tons (Fig. 9.7). It has been found, for example, that a pressure of 42.2 kilograms per square centimeter (600 pounds per square inch) is needed to break the seed coat of a fresh walnut from within, and that water being imbibed by a cocklebur seed develops a force of up to 1,000 times that of normal atmospheric pressure. Yet when water

A.

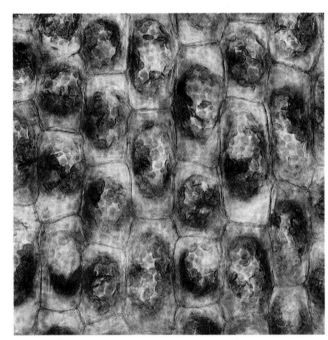

B.

FIGURE 9.5 A portion of a leaf of the water weed *Elodea*. *A*. Normal cells. *B*. Plasmolyzed cells.

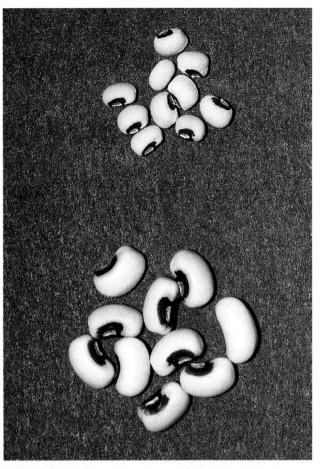

FIGURE 9.6 Black-eyed pea seeds before and after imbibition of water.

and oxygen reach walnut and cocklebur seeds, they germinate readily, as do seeds that fall into the crevices of rocks or have boulders roll over on them.

The huge stone blocks used in the construction of the pyramids of Egypt are believed to have been quarried by hammering rounded wooden stakes into holes made in the face of the stone and then soaking the stakes with water. As the stakes swelled, the force created by imbibition was sufficient to split the rock.

Active Transport

Return for a moment to our two rooms with the tennis balls. Suppose that, in addition to the 100 tennis balls, we drop in 50 slightly underinflated basketballs, also extraordinary in having perpetual motion motors that propel them in any direction at 12 MPH. They should also become randomly distributed throughout the room shortly after they are introduced. Assume, however, that the hole in the wall (which is large enough for the passage of a tennis ball) is not quite large enough to allow a basketball to pass through freely. The basketballs will then remain in the first room. But if we were to install a mechanical arm next to the hole in the second room, and if this arm could grab basketballs that came near the hole and squeeze them through in one direction, basketballs would be transported into the second room *through the expenditure of energy*. The basketballs obviously would gradually accumulate in the second room in greater numbers.

Plants expend energy, too. Plant cells generally contain a larger number of mineral molecules and ions than exist in the soil immediately adjacent to the root hairs. If it were not for the barriers imposed by the differentially permeable membranes, therefore, these molecules and ions would

FIGURE 9.7   A live oak that grew from an acorn lodged in a small crack in the rock. When it rained, the acorn imbibed water and the force of the swelling split the rock. A root is now slowly widening the split.

FIGURE 9.8   A mangrove tree. Mangroves flourish in tropical tidal zones where the salt content of the water is high enough to plasmolyze the cells of most plants. The mangroves still obtain water via osmosis, which takes place because the mangrove cells accumulate an unusually high concentration of organic solutes; some are also able to excrete excess salt.

move from a region of higher concentration in the cells to a region of lower concentration in the soil.

Most molecules needed by cells are polar, and those of solutes may set up an electrical gradient across a differentially permeable membrane of a living cell and require special transport proteins embedded in the membrane (see Fig. 3.7) to pass through. The transport proteins are believed to occur in two forms, one facilitating the transport of specific ions to the outside of the cell and the other facilitating the transport of specific ions into the cell.

The plants absorb and retain these substances against a diffusion (or electrical) gradient *through the expenditure of energy*. This process is called **active transport.** The precise

mechanism of active transport is not fully understood. It apparently involves an enzyme and what has been referred to by some scientists as a proton "pump"—involving the plasma membrane of plant and fungal cells and sodium and potassium ions in animal cells. Both "pumps" are energized by special energy-storing ATP molecules (discussed in Chapter 10).

Some plants such as mangrove and saltbush and also certain algae thrive in areas where the water or soil contains enough salt to kill most vegetation. Such plants accumulate large amounts of organic solutes, including the carbohydrate *mannitol* and the amino acid *proline*. The organic solutes enable the plants to take up water via osmosis (Fig. 9.8), despite the otherwise adverse environment. Some mangroves also have salt glands in their leaves through which they excrete excess salt.

WATER AND ITS MOVEMENT THROUGH THE PLANT

If you were to cover the soil at the base of a plant with foil, place the pot where it would receive light, and then put the potted plant under a glass bell jar, you would notice moisture accumulating on the inside of the jar within an hour or two. Because of the foil barrier, the water could not have come directly from the soil; it had to have come through the plant. More than 90% of the water entering a plant passes through and evaporates—primarily into leaf air spaces and then through the stomata into the atmosphere (see Fig. 9.10), with usually less than 5% of the water escaping through the cuticle. This process of water vapor loss from the internal leaf atmosphere is called *transpiration* (Fig. 9.9).

The amount of water transpired by plants is greater than one might suspect. For example, mature corn plants

FIGURE 9.9 A potted plant sealed under a bell jar. The surface of the soil has been covered with foil. Note the accumulation of moisture on the inside of the glass. The moisture could only have come through the plant by the process of transpiration.

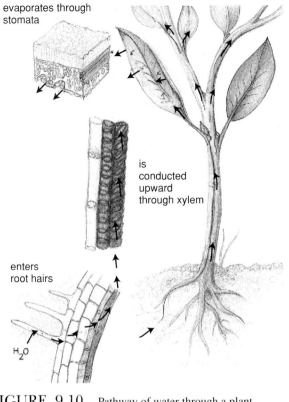

FIGURE 9.10 Pathway of water through a plant.

evaporates through stomata

is conducted upward through xylem

enters root hairs

H_2O

each transpire about 15 liters (4 gallons) of water per week, while four-tenths of a hectare (1 acre) of corn may transpire more than 1,325,000 liters (350,000 gallons) in a 100-day growing season. A hardwood tree utilizes about 450 liters (120 gallons) of water while producing 0.45 kilogram (1 pound) of wood (or 1,800 liters while producing 0.45 kilogram of dry weight substance), and the 200,000 leaves of an average-sized birch tree will transpire from 750 to more than 3,785 liters (200 to 1,000 gallons) per day during the growing season. Humans recycle much of their water via the circulatory system, but if they were to have requirements similar to those of plants, each adult would have to drink well over 38 liters (10 gallons) per day.

Why is so much water involved in the normal processes of living plants? Water constitutes about 90% of the weight of young cells. The thousands of enzyme actions and other chemical activities of cells take place in water, and additional although relatively negligible amounts are used in the process of photosynthesis. The exposed surfaces of the chlorenchyma cells within the leaf have to be moist at all

times, for it is through this film of water that the carbon dioxide molecules needed for the process of photosynthesis enter the cell from the air. Water is also needed for cell turgor, which gives rigidity to herbaceous plants.

Consider also what it must be like in the mesophyll of a flattened leaf that is fully exposed to the midsummer sun in areas where the air temperature soars to well over 38°C (100°F) in the shade. If it were not for the evaporation of water molecules from the moist surfaces, which brings about some cooling, and reradiation of energy by the leaf, the intense heat could damage the plant. Sometimes, the transpiration is so rapid that the loss of water begins to exceed the intake, and the stomata may close, thus preventing wilting (see also in Chapter 11 the discussion of the role of *abscisic acid* relative to excessive water loss).

How does water travel through the roots from 3 to 6 meters (10 to 20 feet) or more beneath the surface and then up the trunk to the topmost leaves of a tree that is more than 90 meters (300 feet) tall? We know that continuous tubular pathways of xylem run throughout a plant, extending from the young roots up through the stem and branches to the tiny veinlets of the leaves; we also know that the water gets to the start of this "plumbing system" by osmosis following a water potential gradient. But water is raised through the columns apparently by a combination of factors, and the process has been the subject of much debate for the past 200 years (Fig. 9.10).

One of the earliest explanations for the rise of water in a living plant was given in 1682 by the English scientist issue. (See the section on the greenhouse effect in Chapter 25 for a discussion of the problem.)

A small increase in carbon dioxide levels could benefit plants, providing the level does not climb beyond 0.2% (which is extremely unlikely). Increases in yields of between 100% and 200% have been obtained by fertilizing plants with carbon dioxide, and some large commercial greenhouses have run pipes over plant beds to supplement their natural supply. The changes in climate that can result from an increase in atmospheric carbon dioxide, however, could have seriously adverse effects that would more than offset positive benefits to plants.

Water

Less than 1% of all the water absorbed by plants is used in photosynthesis; most of the remainder is transpired or incorporated into protoplasm, vacuoles, and other materials. The water used is the source of the oxygen released as a byproduct of photosynthesis, even though carbon dioxide also contains oxygen. This has been demonstrated by using isotopes of oxygen to make both the carbon dioxide and the water used in photosynthesis. When an oxygen isotope is used only in the water, it appears in the oxygen gas released. If, however, it is used only in the carbon dioxide, it is confined to the sugar and water produced and never appears in the oxygen gas, demonstrating clearly that the water is the sole source of the oxygen released.

If water is in short supply, it may indirectly become a limiting factor in photosynthesis; under such circumstances, the stomata usually close and sharply reduce the carbon dioxide supply.

Light

Light exhibits properties of both waves and particles. Energy reaches the earth from the sun in waves of different lengths, the longest waves being radio waves and the shortest being gamma rays. About 40% of the radiant energy we receive is in the form of visible light. If this visible light is passed through a glass prism, it splits up into its component colors. Reds are on the longer wavelength end and violets are on the shorter wave end, with yellows, greens, and blues between (Fig. 10.2). Although nearly all of the visible light colors can be used in photosynthesis, those in the violet to blue and red-orange to red wavelengths are used most extensively. Many in the green range are reflected. Leaves commonly absorb about 80% of the visible light that reaches them.

The intensity of light varies with the time of day, season of the year, altitude, latitude, and atmospheric conditions. On a clear summer day at noon in a temperate zone,

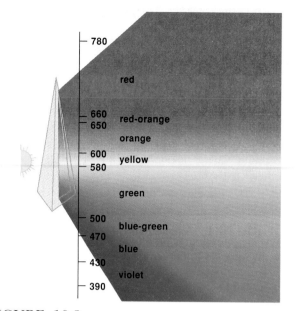

FIGURE 10.2 Visible light that is passed through a prism is broken up into individual colors with wavelengths ranging from 390 nanometers (violet) to 780 nanometers (red).

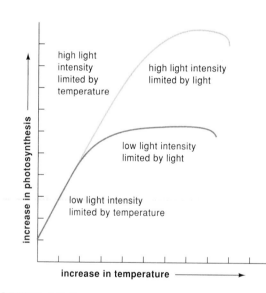

FIGURE 10.3 How temperature and light interact to affect photosynthesis. When temperature is a limiting factor, the rate of photosynthesis is independent of light intensity. When light is a limiting factor, the rate of photosynthesis is independent of temperature.

2. A foot-candle is the light cast by a standard candle at a distance of 1 foot. Foot-candles have been used as units of measurement of light intensity for many years. Note, however, that such units are subject to variables in light perception by human eyes and other variables in candles. They also present problems in conversion to metric measurements. Accordingly, modern scientists have turned to other measurements of light (e.g., radiant energy per square meter). Because these other forms of measurement are not essential to our introductory discussion, the term *foot-candle* is retained here.

sunlight attains an intensity in the vicinity of 10,000 foot-candles.[2] In contrast, consider that a good reading lamp produces only about 25 foot-candles.

Plants vary considerably in the light intensities they need for photosynthesis to occur at optimal (ideal) rates, and other factors, such as the temperature and amount of carbon dioxide available, can be limiting (Fig. 10.3). For example,

CO_2, and hydrogen are removed from the acetyl CoA in the Krebs cycle, which involves enzyme-catalyzed reactions of a series of organic acids.

18. NADH passes the hydrogen gained during glycolysis and the Krebs cycle along an electron transport chain; small increments of energy are released and partially stored in ATP molecules, and the hydrogen is combined with oxygen gas, forming water in the final step of aerobic respiration.

19. Hydrogen removed from glucose during glycolysis is combined with an inorganic ion in anaerobic respiration. The hydrogen is combined with the pyruvic acid or one of its derivatives in fermentation. Both processes occur in the absence of oxygen gas, with only about 7% of the total energy in the glucose molecule being released, for a net gain of two ATP molecules.

20. Two molecules of NADH and two ATP molecules are gained during glycolysis when two 3-carbon pyruvic acid molecules are produced from a single glucose molecule. Another molecule of $NADH_2$ is produced when the pyruvic acid molecule is restructured and becomes acetyl CoA prior to entry into the Krebs cycle.

21. In the Krebs cycle, acetyl CoA combines with 4-carbon oxaloacetic acid, producing first a 6-carbon compound, next a 5-carbon compound, and then several 4-carbon compounds. The last 4-carbon compound is oxaloacetic acid. Two CO_2 molecules are also released during this process.

22. Some hydrogen removed during the Krebs cycle is picked up by FAD and NAD^+; one molecule of ATP, three molecules of NADH, and one molecule of $FADH_2$ are produced during one complete cycle. Energy associated with electrons and/or with hydrogen picked up by NAD^+ and FAD is gradually released as the electrons are passed along the electron transport chain; some of this energy is transferred to ATP molecules during oxidative phosphorylation.

23. Energy used in ATP synthesis during oxidative phosphorylation is believed to be derived from a gradient of protons formed across the inner membrane of a mitochondrion while electrons are moving in the electron transport chain by chemiosmosis.

24. Altogether, 38 ATP molecules are produced during the complete aerobic respiration of one glucose molecule; two are used to prime the process so that there is a net gain of 36 ATP molecules.

25. The conversion of the sugar produced by photosynthesis to fats, proteins, complex carbohydrates, and other substances is termed *assimilation*. Digestion takes place within plant cells with the aid of enzymes. During digestion, large insoluble molecules are broken down by hydrolysis to smaller soluble forms that can be transported to other parts of the plant.

Review Questions

1. What happens in the light reactions of photosynthesis?
2. What roles do water, light, carbon dioxide, and chlorophyll play in photosynthesis?
3. What is glycolysis?
4. What are the differences among aerobic respiration, anaerobic respiration, and fermentation?
5. How do temperature, water, and oxygen affect respiration?
6. Explain digestion and assimilation.

Discussion Questions

1. Since plants apparently benefit from small increases in the carbon dioxide available to them, would you be doing your favorite houseplant a favor by breathing repeatedly on it?
2. Why were the carbon-fixing reactions of photosynthesis formerly called the *dark reactions*?

Additional Reading

Danks, S. M., et al. 1985. *Photosynthetic systems: Structure, function, and assembly.* New York: Wiley-InterScience.

Douce, R., and D. A. Day (Eds.). 1985. *Higher plant cell respiration.* New York: Springer-Verlag.

Fong, F. K. (Ed.). 1982. *Light reaction path of photosynthesis.* New York: Springer-Verlag.

Galston, A. W. 1993. *Life processes of plants: Mechanisms for survival.* New York: W. H. Freeman.

Goodwin, T. W. 1988. *Plant pigments.* San Diego, CA: Academic Press.

Gregory, R. P. 1989. *Photosynthesis.* New York: Routledge, Chapman and Hall, Inc.

Gutteridge, S. 1990. Limitations of the primary events of CO_2 fixation in photosynthetic organisms: The structure and mechanism of rubisco. *Biochimica et Biophysica Acta* 1015: 1–14.

Jones, H. G. 1992. *Plants and microclimate: A quantitative approach to plant physiology.* San Diego, CA: Academic Press.

Lehninger, A. L., D. L. Nelson, and M. M. Cox. 1992. *Principles of biochemistry,* 2d ed. New York: Worth.

Salisbury, F. B., and C. W. Ross. 1992. *Plant physiology,* 4th ed. Belmont, CA: Wadsworth Publishing Co.

Schulze, E. D., and M. M. Caldwell (Eds.). 1994. *Ecophysiology of photosynthesis.*

Ting, I. P. 1985. Crassulacean acid metabolism. *Annual Review of Plant Physiology* 36: 595–622.

Walker, D. 1979. *Energy, plants and man: An introduction to photosynthesis in C_3, C_4 and CAM plants.* Philadelphia: International Ideas, Inc.

A single fiddleneck of a tropical tree fern (Sadleria cyatheoides). As the fern frond (leaf) matures, the fiddleneck uncoils and expands into a beautiful, feathery structure.

11 Growth

From time to time, I grow a few vegetables and berries in my backyard. When Italian squash plants are producing fruit, I have occasionally seen a young squash just beginning to develop, and I have made a mental note to harvest it within a day or two. Then, other matters have distracted me and I have forgotten to follow through, only to discover a few days later that my squash has grown into an enormous "monster" the size of a watermelon. I then have wondered, "How did it grow that big that fast?"

The words **grow, growing,** and **growth** are used in several ways. If, for example, you see a rubber balloon being inflated with gas, you may refer to its gradual increase in diameter as its growth in size. Or if there is a leak in the roof, you might say that the puddle in your living room is growing bigger. In the biological realm, however, growth is always associated with cells. It may be defined simply as an "increase in mass due to the division and enlargement of cells" and may be applied to an organism as a whole or to any of its parts.

Many plants, such as radishes and pumpkins, go through a sequence of growth stages. At first, they grow rapidly, then for a while they show little if any increase in volume, and eventually they stop growing completely. Finally, tissues break down, and the plant dies. Such growth is said to be *determinate*. Parts of other plants, such as the stems of redwood trees, may exhibit *indeterminate* growth and continue to be active for thousands of years.

All living organisms begin as a single cell, which usually divides and keeps dividing until something consisting of possibly billions of cells is formed. As the various cells mature, they usually become larger and then *differentiate*—that is, they develop different forms adapted to specific functions, such as conduction, support, or secretion of special substances. If **differentiation** did not occur, the result would be a shapeless blob with no distinct tissues and little, if any, coordination.

The term **development** is applied to the process of growth and differentiation of cells into tissues, organs, and organisms. What brings about growth and differentiation, and how do they occur? We already know they are controlled by the genes of the chromosomes, which dictate not only the kinds of cells that will develop but also their proportionate numbers and the ultimate size of the organism.

The sizes of the cells themselves are limited by several factors. They are dependent on molecular diffusion within for the transport of oxygen and food and for the removal of wastes, since they have no circulatory organs of their own. As a spherical cell enlarges, the total area of its surface, through which substances enter and leave, does not increase at the same rate as its volume. When a cell's surface area has increased fourfold, for example, its volume has increased roughly six times, obviously putting a great strain on its capacity to admit and exchange materials through its surface.

Most plant cells have become adapted to this problem through various changes in shape that increase the surface area (e.g., development of a convoluted outer surface; flattening of the cell) and also through the development of large vacuoles, with the cytoplasm becoming a thin layer adjacent to the cell wall. These adaptations alone, however, do not increase the strength of the bounding membranes, and most cells divide before their enlargement becomes a significant problem.

PLANT GROWTH REGULATORS AND ENZYMES

The basic units of heredity known as *genes* in the nucleus of each living cell program the synthesis and development of *enzymes,* which control every metabolic step within cells. As previously noted, enzymes are organic catalysts that speed up chemical reactions, usually in the cells where they are produced. They are nearly all proteins and are produced in minute quantities. They generally are not transported from one part of an organism to another.

Genes also dictate the production of **hormones,** which influence many growth phenomena. Hormones, which in plants are produced mostly in growing shoot tips, are chemical substances that differ from enzymes in structure. Like enzymes, they occur in minute quantities, but, unlike enzymes, they ordinarily are transported from their point of origin to another part of the plant where they have specific

Cytokinins

By the close of the 19th century, botanists suspected that there must be something that regulates cell division in plants. Their conjecture led to the discovery in 1913, by Gottlieb Haberlandt of Austria, of an unidentified chemical in the phloem of various plants. This chemical stimulated cell division and initiated the production of cork cambium. In the 1950s, several substances that promote cell division were found in coconut milk (a liquid *endosperm;* endosperm is discussed in Chapter 23), but it was not until 1964 that the identity of such substances was determined in kernels of corn. These various stimulants to cell division came to be known as **cytokinins.**

The several cytokinins now known differ somewhat in their molecular structure and possibly also in origin, but they are similar in composition to adenine. You will recall that adenine is a building block of one of the four nucleotides found in DNA, although none of the cytokinins appears to be derived from DNA. Some cytokinins do, however, occur in certain forms of RNA. Cytokinins are found most abundantly in meristems and other developing tissues, particularly those of young fruit.

If auxin is present during the cell cycle, cytokinins promote cell division by speeding up the progression from the G_2 phase to the mitosis phase, but no such effect takes place in the absence of auxin. Cytokinins also play a role in the enlarging of cells, the differentiation of tissues, the development of chloroplasts, the stimulation of cotyledon growth, and the delay of aging in leaves and in many of the growth phenomena also brought about by auxins and gibberellins. Despite their role in cell division and enlargement, however, there is a total absence of evidence that cytokinins initiate or promote animal cancers or have any other effect on animal cells.

Experiments have shown that cytokinins prolong the life of vegetables in storage. Related synthetic compounds have been used extensively to regulate the height of ornamental shrubs and to keep harvested lettuce and mushrooms fresh. They have also been used to shorten the straw length in wheat, so as to minimize the chances of the plants blowing over in the wind, and to lengthen the life of cut flowers. Many have not yet been approved for general agricultural use.

Abscisic Acid

Although the promotion of growth by auxins and gibberellins had been amply demonstrated by the 1940s, plant physiologists began to suspect that something else produced by plants could have opposite (inhibitory) effects. In 1949, Torsten Hemberg, a Swedish botanist, showed that substances produced in dormant buds blocked the effects of auxins. He called these growth inhibitors *dormins.*

In 1963, three groups of investigators working independently in the United States, Great Britain, and New Zealand discovered a growth-inhibiting hormone, which was in 1967 officially called **abscisic acid (ABA).** Later, it was shown that ABA and dormins were one and the same.

ABA is synthesized in plastids, apparently from carotenoid pigments. It is found in many plant materials but is particularly common in fleshy fruits, where it evidently prevents seeds from germinating while they are still on the plant. When ABA is applied to seeds outside of the fruit, germination usually is delayed. Because the stimulatory effects of other hormones are inhibited by ABA, cell growth is usually also inhibited.

ABA was originally believed to promote the formation of abscission layers in leaves and fruits. However, the evidence suggesting that ethylene (discussed next) is far more important than ABA in abscission now is overwhelming, and, despite its name, ABA has little, if any, influence on the process.

When ABA is applied to active plant buds, the leaf primordia become bud scales and the bud goes into a dormant winter condition. Since nursery plants being shipped stand to suffer less damage in a dormant condition than in an active one, there is practical value to the application of ABA and similar growth inhibitors to such nursery stock. The inhibiting effects of ABA can be reversed by the application of gibberellins.

ABA apparently helps leaves respond to excessive water loss. When the leaves wilt, ABA is produced in amounts several times greater than usual. This interferes with the transport or retention of potassium ions in the guard cells, causing the stomata to close. When the uptake of water again becomes sufficient for the leaf's needs, the ABA breaks down and the stomata reopen. Despite the growth-inhibiting effects of ABA, there is no evidence that it is in any way toxic to plants.

Ethylene

In 1934, R. Gane discovered that the simple gas **ethylene** was produced naturally by fruits; it had been known for some time before that the ripening of green fruits could be accelerated artificially by placing them in ethylene. It is now known that ethylene is produced not only by fruits but also by flowers, seeds, leaves, and even roots. Several fungi and a few bacteria are also known to produce it, and its regulatory effects make it a hormone in the broad sense of the word. Ethylene is produced from the amino acid methionine; oxygen is required for its formation.

The production of ethylene by plant tissues varies considerably under different conditions. A surge of ethylene lasting for several hours becomes evident after various tissues, including those of fruits, are bruised or cut, and applications of auxin can cause an increase in ethylene production of two to 10 times. As pea seeds germinate, the seedlings produce a surge of ethylene when they meet interference with their growth through the soil. This apparently causes the stem tip to form a tighter crook, which may aid the seedling in pushing to the surface.

When a storm stirs up a green field or animals pass across it, plants respond to these and other mechanical stresses with an increase in the production of fibers and

FIGURE 11.5   Effect of ethylene on holly twigs. Two similar twigs were placed under glass jars for a week; at the same time, a ripe apple was placed under the jar on the right. Ethylene produced by the apple caused abscission of the leaves.

collenchyma tissue. Cell elongation is also inhibited, resulting in shorter, sturdier plants. The responses, called *thigmomorphogenesis,* are under the control of genes that are activated by touch. Several substances, including enzymes, ethylene, and a protein called *calmodulin,* are involved. Calmodulin, which requires calcium for its actions, plays a role in several kinds of plant reactions to stimuli.

Ethylene apparently can trigger its own production. If minute amounts are introduced to the tissues that produce the gas, a tremendous response by the tissues often results. These tissues may then produce so much ethylene that the part concerned can be adversely affected. Flowers, for example may fade in the presence of excessive amounts, and leaves may abscise (Fig. 11.5).

In ancient China, growers used to ripen fruits in rooms where incense was being burned, and citrus growers used to ripen their fruits in rooms equipped with a kerosene stove. In houses where gas heating is used, occupants often experience great difficulty in growing houseplants, and greenhouse heaters using such fuel create a dilemma for owners attempting to promote plant growth.

In the days before electric street lights became commonplace, gas lights were used, and in some cities in Germany, leaves fell from the shade trees if they were located near a gas line that leaked. In all of these instances, minute amounts of ethylene gas resulting from the fuel combustion or leakage brought about the results, both good and bad. In fact, as little as one part of ethylene per 10 million parts of air may be sufficient to trigger responses.

Today, commercial use of ethylene is extensive. It has been used for many years to ripen harvested green fruits, such as bananas, mangoes, and honeydew melons, and to cause citrus fruits to color up before marketing. Since ethylene production almost ceases in the absence of oxygen, apple and other fruit growers have found that if they place unripe fruit in sealed warehouses after harvest, pump out the air, reduce temperatures to just above freezing, and replace the air with inert nitrogen gas or carbon dioxide, the fruit will remain metabolically inactive for long periods. The growers can then remove the fruit in batches throughout the year, add as little as one part per million of ethylene, and have ripe fruit at any time there is a demand. This is why apples are always available in supermarkets, even though harvesting is usually confined to a few weeks in the fall.

Fruits that respond to ethylene usually have a major increase in respiration just before ripening occurs. The increase in ethylene production at that time is often up to 100 times greater than it was a day or two earlier. The accompanying major increase in respiration is called a *climateric,* and fruits that exhibit such phenomena are called *climateric fruits.* Some fruits such as grapes are *nonclimateric* and do not respond in this way to ethylene.

A few growers still use natural ethylene to ripen pears and peaches by wrapping each fruit individually in tissue paper. The paper retards the escape of ethylene from the fruit and hastens ripening. "Resting" potato tubers will sprout following brief applications of ethylene, and seeds may be stimulated to germinate if given a short exposure to the gas just before sprouting, although treatment after sprouting inhibits growth.

Ethylene is used in Hawaii to promote flowering in pineapples, and it causes members of the Pumpkin Family to produce more female flowers and thus more fruit. Because trees grown in containers in nurseries are usually crowded, they are often tall and spindly, but applications of small amounts of ethylene to the trunks while they are enclosed in plastic tubes causes a marked thickening, making the trees sturdier and less likely to break.

Other Hormones or Related Compounds

A number of compounds called *oligosaccharins,* which are released from cell walls by enzymes, influence cell differentiation, reproduction, and growth in plants and therefore must be considered hormones. However, oligosaccharins produce their effects at concentrations of up to 1,000 times less than those of auxins, and the effects (e.g., growth promotion, inhibition of flowering) are not only highly specific but the responses to them are essentially identical in all species. *Brassosteroids,* which have a gibberellinlike effect on plant stem elongation, are known from legumes and a few other plants. Yams, which incidentally are a source of DHEA—a hormone whose production by humans tends to decrease with age—are also the source of *batasins.* Batasins promote dormancy in *bulbils,* which are produced from axillary buds in lilies and a few other plants.

HORMONAL INTERACTIONS

Apical Dominance

For centuries, gardeners and nursery workers have often deliberately removed terminal buds of plants to promote bushier growth, knowing that the buds are involved in *apical dominance.* Apical dominance is the suppression of the growth of the lateral or axillary buds, believed to be brought about by an auxinlike inhibitor in a terminal bud. It is strong in trees with conical shapes and little branching toward the top (e.g., many pines, spruces, firs) (Fig. 11.6) and weak in trees with stronger branching (e.g., elms, ashes, willows).

There is evidence that, as the tissues produced by the apical meristem of the terminal bud increase in length, the source of the inhibitor moves farther away and the concentration of the inhibitor gradually drops in the vicinity of the lateral buds. We presume that the inhibitory effect eventually falls below a threshold, allowing the lateral buds to grow. Each lateral bud then itself becomes a terminal bud that produces its own inhibitor, which in turn suppresses the new lateral buds developed beneath it. These presumptions, however, have not yet been confirmed and some studies suggest that a cytokinin deficiency in the lateral buds actually plays a greater role in apical dominance than auxinlike in-

FIGURE 11.6　A Jeffrey pine tree. The trunk of this tree, which would normally be single, is forked because of the earlier removal of the terminal bud.

hibitors. Further studies are needed before the precise causes of apical dominance are fully understood.

Senescence

The breakdown of cell components and membranes that eventually leads to the death of the cell is called **senescence.** As mentioned in Chapter 8, the leaves of deciduous trees and shrubs senesce and drop through a process of abscission every year. Even evergreen species often retain their leaves for only two or three years (with the notable exception of bristlecone pines, which hold their leaves for up to 30 years), and the aboveground parts of many herbaceous perennials senesce and die at the close of each growing season.

Why do plant parts senesce? Some studies have suggested that certain plants produce a senescence "factor" that behaves like, or is actually, a hormone, but we are not yet certain of the precise mechanisms involved. We do know, however, that both ABA and ethylene promote senescence, while auxins, gibberellins, and cytokinins delay senescence in a number of plants that have been studied. Other factors, such as nitrogen deficiency and drought, also hasten senescence.

OTHER INTERACTIONS

The development of roots and shoots in both tissue culture and in girdled stems is also regulated by a combination of auxins and cytokinins. Experiments with living pith cells of tobacco plants have shown that such cells will enlarge when supplied with auxin and nutrients, but they will not divide unless small amounts of cytokinin are added. By varying the amounts of cytokinin, it is possible to stimulate the pith cells to differentiate into roots or into buds from which stems will develop. This last regulatory effect can be used to offset apical dominance. The extent of the suppression depends on how close the axillary buds are to the terminal bud and on the amount of inhibitor involved. Experiments with pea plants have shown that the axillary buds will begin to grow as little as four hours after a terminal bud has been removed. If cytokinins are applied in appropriate concentration to axillary buds, however, they will begin to grow, even in the presence of a terminal bud.

Gibberellins, cytokinins, and enzymes all play a role in the germination of seeds. After water has been imbibed by cereal seeds, gibberellins are released by the embryo and stimulate the secretion of enzymes that digest endosperm. In many dicot seeds, cytokinins stimulate the production of starch-digesting enzymes.

PLANT MOVEMENTS

We noted in Chapter 2 that all living organisms exhibit movement but that most plant movements are relatively slow and imperceptible unless seen in time-lapse photography or demonstrated experimentally. Now we need to examine how and why plant movements occur.

Growth Movements

Growth movements result from varying growth rates in different parts of an organ. They are mainly related to young parts of a plant and, as a rule, are quite slow, usually taking at least two hours to become apparent although the plant may have begun microscopic changes within minutes of receiving a stimulus. The stimulus may be either internal or external.

Movements Resulting Primarily from Internal Stimuli

Helical (Spiraling) Movements Charles Darwin once attached a tiny sliver of glass to the tip of a plant growing in a pot. Then he suspended a piece of paper blackened with carbon over the tip, and, as the plant grew, he raised

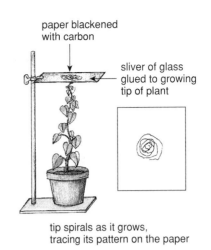

paper blackened with carbon

sliver of glass glued to growing tip of plant

tip spirals as it grows, tracing its pattern on the paper

FIGURE 11.7 Charles Darwin's demonstration of spiraling growth.

the paper just enough to allow the tip to touch the paper without hurting the plant. He found that the growing point traced a spiral pattern in the blackened paper. We know now that such helical movements of growth are common to many plants (Fig. 11.7).

Nodding Movements Members of the Legume Family, such as garden beans, whose ethylene production upon germinating causes the formation of a thickened crook in the hypocotyl, exhibit a slow oscillating movement (i.e., the bent hypocotyl "nods" from side to side like an upside-down pendulum) as the seedling pushes up through the soil. This nodding movement apparently facilitates the progress of the growing plant tip through the soil.

Twining Movements Although twining movements are mostly stimulated internally, external forces, such as gravity and contact, may also play a role in these movements. They occur when cells in the stems of climbing plants, such as morning glory, elongate to differing extents, causing visible spiraling in growth (in contrast with the spiraling movements previously mentioned, which are not visible to the eye). Tendril twining, which is initiated by contact, results from an elongation of cells on one side of the stem and a shrinkage of cells on the opposite side, followed by differences in growth rates (Fig. 11.8). Some tendrils are stimulated to coil by auxin, while others are stimulated by ethylene.

Contraction Movements In Chapter 6, we noted that the bulbs of a number of dicots and monocots have contractile roots, which pull them deeper into the ground. In lilies, for example, seeds germinating at the surface ultimately produce bulbs that end up 10 to 15 centimeters (4 to 6 inches) below ground level because of the activities of contractile roots. There is some evidence that temperature fluctuations at the surface determine how long the contracting will continue. When the bulb gets deep enough that the differences between daytime and nighttime

FIGURE 11.8 Typical twining of a tendril produced by a manroot plant. Note that the direction of coiling reverses near the midpoint.

direction of a stimulus) are called **nastic.** In prayer plants (see Fig. 11.15) and others where the leaves or flowers fold up and reopen daily, turgor movements are also involved, as are the external stimuli of light and temperature. These movements are controlled by a biological "clock" on cycles of approximately 24 hours and are referred to as *circadian rhythms* (discussed under *sleep movements* on page 192). In a few instances, nastic movements apparently are initiated by external stimuli alone.

Movements Resulting from External Stimuli

Permanent, directed movements resulting from external stimuli are commonly referred to as **tropisms.**

Phototropism The main shoots of most plants growing in the open tend to develop vertically, although the branches often grow horizontally. If a box is placed over a plant growing vertically and a hole is cut to admit light from one side, the tip of the plant will begin to bend toward the light within a few hours. If the box is later removed, a compensating bend develops, causing the tip to grow vertically again. Such a growth movement toward light is called a *positive phototropism* (Fig. 11.9). A similar bending away from light is called a *negative phototropism.* The shoot tips of most plants are positively phototropic, while roots are either insensitive to light or negatively phototropic.

Leaves often twist on their petioles and, in response to illumination, become perpendicularly oriented to a light source. In fact, many plants have solar-tracking leaves with the blades oriented at right angles to the sun throughout the day. Some botanists have referred to solar-tracking movements as *heliotropisms* but, unlike in the phototropic responses of stems and roots, growth is not involved.

It has been widely reported that sunflowers exhibit heliotropic movements and face the sun throughout the day. This is not true, except, perhaps, when the plants are very young; sunflowers face east as they develop and remain facing in that direction until the seeds are mature and the plant dies.

Strictly speaking, the twisting of petioles that facilitates heliotropic movements should be called *phototorsion,* because *motor cells* (in *pulvini,* which are special swellings at the bases of leaves or leaflets) at the junction of the blade and the petiole control the movement (see *turgor movements* on page 192). When viewing a tree from directly overhead or observing a vine growing on a fence, it is possible to note how surprisingly little overlap of the leaves occurs. Each leaf is oriented so that it receives the maximum amount of light available (Fig. 11.10).

We have already noted that if the tip of a coleoptile is covered or removed, the structure will not bend toward light and that auxin is produced in the tip (see Fig. 11.1). We have also noted that auxin promotes the elongation of cells, at least in certain concentrations. For some time, it was believed that stem tips bent toward light because auxin was

temperatures are slight, the contractions cease. The aerial roots of some banyan trees straighten out by contraction after the roots have made contact with the ground. The "shrinking" of roots has been shown to take place at the rate of 2.2 millimeters (0.1 inch) a day in sorrel.

Nastic Movements When flattened plant organs, such as leaves or flower petals, first expand from buds, they characteristically alternate in bending down and then up as the cells in the upper and lower parts of the leaf alternate in enlarging faster than those in the opposite parts. Such nondirectional movements (i.e., movements that do not result in an organ being oriented toward or away from the

FIGURE 11.9   A cyclamen plant that received light from one direction for several weeks. Note how all visible plant parts are oriented on the side that received light.

FIGURE 11.10   Wild grape vines in the fall. Note how little overlap of leaves has occurred.

destroyed or inactivated on the exposed side, leaving more growth-promoting hormone on the side away from the light, causing the cells there to elongate more and produce a bend. Careful experiments have shown, however, that stem tips growing in the open have the same total amount of auxin present as do stem tips from the same species receiving light from one side. Other experiments indicate that the auxin migrates away from the light, accumulating in greater amounts on the opposite side, promoting greater elongation of cells on the "dark" side. Apparently, an active transport system enables the auxin to migrate against a diffusion gradient.

Different intensities of light may bring about different phototropic responses. In Bermuda grass, for example, the stems tend to grow upright in the shade and parallel with the ground in the sun. In other plants, such as the European rock rose, which grows among rocks or on walls, the flowers are positively phototropic, but, once they are fertilized, they become negatively phototropic. As the pedicels (stalks) elongate, the developing fruits are buried in cracks and crevices, where the seeds then may germinate.

Phototropic responses are not confined to flowering plants. A number of mushrooms, for example, show marked positive responses to light, and certain fungi that grow on horse dung have striking phototropic movements, which are discussed in Chapter 19.

Gravitropism   Growth responses to the stimulus of gravity are called **gravitropisms.** The primary roots of plants are positively gravitropic, while shoots forming the main axis of plants are negatively gravitropic (Fig. 11.11). There is evidence that plant organs perceive gravity through the movement of large starch grains in special cells of the root cap (discussed in Chapter 6). The starch grains, called *statoliths*, are also found in coleoptile tips and in the endodermis. When a potted plant is placed on its side, the statoliths will, within a few minutes, begin to float or tumble down until they come to rest on the side of the cells closest to the gravity stimulus. In roots, the cells on the side opposite the stimulus begin elongating within 10 seconds to an hour or two, bringing about a downward bend, while the opposite occurs in stems (figs. 11.12 and 11.13).

Some cell biologists have suggested that mitochondria and Golgi bodies (dictyosomes) also respond to gravity, but precisely how they or the starch grains bring about the response is the subject of conjecture. It may be that auxin and ABA, along with calcium ions, are agents

FIGURE 11.11 This *Coleus* plant was placed on its side the day before the photograph was taken. The stems bent upward within 24 hours of the pot being tipped over.

involved in modifying cell elongation so as to produce these gravitropic bendings.

The stimulus of gravity can be negated by rotating a plant placed in a horizontal position. A simple device called a *clinostat* uses a motor and a wheel to rotate a potted plant slowly about a horizontal axis (Fig. 11.14). As the plant rotates, both the stem and roots continue to grow horizontally instead of exhibiting characteristic gravitropic responses. Obviously, neither the starch grains in the statoliths nor other organelles can settle while the plant is moving, and this apparently prevents transport of auxin, which would bring about cell elongations that produce curvatures of root and stem.

Other Tropisms   A plant or plant part response to contact with a solid object is called a *thigmotropism*. One of the most common thigmotropic responses is seen in the coiling of tendrils and in the twining of climbing plant stems (see figs. 7.13 and 11.8). Such responses can be relatively rapid, with some tendrils wrapping around a support two or more times within an hour. The coiling results from cells in contact becoming slightly shorter while those on the opposite side elongate.

Roots often enter cracked water pipes and sewers. In fact, roots have been known to grow upward for considerable distances in response to water leaks. Some researchers have called such growth movements *hydrotropisms,* but most plant physiologists today doubt that responses to water and several other "stimuli" are true tropisms. Other external stimuli that produce responses designated tropic by some scientists are chemicals (*chemotropism*), temperature (*thermotropism*), wounding (*traumotropism*), electricity (*electrotropism*), dark (*skototropism*), and oxygen (*aerotropism*).

It has been noted that greater concentrations of roots occur on the north and south sides of wheat seedlings, and it has been suggested that magnetic forces may be involved; the term *geomagnetotropism* has been proposed for this phenomenon. Some of these tropic or tropiclike responses have been artificially induced, but others take place regularly in nature. For example, germinating pollen grains produce a long tube that follows a diffusion gradient of a chemical released within a flower; this is considered a chemotropism. Thermotropic responses to cold temperatures may be seen in the shoots of many common weeds, which grow horizontally when cold temperatures prevail and return to erect growth when temperatures become warmer.

Turgor Movements

Turgor movements result from changes in internal water pressures. The cells concerned may be in normal parenchyma tissue of the cortex, or they may be in special swellings called *pulvini* located at the bases of leaves or leaflets. Some turgor movements may be quite dramatic, taking place in a fraction of a second. Others may require up to 45 minutes to become visible.

"Sleep" Movements (Circadian Rhythms)
Members of several flowering plant families, particularly the Legume Family (which includes peas and beans), exhibit movements in which either leaves or petals fold as though "going to sleep" (Fig. 11.15). The folding and unfolding usually takes place in regular daily cycles, with folding most frequently taking place at dusk and unfolding occurring in

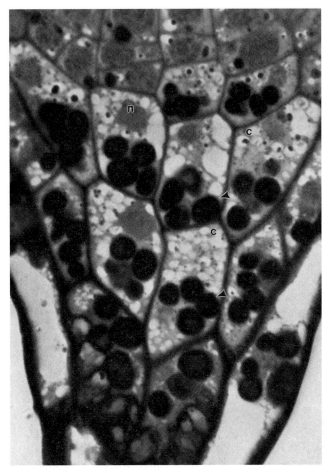

FIGURE 11.12   A root cap of a tobacco plant. The force of gravity is at the bottom of the picture. Note that the amyloplasts (more or less spherical dark objects) are toward the bottom of each cell. The amyloplasts are believed to function as statoliths in the perception of gravity by roots, ×2,000.

(Light micrograph courtesy John Z. Kiss)

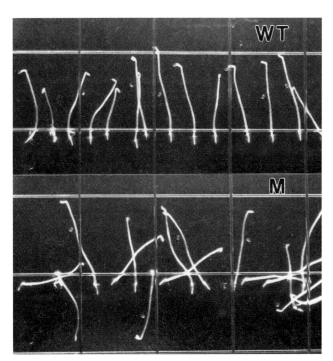

FIGURE 11.13   Tobacco seedlings grown in the dark. The source of gravity is at the bottom of the pictures. Normal "wild type" seedlings in the top row are all more or less perpendicular to the gravity. The seedlings in the bottom row are mutants with much less starch than normal plants. The mutant seedlings are disoriented, suggesting that any amyloplasts of typical mass function as statoliths in the perception of gravity, ×0.5.

(Courtesy John Z. Kiss)

the morning. Such cycles, which have been more extensively documented in the Animal Kingdom, have come to be known as **circadian rhythms.**

Circadian rhythms appear to be controlled internally by the plants, although they are also geared to changing day lengths and seasons. They do not generally accelerate when temperatures increase. The actual timing of circadian rhythms varies with the species, although most plants exhibiting sleep movements do so at dusk and at dawn. The flowers of several species, however, open their flowers at different hours of the night.

About 200 years ago, the famous Swedish botanist Linnaeus planted wedge-shaped segments of a circular garden with plants that exhibited sleep movements. The plants were arranged in successive order of their sleep movements throughout a 24-hour day. One could tell the approximate time simply by noting which part of the garden was "asleep" and which was "awake." A few others copied the garden clock idea, but the expense and difficulty of obtaining all the plants

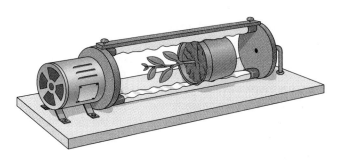

FIGURE 11.14   A clinostat, which is a tool used by plant biologists to negate the effects of gravity. Growing plants or seedlings are slowly rotated so that the statoliths in cells that perceive gravity do not settle to the bottom, and typical growth or bending of stems or roots away from gravity does not occur.

A.   B.

FIGURE 11.15   A prayer plant. *A*. The plant at noon. *B*. The same plant at 10 P.M., after "sleep" movements of its leaves have occurred.

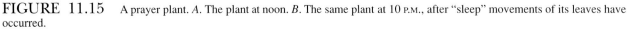

from different parts of the world and replanting them each year proved to be too great for the practice to be continued.

The movements are produced by turgor changes caused by the passage of water in and out of cells at the bases of the leaves or leaflets. The function of these movements is not clear, and the rhythms are also not confined to sleep movements. Species of certain warmer marine water algae called *dinoflagellates* (discussed in Chapter 18) glow in the dark through *bioluminescence*, a process by which chemical energy is reconverted to light energy. One species always glows brightly within two or three minutes of midnight, even if it is maintained in culture in continuous dim light. This particular dinoflagellate also glows when culture containers in which it is suspended are jarred, and it displays two other types of circadian rhythms. One rhythm involves cell division, with peak mitotic activity occurring just before dawn, and the other rhythm pertains to photosynthesis, which reaches a maximum around noon.

Another type of rhythm is seen in the giant bamboos of Asia, which send up huge flowering stalks every 33 or 66 years, even if the plants have been transplanted to other continents and are growing under different conditions. These flowering stalks use up all the energy reserves of the bamboo, and they die shortly after appearing. This has especially been a problem in cities where nearly all of the bamboo plantings were propagated from a single source and then all died simultaneously after flowering.

Water Conservation Movements

Leaves of many grasses have special thin-walled cells (*bulliform cells*) below parallel, lengthwise grooves in their sur-faces. During periods when sufficient water is not available, these cells lose their turgor and the leaf rolls up or folds (Fig. 11.16). Experiments with certain prairie grasses have shown that the rolling effectively reduces transpiration to as low as 5% to 10% of normal.

Contact Movements

The sudden movements of bladderworts (discussed in Chapter 7) involve turgor changes apparently triggered by electric charges released upon contact or as a result of variations in light or temperature. The springing of the trap of the Venus flytrap (also discussed in Chapter 7) was thought to be brought about in similar fashion, but recent research has shown that the trap closes when its outer epidermal cells expand rapidly, and it reopens when the inner epidermal cells expand in the same way. About one-third of the ATP available in the cells is used in each movement, so repeated stimulation of the trap by touching the trigger hairs readily fatigues the trap if sufficient time for ATP replenishment is not given between stimulations.

The sensitive plant (*Mimosa pudica*) has well-developed swellings (*pulvini*) at the bases of its many leaflets and a large pulvinus at the base of each leaf petiole. When the leaf is stimulated by touch, heat, or wind, there is a type of chain reaction in which potassium ions migrate from one-half of each pulvinus to the other half. This is followed by a rapid shuttling of water from the pulvinar parenchyma cells of one-half to those of the other half. The loss of turgor results in the folding of both the leaflets and the leaf as a whole (Fig. 11.17).

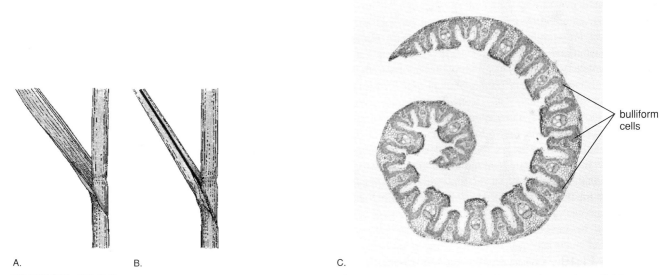

A.   B.   C.

bulliform
cells

FIGURE 11.16   Water conservation movement in a grass leaf when insufficient water to maintain normal turgor is available. *A*. The leaf when adequate water is available. *B*. The leaf after it has rolled up. *C*. Enlargement of a cross section of a rolled leaf showing the large, thin-walled *bulliform* cells (arrows), which partially collapse under dry conditions and thus bring about the rolling of the leaf blade.

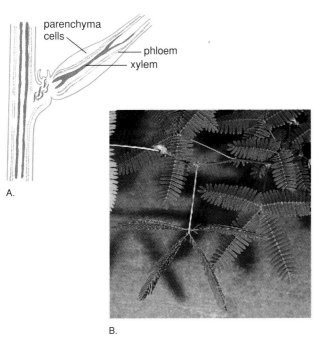

parenchyma
cells
phloem
xylem

A.

B.

FIGURE 11.17   *A*. A longitudinal section through the pulvinus of a sensitive plant (*Mimosa pudica*). *B*. The leaflets of the leaf toward the bottom of this picture have folded upward in response to being bumped. The other leaves of this sensitive plant have remained fully expanded.

A.

B.

FIGURE 11.18   A bush monkey flower. The white, two-lobed structure in the center is the stigma. *A*. The stigma as it appears before pollination. *B*. The stigma 2 seconds after being touched by a pollinator; the lobes have rapidly folded together.

If a part of the stem of a sensitive plant is cut off and then immediately reattached with a water-filled piece of rubber tubing, it can be shown that something is transmitted through the water from above the cut. Within a few minutes after a leaf above the cut is stimulated to fold, a leaf or two below the cut will also fold. Although potassium ions have

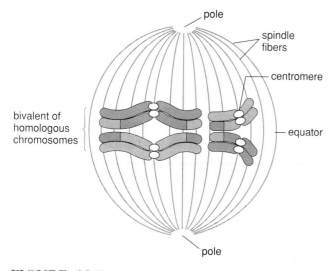

pole
spindle fibers
centromere
equator
pole
bivalent of homologous chromosomes

FIGURE 12.5   A diagram of homologous chromosomes at metaphase I of meiosis.

The diploid body itself is called a **sporophyte.** It develops from a zygote and eventually produces **spore mother cells (meiocytes),** each of which undergoes meiosis, producing four **spores.** The haploid bodies that develop from these spores are called **gametophytes.** These eventually form sex structures, or cells, in which gametes are produced by mitosis.

As becomes evident in chapters to follow, the gametophytes of many primitive forms constitute a large part of the visible organism, but as we progress up through the Plant Kingdom to more complex plants, they become proportionately reduced in size until they may be only microscopic. The switch from one generation to the next takes place as spores are produced when spore mother cells undergo *meiosis* and again when a zygote is produced through fusion of gametes, or **fertilization** (also called **syngamy**).

Although the sporophyte generation of some primitive organisms may consist of a single cell (the zygote), the basic plan of Alternation of Generations can be seen in the Protoctistan, Fungal, and Plant Kingdoms. It becomes most conspicuous, however, in the Plant Kingdom, and it differs from one organism to the next in the forms of the various bodies and cells.

The accompanying diagram (Fig. 12.6) and the following six rules pertaining to Alternation of Generations in the majority of plants and other green organisms, apply to the life cycle of any sexually reproducing organism discussed in this book.

1. The first cell of any *gametophyte generation* is normally a *spore* (*sexual spore* or *meiospore*), and the last cell is normally a *gamete.*

2. Any cell of a gametophyte generation is usually *haploid* (*n*).

3. The first cell of any *sporophyte generation* is normally a *zygote,* and the last cell is normally a *spore mother cell* (*meiocyte*).

4. Any cell of a sporophyte generation is usually *diploid* (2*n*).

5. The change from a sporophyte to a gametophyte generation usually occurs as a result of *meiosis.*

6. The change from a gametophyte to a sporophyte generation usually occurs as a result of *fertilization* (fusion of gametes), which is also called *syngamy.*

The word *generation* as used in *Alternation of Generations* simply means *phase of a life cycle* and should not be confused with the more widespread use of the word pertaining to time or offspring.

Summary

1. Reproduction may take place through natural vegetative propagation or by spores (asexual reproduction) or by sexual processes (sexual reproduction). In sexual reproduction, two gametes unite, forming a zygote, which is the first cell of a new individual.

2. The process of meiosis ensures that gametes will have half the chromosome number of the zygotes and also usually ensures that offspring will not be identical to the parents in every respect.

3. Meiosis takes place by means of two successive divisions, each of which, like mitosis, is divided into arbitrary phases even though the process is continuous. In prophase I, the chromosomes become paired, often exchange parts, and then separate. The similar chromosomes of each pair are referred to as being homologous. Exchange of parts of chromatids may be affected by the parts initially crossing over to and from one another, forming chiasmata, and then tearing apart.

4. In metaphase I, the chromosomes become aligned at the equator in pairs, and in anaphase I, whole chromosomes from each pair migrate to opposite poles. In telophase I, the chromosomes either partially revert to their interphase state or initiate the second division, which is essentially similar to mitosis.

5. In prophase II, the chromosomes of each of the two groups become shorter and thicker again, with both groups becoming aligned at their respective equators in metaphase II. In anaphase II, the chromatids of each chromosome separate and migrate to opposite poles, and in telophase II, the chromatids (now called *chromosomes* again) lengthen, and new nuclear envelopes and nucleoli appear for all four groups. New cell walls are produced between each of the four groups.

6. If the chromatids have exchanged parts earlier, none of the four groups may have identical combinations of DNA, and each group has half the original number of chromosomes. Each of the four cells constitutes a

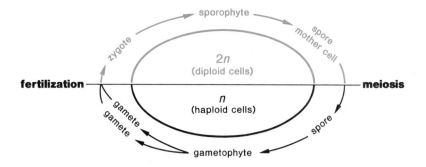

FIGURE 12.6   A typical life cycle of plants or related organisms that undergo sexual reproduction.

spore (sexual spore or meiospore), which may develop into a body or structure within which gametes may be produced by mitosis.

7. Any cell having one set of chromosomes is said to be haploid or to have *n* chromosomes; any cell having two sets of chromosomes is said to be diploid or to have *2n* chromosomes.

8. In the life cycle of an organism that undergoes sexual reproduction, there is an alternation between a haploid phase and a diploid phase. The haploid body is called a *gametophyte* and the diploid body is called a *sporophyte*.

9. The change from the haploid phase to the diploid phase occurs when two gametes (each *n*) unite, forming a zygote (*2n*) in the process of fertilization (syngamy). The change from the diploid to the haploid phase occurs as a result of a spore mother cell (meiocyte) undergoing meiosis, when a *2n* cell becomes four *n* cells. This switching of phases is referred to as Alternation of Generations.

Review Questions

1. Indicate when during meiosis each of the following events occurs: (a) crossing-over, (b) chromatids separating at their centromeres and migrating to opposite poles, and (c) chromosomes aligning themselves in pairs at the equator.

2. Is there any difference between mitosis and the second division phases of meiosis?

3. What is the significance of meiosis with respect to sexual reproduction?

4. What is meant by saying that the cells of a sporophyte phase are diploid or *2n*?

5. At what stage of a life cycle does the chromosome number of cells switch from *n* to *2n*?

Discussion Questions

1. If an organism reproduces very freely by asexual means, is there any advantage to its also reproducing sexually?

2. Would it make any difference if the events of the second division of meiosis occurred before the events of the first division?

3. Would it be correct to say that a bivalent has four times the amount of DNA present in the chromosomes of a cell in anaphase II?

4. Would you assume that the length of a chromosome might have something to do with the number of crossovers it might form with its homologue?

5. Two mitotic divisions may take place in as little as a few hours, while meiosis may take much longer. What reasons can you suggest for this?

Additional Reading

Becker, W. M., and D. W. Deamer. 1991. *The world of the cell*. Redwood City, CA: Benjamin/Cummings Publishing Co., Inc.

Darnell, L. 1991. *Cell biology and composition*. New York: W. H. Freeman.

DeRobertis, E. D. P., and E. M. DeRobertis Jr. 1987. *Cell and molecular biology*, 8th ed. Philadelphia: Lea and Febiger.

Herrmann, H. 1990. *Cell biology*. New York: HarperCollins Pubs., Inc.

John, B. 1990. *Meiosis*. New York: Cambridge University Press.

Raven, P. H., R. F. Evert, and S. E. Eichorn. 1992. *Biology of plants*, 5th ed. New York: Worth Publishers.

Wolfe, S. L. 1993. *Cell and molecular biology*. Belmont, CA: Wadsworth Publishing Co.

Narrow-tubed monkey flowers (Mimulus angustatus), *a vernal pool species of northern California. (Courtesy Robert A. Schlising)*

13 Genetics

Overview————————————

In this chapter, the story of Karpechenko's crossing of a radish and a cabbage introduces the topic of genetics. This is followed by a brief history of Mendel and his classic experiments with pea plants. Each of Mendel's principles or laws is discussed, and examples of monohybrid and dihybrid crosses are given. Next, the backcross, linkage, chromosomal mapping, the Hardy-Weinberg law, and interactions between genes are introduced. The chapter concludes with a discussion of DNA structure and protein synthesis.

Some Learning Goals

1. Understand the significance of Mendel's experiments with peas.
2. Understand the following terms: allele, dominance, phenotype, genotype, homozygous, heterozygous, monohybrid cross, dihybrid cross, backcross, linkage, chromosomal mapping, and the Hardy-Weinberg law.
3. Give the ratios of the offspring in the first two generations from a monohybrid and a dihybrid cross. Describe the genotypes involved.
4. Be able to solve simple genetic problems.
5. Understand how genes interact.
6. Know the kinds and functions of RNA.
7. Be able to discuss the nature of DNA and protein synthesis.

When a plant is being grown as a crop, the more parts of the plant that are edible or otherwise useful, the more efficient the use of the land on which the crop is grown. Some parts of many crop plants are not used, however, except possibly as ingredients in compost. The roots of lettuce, for example, are not useful for either food or fuel. In tomatoes and grapes, only the fruits are edible; in potatoes, only the tubers are used, and in cotton, the fibers and the seeds constitute the useful parts.

In 1928, the Russian biologist G. D. Karpechenko reported on an experiment in which he tried to cross two food plants in the Cabbage Family: the radish, with an edible root, and the cabbage, with edible leaves. One might assume that such a cross, if successful, would produce an ideal crop plant. Karpechenko found that these two plants, although related, did not cross readily, but he did succeed in obtaining a few seeds from such crosses. When these seeds were planted, the results were dramatic. Huge plants grew and produced fertile seeds of their own. From a crop point of view, however, the cross was a dismal failure because these large plants had leaves mostly like those of a radish and roots like those of a cabbage—they were virtually useless!

Such crossing experiments are a part of **genetics,** the science that deals with heredity, or natural inheritance. Genetics is one of the youngest of the biological sciences, having developed after the details of meiosis and mitosis became understood at the turn of the century. It has since be-

come a vast and very important field of study, having the potential not only for improving food production in a hungry world but also for solving various problems related to the inheritable human diseases, the effects of radiation, the control of population growth, and many other matters of direct human interest and concern.

MENDEL AND HIS WORK

Genetics as a science originated about a generation before its significance became appreciated in the scientific community as a whole. An Austrian monk, Gregor Mendel (Fig. 13.1), taught between 1853 and 1868 in what became the Czechoslovakian city of Brno. In the monastery there, he carried on a wide range of studies and experiments in physics, mathematics, and natural history. He also became an authority on bees and meteorology and kept notes on experiments involving two dozen different kinds of plants. Today, he is best known for the studies he conducted with a number of varieties of pea plants.

Mendel published the results of his studies on pea plants in a biological journal in 1866, and he sent copies of his paper to leading European and American libraries. He also sent copies to at least two eminent botanists of the time.

FIGURE 13.1   Gregor Mendel.
(Photo ca. 1870s, Brünn, Moravia. Courtesy, Brooklyn Botanic Gardens, New York. Print at Hunt Institute, Pittsburgh, PA)

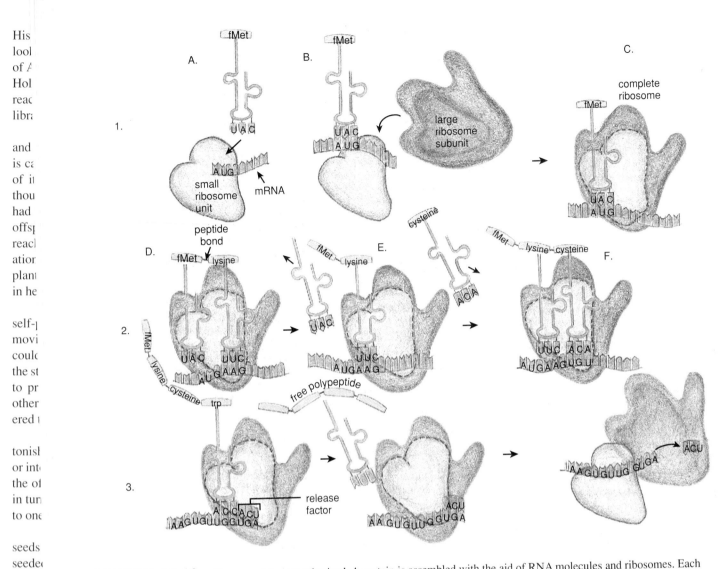

FIGURE 13.10 How a protein is synthesized. A protein is assembled with the aid of RNA molecules and ribosomes. Each ribosome consists of two subunits, one larger than the other. The process can be divided into three phases: 1. *Initiation. A.* An mRNA molecule from chromosomal DNA becomes attached to a small ribosomal subunit. *B.* A tRNA molecule bearing formylated methionine (Met) hydrogen bonds to the initiation codon (AUG) on the mRNA. *C.* The large ribosomal subunit locks into place. 2. *Elongation. D.* A second tRNA molecule, with an amino acid attached, moves into the ribosome, and its anticodon bonds to the mRNA. *E.* A peptide bond forms between the two amino acids that are now side by side. At the same time, the bond between the first amino acid and its tRNA molecule is broken, and the tRNA is released. *F.* As the first tRNA molecule is released, the second tRNA molecule moves into its place, and a third tRNA moves in behind it while another peptide bond is formed. A polypeptide chain builds as the process is repeated. 3. *Termination.* When a *termination codon* is reached, the polypeptide chain is separated from the last tRNA molecule, which is then released. Finally, a *release factor* triggers the separation of the two ribosomal subunits.

Summary

1. Genetics, the science of heredity and natural inheritance, includes crossing experiments, such as Karpechenko's cross of a cabbage and a radish.

2. The science of genetics originated with Gregor Mendel, an Austrian monk, who performed experiments with peas. Mendel's crosses involved pea varieties that had pairs of contrasting characteristics carried on a separate pair of homologous chromosomes. The offspring of the parental plants crossed were designated the F_1 generation, and the offspring of crosses between members of the F_1 generation were designated the F_2 generation. F_1 plants were also called hybrids.

3. Mendel, who was exceptionally fortunate in his choice of characteristics to study, meticulously counted the number of all offspring. He found that the F_1 hybrids

His
lool
of *F*
Hol
read
libra

and
is ca
of it
thou
had
offs
reacl
atior
plan
in he

self-
movi
coulc
the st
to pr
other
ered t

tonisl
or int
the of
in tur
to on

seeds
seede
seven

were
and th
tion (f
ally ab
the *se*

cross,
dition,
ing ch
becaus
whole

Mende
charact
other c
do not
was als
perimei
just sev
characte
chromo

of his pea crosses all resembled their parents in appearance, but the F_2 offspring were produced in a ratio of three plants resembling one parent to one plant resembling the other parent.

4. Mendel called the agents for the characteristics inside the plants *factors* and deduced that each plant must have two factors for each characteristic. His law of unit characters states that "factors, which always occur in pairs, control the inheritance of various characteristics." The factors later became known as *alleles,* which are pairs of genes.

5. Mendel called the suppressing member of a pair of factors a *dominant* and its counterpart a *recessive.*

6. Phenotypes, which are described with words, denote appearance, and genotypes, for which letters are used, designate genetic makeup. Dominants are shown with capital letters and recessives with lowercase letters.

7. A homozygous plant has identical members of a pair of alleles. A heterozygous plant has contrasting members of a pair of alleles. The F_1 generation phenotypes may be intermediate between the parents as a result of absence of dominance (incomplete dominance).

8. The offspring of a monohybrid cross are produced in a ratio of 3:1, when dominants and recessives are involved. The offspring of a dihybrid cross involving dominance and two pairs of genes carried on separate pairs of chromosomes usually are produced in a phenotypic ratio of 9:3:3:1.

9. Backcrosses, which involve $F_1 \times$ the recessive parent, are used to check the results of genetic experiments. Backcrosses normally produce phenotypic ratios of 1:1 (monohybrid) or 1:1:1:1 (dihybrid).

10. Linked genes are inherited together. Bateson and Punnett discovered that expected 9:3:3:1 ratios in the F_2 phenotypic generation of certain sweet pea crosses did not occur. The 7:2:2:7 ratio they obtained was postulated to be the result of linkage and occasional crossing-over.

11. Chromosomal mapping involves calculating crossover percentages to determine the relative position of genes on chromosomes. The closer the genes are to each other on a chromosome, the less likely they are to be involved in a crossover and vice versa.

12. The Hardy-Weinberg law explains why recessive characteristics in a population do not eventually disappear. Selection is the most significant cause of changes in the proportions of dominant and recessive genes in a population.

13. Genes interact with one another, and so a phenotype is seldom the direct expression of its alleles.

14. Watson, Crick, and Wilkins were awarded a Nobel Prize in 1953 for developing a model of DNA that is now accepted as authentic. DNA replicates (duplicates itself) by "unzipping" down its nucleotide "ladder rungs" and having new nucleotides replace the old ones until two new chains of DNA have been formed.

15. New proteins are formed precisely according to coded information contained in the DNA of the nucleus by means of RNA, which occurs in three forms. Messenger RNA forms along part of a DNA molecule and is carried out of the nucleus into the cytoplasm, where it becomes attached to a ribosome. Ribosomes "read" the coded information and participate in assembling amino acids together to form a protein, with the aid of transfer RNA molecules. Ribosomes themselves are largely composed of ribosomal RNA.

Review Questions

1. When did genetics originate as a science?

2. Why did Mendel succeed where others had failed?

3. Who were Tschermak, Correns, and de Vries?

4. Define *hybrid, F_1 generation, phenotype, genotype, homozygous, heterozygous, dominant,* and *recessive.*

5. What are Mendel's laws?

6. In a variety of flowering plants, blue flowers and dwarf habit are controlled by dominant genes; white flowers and tall habit are the recessives. The gene pairs are carried on separate pairs of chromosomes. If a homozygous blue, dwarf plant is crossed with a homozygous white, tall plant,

   a. what are the phenotypes and genotypes of the F_1 generation?

   b. how many different *kinds* of genotypes could occur in the F_2 generation?

   c. what phenotypic ratio would you expect to occur in the F_2 generation?

   d. what genotypes could result from the cross $BBDD \times BbDd$?

   e. what phenotypes could result from the cross $bbDd \times BBDd$?

7. In summer squashes, disk-shaped fruit is dominant over spherical fruit and white fruit color is dominant over yellow fruit color. If a homozygous white, disk-shaped variety is crossed with a homozygous yellow, spherical variety, how many different homozygous genotypes are possible in the F_2 generation?

8. A cross is made between a white-flowered plant and a red-flowered plant of the same variety. The F_2 generation consists of 32 white-flowered plants, 64 pink-flowered plants, and 29 red-flowered plants. Explain.

9. In tomatoes, smooth skin is dominant and fuzzy skin is recessive; also, tallness is dominant and dwarfism is recessive. A homozygous smooth, tall variety is

crossed with a homozygous fuzzy, dwarf variety, and the F_1 is backcrossed to a homozygous fuzzy, dwarf variety. The offspring of the backcross are 95 smooth, tall; 3 fuzzy, tall; 4 smooth, dwarf; 96 fuzzy, dwarf. Is linkage involved? If so, what is the percentage of crossing-over?

10. In squashes, white color, which is controlled by a single gene, is dominant, and yellow is recessive. Give the phenotypes and the genotypes for each of the following crosses: $Ww \times Ww$; $Ww \times ww$; $WW \times ww$.

11. If red flowers are dominant and white flowers are recessive in an *annual* variety of *self-pollinating* beans, assume that you plant one heterozygous bean seed on an island, and it germinates and thrives. What will the colors and ratios of the phenotypes be in the F_4 generation?

12. In snapdragons, red flower color shows incomplete (lack of) dominance over white flower color. If you wanted to produce seeds that would produce only pink-flowered plants when sown, how would you go about it?

13. If one plant is homozygous for three dominant characteristics and another plant is correspondingly recessive for the same characteristics, what proportion of the F_2 generation will resemble each parent if the two plants are crossed?

14. What would happen in the production of gametes if a species had an odd number of chromosomes in each of its cells?

15. What is meant when genes are said to be linked?

16. What is chromosomal mapping and how is it done?

17. What is the Hardy-Weinberg law?

18. Describe the structure and duplication of DNA.

19. How are proteins synthesized?

20. What is the nature and function of each form of RNA?

Discussion Questions

1. Mendel is said to have been exceptionally lucky in his discoveries. Do you think most scientists who make significant discoveries are lucky? Explain.

2. The peas Mendel worked with were largely self-fertile. Of what advantage or disadvantage to a species is such a phenomenon?

3. Does DNA really control the synthesis of proteins? Explain.

Additional Reading

Brown, T. A. 1990. *Genetics: A molecular approach.* New York: Chapman and Hall.

Gardner, E. J., et al. 1991. *Principles of genetics,* 8th ed. New York: Wiley.

Grant, V. 1991. *The evolutionary process,* rev. ed. New York: Columbia University Press.

Klug, W. S. 1994. *Concepts of genetics,* 4th ed. New York: Macmillan.

Maynard-Smith, J. 1989. *Evolutionary genetics.* New York: Oxford University Press.

Olby, R. C. 1985. *The origins of Mendelism,* 2d ed. Chicago: University of Chicago Press.

Otte, D., and J. A. Endler (Eds.). 1989. *Speciation and its consequences.* Sunderland, MA: Sinauer Associates, Inc.

Russell, P. J. 1992. *Genetics,* 3d ed. New York: HarperCollins College Pubs., Inc.

Tamarin, R. 1991. *Principles of genetics,* 3d ed. Dubuque, IA: Wm. C. Brown Publishers.

Chapter Outline

A flower of butterfly pea (Clitoria ternata), *a tropical vine whose flower construction ensures that both the anthers and pistil touch the backs of visiting insects. The seeds are believed to be toxic to livestock.*

Plant Biotechnology and Propagation

14

Overview———————————

After a brief introduction to genetic engineering, the techniques and applications of recombinant DNA technology are discussed. Then some uses of hybridization, polyploidy, and mutations in plant breeding are explored, and the basics of tissue culture and mericloning are reviewed. The chapter closes with an overview of the principal types of vegetative propagation and grafting.

Some Learning Goals

1. Understand the functions of enzymes involved in the development of recombinant DNA.
2. Be able to delineate the basic steps involved in gene splicing.
3. Know several applications of genetic engineering.
4. Learn the roles of hybridization, polyploidy, and mutations in traditional plant breeding.
5. Understand the positive and negative aspects of the "Green Revolution."
6. Be able to explain tissue culture, mericloning, and related techniques to others.
7. Learn the basic types of vegetative propagation.

S ome institutions with important around-the-clock services, such as hospitals and hotels, often have backup generators, which function during power failures. Backup generators probably will be needed for many types of power equipment throughout the foreseeable future, but a scientist at the University of California at Davis has predicted that, within a few years, backup generators for some types of lighting may become obsolete. He believes shrubs that glow in the dark will line airport runways, freeways, and sidewalks. The glowing shrubs are expected to be developed, through the use of *genetic engineering* techniques, by introducing into the shrubs genes from luminescent bacteria (Fig. 14.1). The initial glow is expected to be intensified, also through genetic engineering.

The genetic engineering techniques involved in the development of new crops, medicines, disease control or elimination, waste management, oil spill cleanups, and shrubs that glow, as well as a varied assortment of manipulations or uses of plants or plant cells and tissues, are all part of the field of **biotechnology.**

Prior to the 1970s, most plant biotechnology involved traditional forms of plant breeding, such as hybridization (crossing of different varieties) and selection for desirable traits. Such breeding involved transfer of pollen from anthers to stigmas, which, in turn, normally brings about a sexual blending of characteristics. Modern biotechnological techniques, however, are largely asexual, providing much more precise control over the features expressed in the offspring. A brief look at both traditional and modern biotechnology forms the body of this chapter.

GENETIC ENGINEERING OR RECOMBINANT DNA TECHNOLOGY

If genes for desirable traits in plants (e.g., superior nutritional quality, metabolic efficiency, capacity for nitrogen fixation) are isolated from selected plants and can be incorporated into bacteria, the bacteria may produce large amounts of recombinant DNA for introduction into other plants. If the desirable trait has been successfully incorporated, the plants, in turn, can be propagated asexually.

The process of genetic engineering, or gene "splicing," involves removing a gene from its normal location and inserting it into a circular strand of DNA; the restructured DNA is then transferred into cells of another species.

The procedure begins with the isolation of pure bacterial plasmid DNA. A **plasmid** is a small, circular DNA fragment. As indicated in Chapter 17, bacterial cells differ from other cells in having no nucleus or membrane-bound organelles, such as plastids or mitochondria. Each bacterial cell does, however, have a long strand of DNA and also usually up to 40 or more plasmids.

Isolation of Plasmid DNA

Isolation of DNA is accomplished by breaking up the bacterial cells and extracting the DNA material with solvents such as phenol. The chromosomes can be separated from the plasmids by placing the extracted material in salt solutions that precipitate the chromosomes but not the plasmids.

Restriction Enzymes

Next, the linkages between adjoining nucleotides of plasmids are broken by special bacterial enzymes known as *restriction enzymes,* which break a circular plasmid at a specific nucleotide sequence, leaving a linear strand with "sticky" ends where the nucleotides are unpaired. The ends are called *sticky* because they attract complementary nucleotide sequences.

Repair Enzymes

Recombinant plasmids are made by mixing large numbers of plasmid DNA segments with fragments of the desired DNA, which has been broken by the same restriction enzyme so that the sticky ends are complementary to each other. When so mixed, a fragment from one source may associate with a fragment from the other source, and with the aid of a *repair enzyme* (DNA ligase), the two fragments may link, yielding a circular recombinant DNA molecule with parts of both original molecules. The recombinant plasmids are then inserted into bacterial cells where they replicate, thus introducing new traits into a particular strain of bacteria. The new inserted gene can make a new protein (Fig. 14.2).

A.

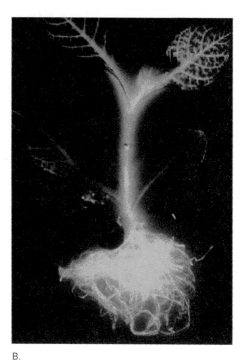

B.

FIGURE 14.1 A tobacco plant that has been genetically engineered to contain genes from a firefly. The plant glows when it is bathed in ATP, which provides the energy that activates the enzyme luciferase. The luciferase "turns on the glow" of luciferin.

(*A.* Courtesy Keith V. Wood; *B.* From D.W. Ow et al. 1986. "Transient and stable expression of the firefly luciferase gene in plant cells and transgenic plants." *Science* 234 (November 14, 1986): 856–59. Copyright © 1986 by the AAAS.)

Protein Sequencers and Gene Synthesizers

Bacteria are easily propagated, and it is relatively simple to obtain large amounts of DNA with enormous potential for scientific studies. The simplest bacterium, however, has at least 600 different proteins, and in the past it has taken several months and considerable expense to isolate a desirable protein from all the others.

The development during the 1980s of two machines, *protein sequencers* and *gene synthesizers,* has dramatically reduced the costs in time and funds for *cloning,* or producing genetically identical genes. Protein sequencers can reveal the precise sequence of amino acids in a protein. With that information, one can produce synthetic gene fragments that code for the protein. The fragments can then be introduced into the DNA of bacteria. Modified bacteria can be cultured to produce large quantities of the desired protein.

Cell Bombardment and Electroporation

Cell bombardment, an increasingly used method of gene transfer, relies on shooting DNA-coated particles into plant cells. The new DNA becomes incorporated in about 2% of the cells. Monsanto Company has used cell bombardment to introduce into corn a gene from BT (*Bacillus thuringiensis*) bacteria that makes the corn resistant to the European corn borer—a major pest that annually damages about 20% of the U.S. corn crop at a cost of nearly $1 billion.

In 1995, approval was also given for the release of a potato that was genetically altered to produce in its leaves the BT gene that kills Colorado potato beetles. BT genes are effective against a wide variety of caterpillars but harmless to almost all other organisms. It is anticipated that attempts will also be made to control in this fashion tomato hornworms, cabbage loopers, and similar pest larvae of many other food plants.

Electroporation involves the incubation of cells that are placed in an electrical field. Short electrical pulses cause the cell walls to become sufficiently permeable to admit foreign DNA, which is introduced without force into the cells.

These methods of gene transfer have the advantages of reliability and broad application; the disadvantages are that complex DNA insertions can be time-consuming and expensive; copies of DNA also can become rearranged in an undesirable manner.

Other Applications of Genetic Engineering

A common soil bacterium, *Agrobacterium tumefaciens,* which causes slow-growing tumors known as *crown gall* in woody dicots, has the capacity to use tDNA to transfer DNA into plant cells. The tumor-inducing plasmid can be altered by the deletion of the disease-causing parts and their

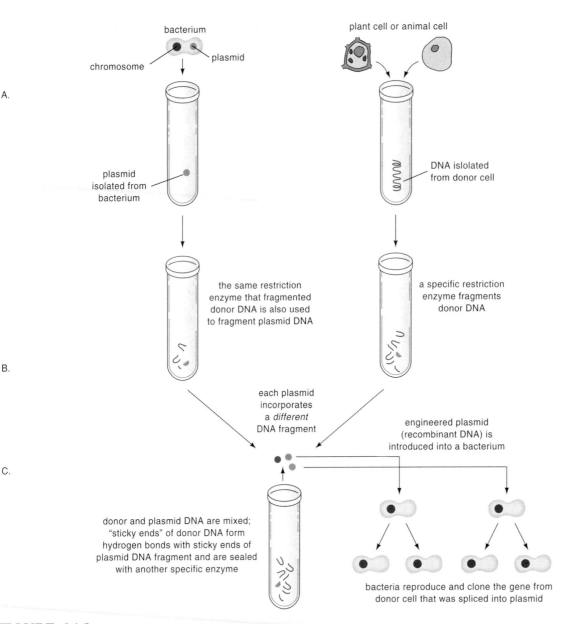

FIGURE 14.2   How recombinant bacteria are produced. *A*. DNA isolated from a plant of animal cell, and a plasmid isolated from a bacterial cell, are fragmented with the same restriction enzyme and then mixed. *B*. Some "sticky ends" of DNA fragments from both cells bond together, forming recombinant DNA. *C*. The recombinant DNA is inserted into a bacterium, which multiplies the recombinant DNA as it divides.

(After Lewis, Ricki, *Life*. 1992. Wm. C. Brown Publishers, Dubuque, IA.)

replacement with desirable genes. Use of crown gall bacteria for gene transfer has the advantage, in many instances, of incorporating only one precisely defined segment of DNA into a plant cell. An important goal of agricultural genetic engineers is the adaptation of the crown gall bacterium gene transfer system to the improvement of cereal crops.

Bacteria have been used for many years in the production of industrial chemicals, but genetic engineering may greatly increase such use. More complex organisms have more complex genetic complements, however, and much additional research is needed before the elimination of diseases and defects in higher plants and animals becomes commonplace as a result of genetic engineering.

Viruses, some of which naturally infect plants, are also easily propagated and are subject to similar engineering techniques. It is possible that viruses, discussed in Chapter 17, will play a major role in the genetic manipulation of higher plants in the future.

Early Experiments

Pioneering experiments with higher animals and plants were already under way by the late 1970s and early 1980s. In 1980, for example, J. Gordon of Yale University injected thousands of recombinant plasmids into newly fertilized mouse eggs that were in a dish and then implanted the eggs in the oviducts of about 10 female mice. The mice produced 78 offspring, two of which proved to have traces of the new DNA in their cells.

In 1981, scientists of the University of Wisconsin and the U.S. Department of Agriculture reported transferring a gene from a French bean seed into a cell of a sunflower plant and called the tissue that developed after the transfer "sunbean." The gene, which directs major protein production, is stable in the sunflower cell, and high levels of bean protein production are anticipated when technology is developed to regenerate a sunflower plant from the "sunbean" cells.

Among other genetically engineered organisms developed during the 1980s are the so-called ice-minus bacteria, which produce a protein coat that deters the formation of ice crystals and reduces frost damage to crops. Although this organism has enormous potential for extending growing seasons and ranges for a variety of crops, its developers experienced strong opposition to field testing, even though they received permission to do so. Much of the opposition was based on a fear of the unknown and the belief that the release of genetically engineered organisms outside the laboratory could be potentially disastrous if unanticipated effects, such as uncontrollable growth or virulent mutations, were to occur.

Later Developments

The United States in 1986 withdrew and then reinstated the first license for a genetically engineered vaccine that contained a herpes virus that could not duplicate itself. To date, there is no cure for herpes, and a vaccine to prevent it would be a major step in the anticipated eventual suppression or elimination of it and numerous other diseases.

In 1995, a researcher at Texas A&M University reported transferring into tobacco and potato plants a gene that in mice creates antibodies that protect against hepatis B. The hepatitis B vaccine series currently costs $120, but with genetically engineered plants doing all the work, scientists believe the cost of the vaccine can be reduced to about $1. Such cost reduction is particularly significant for countries such as China, where 150 million mostly poor people carry the disease.

In 1986, the Environmental Protection Agency granted the Mycogen Corporation of San Diego, California, permission to conduct small-scale field tests of a genetically engineered pesticide. In 1989, a Cornell University entomologist sprayed cabbage plants with a genetically disabled virus specific for cabbage loopers. The purpose of the Cornell test was to see if the genetically disabled virus would disappear quickly after attacking insects. If this proves to be the case, scientists later hope to splice into the cabbage looper virus a new gene that will be more deadly to insect pests but will disappear after spreading quickly throughout the insects. If these experiments are successful, dependence on the roughly three million pounds of dangerous pesticides used annually in the United States could be significantly reduced.

In 1995, the Canadian Department of Agriculture and Agri-Food approved for spring planting in Saskatchewan two genetically altered varieties of canola seed that are capable of surviving otherwise fatal herbicides. As a result, it is believed one application of a biodegradable herbicide may effectively replace several herbicide sprays per season.

Scientists from the United States and Australia have produced insect-resistant seeds by introducing into common peas a gene that prevents weevils from digesting starch. The gene is active only while the seed is dormant and becomes inactive after germination. In 1995, second generation plants that are seed weevil-resistant were produced. In these plants, the weevils are not killed but essentially starve to death. It is hoped that the gene can now be introduced into the seeds of many other crops.

Doctors at the National Institutes of Health in Bethesda, Maryland, received federal approval in 1989 to begin injecting terminal cancer patient volunteers with billions of their own white blood cells that had been engineered with bacterial genes that function as a tracking device. Although the "tracking" genes are not therapeutic, they do serve to aid doctors in monitoring how well the cells are fighting cancer.

Earlier experiments of a slightly different nature with 20 victims of a rapidly growing skin cancer used white blood cells, known as tumor-infiltrating lymphocytes, that had been cultured outside the body. A remarkable shrinkage of the tumors was demonstrated in 12 of the patients.

In California, a genetically engineered sun lotion that promises to protect humans from three types of cancer-causing radiation is being developed.

A list of some of the medicinal drugs already produced with the aid of genetic engineering is shown in Table 14.1.

Calgene has also successfully engineered a tobacco plant that is highly resistant to Roundup, a common commercial herbicide. If most crop plants could be engineered to become that resistant, herbicides selective for only undesirable weed plants could be sprayed on crop fields without affecting the desirable plants. In 1994, a Roundup-resistant soybean was developed that first was released to farmers in 1996.

Plant pathologists at the U.S. Department of Agriculture have isolated genes that can be bred into popular varieties of beans to make them resistant to 33 kinds of rust fungus. This is especially important because rust fungi can mutate and become resistant to fungicides in a single season. It takes numerous breeding steps to introduce the full complement of rust resistance genes into the beans, but once the process has been completed, spraying for fungi may be eliminated or at least significantly reduced.

Certain fish that inhabit Arctic and Antarctic regions produce an organic form of antifreeze that prevents water molecules in their bodies from freezing. Genetic engineers have successfully cloned the gene responsible for the antifreeze, which they expect soon to be able to produce and introduce into ice cream to prevent it from forming ice crystals after it has slightly thawed and then been refrozen.

FIGURE 14.3   Hybrid corn plants.

solution at just the right strength (usually 0.1% to 0.3%) to kill most of them and then checking for the small percentage (usually about 5%) of surviving seedlings whose cells have twice the original number of chromosomes.

Some tetraploids are artificially induced without the use of colchicine. In tomatoes, for example, wounding of the plant causes it to produce *callus* tissue whose cells may have double the basic complement of chromosomes. In corn, tetraploids may be produced by applying heat to seeds. Polyploidy is further discussed under the "Role of Hybridization in Evolution" in Chapter 15.

Mutations

Mutations, which involve a change in a gene or chromosome and occur naturally in all living organisms, can be artificially induced by radiation and chemicals. Like those occurring naturally, artificially induced mutations are largely harmful, and the expense of screening large populations for a rare desirable mutant form generally is not practical. A few useful modifications have been obtained, however, through induced mutation. These include higher yielding strains of *Penicillium* mold, blight resistant varieties of oats, and an improved grape from a variety whose individual fruits previously were bunched too compactly. Additional discussion of the role of mutations in evolution is given in Chapter 15.

TISSUE CULTURE AND MERICLONING

Tissue Culture

A *tissue culture* is essentially a mass of **callus** tissue (an undifferentiated group of cells) growing on an artificial medium. It can be started from almost any part of the plant, although tissues taken from the vicinity of meristems usually produce the best results. With the proper media, the callus tissue eventually differentiates into shoots and roots or sometimes into plant embryos. The technique was originally used in the culture of cells from carrot, tobacco, and citrus, but it is now used with a widespread variety of plants.

Using aseptic practices (i.e., practices that seek to eliminate bacteria and other disease organisms), the technique is designed to isolate unique tissues with the potential to become new disease-free plants; it also permits rapid multiplication of the materials. The best results are obtained when the procedures are conducted in a room or chamber that has been sterilized with ultraviolet light or has been steam-heated before work begins. All instruments, glassware, and media need to be sterilized in advance. An *autoclave*, which is something like a large pressure cooker, is generally used for this purpose (Fig. 14.4). The ingredients for the media are measured very precisely and usually include distilled water, various inorganic salts, and organic substances, such as yeast extracts, vitamins, sugar, and hormones.

many plants, including about half of all flowering plants, have more than two sets of chromosomes in their cells and are therefore **polyploid.** Polyploids that have three, four, six, or eight sets of chromosomes are said to be *triploid, tetraploid, hexaploid,* or *octoploid,* respectively. Polyploidy can arise in various ways. Hybridization followed by a doubling of the chromosome number, as exemplified by the bread wheat just mentioned, results in a polyploid called an *alloploid.* The chromosome sets can also be doubled when pairs of chromosomes fail to separate during mitosis or meiosis. The resulting polyploid is called an *autoploid;* a diploid plant could become a tetraploid by either alloploidy or autoploidy.

Since tetraploids are often bigger or more vigorous than their diploid counterparts, plant breeders sometimes try to develop tetraploids artificially by treating seeds with *colchicine,* an alkaloidal drug derived from the corms of the autumn crocus. The colchicine interferes with spindle formation during mitosis, so pairs of chromosomes do not separate during anaphase, and if the cell with the double set of chromosomes survives and becomes functional, a tetraploid plant may be the ultimate result. Developing tetraploids with colchicine usually involves treating seeds with a colchicine

FIGURE 14.4 An autoclave, in which pressurized steam is used to sterilize media, glassware, and instruments.

The considerable potential for crop improvement offered by tissue culture lies in the relative ease with which selection for desirable traits can be accomplished in large populations of cells cultured in a few glass containers instead of in populations of plants on large acreages of land and in varied environmental conditions. Manipulation of the cells often involves subjecting them to a variety of stresses, such as herbicide and other poisons, heat, cold, disease bacteria and fungi, and radiation, and then isolating cells that survive from among those that are susceptible.

Shoot Meristem Culture (Mericloning)

Orchids are among the most prized of all cultivated plants. They also produce the smallest seeds known, and it often takes seven or more years from seed until the first flower appears. Normally, the seeds have to become associated with a specific fungus before they will germinate, and such fungi may not occur outside of the orchid's natural habitat. In ad-

dition, because of variables inherited from the parents, orchid hybridizers do not know if they have been successful in developing plants with superior characteristics until the orchids flower several years later. The growers have partially overcome both problems, however, with the use of the shoot meristem, or *mericloning,* technique.

In mericloning, certain hormones and other substances are substituted for the fungus in artificial growing media. The technique usually involves the removal of the apical meristem and the youngest one or two leaf primordia, with the use of a dissecting microscope under aseptic conditions (Figs. 14.5 and 14.6). The meristem is then placed on a sterile medium and kept at room temperature under lights (Fig. 14.7). Adventitious roots usually develop within a month. The tissue that develops prior to the appearance of shoots and roots is subdivided as it grows, and each subdivision is then capable of becoming a new plant. By this process, commercial growers produce in a relatively short time thousands of plants, all genetically identical to the plant from which the meristem originated.

Artificial Seeds

Scientists working with tissue cultures are aware that, with the addition of certain hormones and nutrients to the culture medium, embryos will develop from the callus of some (but apparently not all) species of seed plants. Genetically identical embryos mass-produced through tissue culture can then be packaged with a food supply, hormones, and a biodegradable protective coat. The need for fertilizers and pesticides can also be reduced by adding nitrogen-fixing bacteria, fertilizers, pesticides, and other enhancements to the package. At present, the production of such artificial seeds is more costly than conventional seed production, but several companies are exploring more efficient production techniques. Already, however, the costs are partially offset by the fact that the artificial seeds produce a crop of identical plants, which significantly lessens the expense of harvesting.

Protoplast Fusion

Tissue culture also lends itself to various genetic engineering techniques, including the transfer of organelles from one cell to another and the hybridization of cells through *protoplast fusion*. This latter technique involves a culture of two different kinds of plant cells; the walls of the cells are digested with enzymes, leaving naked protoplasts. A small percentage of the protoplasts may, on their own, undergo fusion in the culture, resulting in hybrid cells with genes from both parent cells.

Other protoplasts can be stimulated by lasers or electrical charges to unite or be induced to do so with chemicals, such as diluted polyethylene glycol (antifreeze). The hybrid cells are isolated from those with identical genotypes by culturing them on a medium on which only the hybrid cells can grow. Regeneration of plants from fused protoplasts has

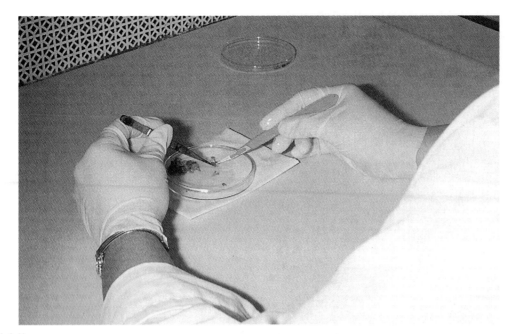

FIGURE 14.5   A skilled technician removing the apical meristem from an orchid plant.

proved to be difficult for monocots such as corn, wheat, and rice but seems promising for dicots.

If the organelles (e.g., chloroplasts) of one cell are more desirable than those of the other cell, the nucleus of the cell with less desirable organelles can be destroyed by radiation. A plant resulting from the fusion of two different protoplasts is called a *somatic hybrid;* those with the cytoplasm of two cells but only one nucleus are referred to as *cybrids*. Somatic hybrids frequently are less desirable than the parents, but several exceptions produced by protoplast fusion have resulted in potentially valuable grain crop plants.

Clonal Variants

Since tissue culture techniques came into widespread use, scientists have occasionally been surprised to find in their cell cultures cells that differ from others in appearance. For reasons that are not clear, such variants appear much more frequently than mutants do in field-grown crops. Some of the variants have superior color or other desirable vegetative characteristics, and scientists are now developing new varieties of vegetables and other food plants by searching for such variants in cell cultures.

Some cell cultures are derived from pollen grains or other haploid tissues or cells. Each cell in such cultures has only one set of chromosomes and therefore cannot develop spore mother cells that could undergo meiosis and develop gametes. However, the drug colchicine can be used to interfere with spindle formation during mitosis of such cells. This can result in two sets of chromosomes remaining in a single cell, and if this diploid cell divides and the tissue differentiates, a homozygous plant is the ultimate result. Producing a homozygous plant in this manner can be accomplished

within a year or two, which is about four years less time than it would take to produce a similar plant by ordinary breeding and selection in the field.

TRADITIONAL VEGETATIVE PROPAGATION

The value of being able to take a piece of stem of a desirable plant from one locality to another and induce it to grow into a new plant identical to its parent or of being able to divide rhizomes, tubers, and other parts of plants and have them each develop into new plants has been recognized since ancient times. *Grafting,* which involves the permanent union of parts of two plants, also has its origins in antiquity. Plants with desirable features, such as superior fruits or flowers or attractive leaves, may be grafted onto related plants whose chief attraction may lie in a vigorous root system or one that is resistant to disease.

The Chinese were apparently practicing grafting by the year 1000 B.C., and Theophrastus discussed grafting and other forms of vegetative propagation in his book *Causes of Plants,* written in the third century B.C. During the 14th and 15th centuries A.D., large numbers of plants were imported into European gardens and maintained by grafting. Even though the internal nature of the graft union was not fully understood, many forms of grafting were developed; no fewer than 119 different methods of grafting were described in 1821 by A. Thouin. Today, the practice of various forms of vegetative propagation is almost universal and is an important part of the economies of the world. Some of the more widespread techniques of vegetative propagation are

FIGURE 14.6 Flasks of sterile growing medium to which meristematic orchid tissue has been added. The flasks are rotated under lights. Roots and shoots appear within a few weeks.

FIGURE 14.7 The plants that have developed from the cultured meristematic tissue are separated and further cultured to maturity.

discussed in the following sections; details of grafting are discussed in Appendix 4.

Stem Cuttings

Occasionally, a basic change may occur in the chromosomes of a cell, and the result is a bud or a whole plant that develops differently from the rest of the tree or variety. Such a *mutation* is likely to be undesirable or to result in the death of the cell or tissue. In a few instances, however, the mutation may produce superior characteristics, such as better flowers or fruit, dwarf forms, or disease resistance. If you were to plant seeds from such a mutant form, the plant that would develop could differ as much from the mutant parent as you do from either your mother or your father. Since pollen from one parent has to be transferred to a flower of another parent for fruits and seeds to form, the mutant plant carries characteristics from both parents and thus may pass on to future generations a variety of combinations of characteristics.

When a mutation is desirable, the element of chance inheritance can be eliminated by cultivating tissue directly from the mutant plant, so that each of the offspring has characteristics (phenotypes) identical to those of the single parent; the desirable feature can thus be perpetuated. Wash-

ington navel oranges, for example, have come from a mutation that appeared on a sweet orange tree growing near Bahia, Brazil, around 1820. The grower, whose identity is not known, apparently recognized the superior nature of the fruit on a branch of one of the trees and, either by bud grafting or by making cuttings, multiplied the new form. Fifty years later, a missionary sent a dozen budded navel orange trees from Brazil to the U.S. Commissioner of Agriculture in Washington, D.C., and in 1873, two trees were shipped from Washington to Riverside, California. From these California trees, one of which is still living, almost the entire navel orange industry of the world has been developed through vegetative propagation.

In propagation by *cuttings,* which may have been used by the Brazilian orange grower, pieces of a plant are induced to produce roots and are then planted to grow on their own.

A large number of plants can be started with cuttings from a few trees or shrubs in a relatively small amount of space and time, and little skill is required. Cuttings also do not have the problem of incompatibility, which is found with grafting. It is not always desirable to propagate by cuttings, however, particularly if the variety concerned does not have a sturdy, disease-resistant root system. Some plants are also very difficult to start from cuttings because they do not readily form roots. Next to planting seeds, however, cuttings have become the most widely used means of multiplying plants in the world today.

Most cuttings are made from stems that are 0.6 to 2.5 centimeters (0.25 to 1.00 inch) thick and contain at least two nodes. They may vary in length from 7.5 centimeters to

The New Harvest: Bioengineered Foods

While George Washington probably never cut down a cherry tree, he certainly grew tomatoes in his Mt. Vernon garden. Tomatoes were a favorite gardening plant in colonial America and remain popular today as Americans spend about $3.5 billion a year to buy fresh tomatoes. But the fresh summer tomatoes grown by so many backyard gardeners have a new cousin—a genetically modified variety called the Flavr Savr™. Introduced to markets in 1994, the Flavr Savr™ tomato was the first genetically engineered food to reach supermarket shelves. Calgene, a biotechnology company in Davis, California, took 12 years to develop and receive FDA approval for the Flavr Savr™ tomato.

Why would Calgene spend this amount of time and money to create the Flavr Savr™? It will come as no surprise that of the most common items bought fresh in supermarkets, tomatoes rank last in customer satisfaction according to a survey conducted by the U.S. Department of Agriculture. So if a company could produce a better tasting tomato, it stands to make a good profit. The problem with commercially grown tomatoes is this— most fresh market tomatoes are harvested green, prior to the development of the tomatoes' full flavor potential. They are then transported in refrigerated trucks to storage houses where ethylene gas is used to induce the red coloration natural to the ripening process. Only then are

Vine ripened Flavr Savr™ tomatoes.
(Photo courtesy Calgene, Inc., Davis, CA)

tomatoes shipped to market. The typical result is a tomato that "looks red but eats green," with a taste far removed from home-grown, vine-ripened tomatoes.

Picking "green" is done because growers are in a race with nature. They need to get the tomatoes to market before spoilage occurs. If tomatoes are allowed to ripen on the vine, they begin to get soft because tomatoes slowly self-destruct for dispersal of their seeds.

Calgene reasoned that they could give growers an additional three to five days of vine-ripening time before harvesting—and flavor could be built on the vine resulting in a more flavorful tomato—if this softening process

---

60.0 centimeters (3 inches to 2 feet). With deciduous plants, the cuttings are made—while the buds are dormant—from healthy wood of the previous season that has been growing in the sun. Pieces may be cut transversely, or a small part of wood from an older branch may be left attached at the bottom end. If a number of cuttings are being made at one time, they may be tied together with rubber bands, placed bottom ends down in cool, damp sand or sawdust from which water will drain easily, covered, and stored until spring. They should be checked from time to time to see if buds are developing. If they are, the cuttings should either be planted or refrigerated, as development of buds without corresponding root growth will kill the material.

In the spring, the bottom ends of the cuttings can be inserted 4 to 6 centimeters (3.0 to 4.5 inches) in damp sand (to minimize problems involving soil fungi) or directly into the soil. Here adventitious roots develop, and normal growth

should occur. Evergreen plants tend to produce adventitious roots much more slowly, taking from several months to a year to produce sufficient growth to become established. Since cuttings may dry out, they need to be misted with water frequently, and they also need to be kept out of the sun and handled rapidly while they are being prepared. Leaves from the lower part of the cutting should be removed. If bottom heating is available, the additional warmth usually accelerates the development of the roots as long as high humidity can be maintained.

Leaf Cuttings

Houseplants, such as African violets, peperomias, begonias, sansevierias, and others with somewhat succulent leaves can be propagated in various ways from their leaves (Fig. 14.8). One commonly used method is to cover a jar or pan

---

could be retarded. Their solution was to slow down softening by inserting a new gene into the tomato plant.

Calgene scientists knew that the cells of tomatoes are held together by a natural chemical called pectin that acts like a cement. During ripening, an enzyme called polygalacturonase (PG) begins to dissolve the pectin, creating the softening. Their idea was to take the PG gene that specifies the enzyme polygalacturonase, cut it out of the chromosome with molecular scissors called restriction enzymes, and reinsert it **backwards** into the chromosome. This backward gene, called an **antisense gene,** produces a product (messenger RNA) that interferes with the naturally occurring PG messenger RNA. As a result, the amount of the PG enzyme produced is reduced significantly in the ripening tomato—and softening is delayed.

The process of inserting new genes into plants is these days relatively simple. Once Calgene researchers identified and cloned the PG gene, they reversed the gene and spliced this reversed, antisense PG gene into the DNA of a normal soil-inhabiting bacterium called *Agrobacterium.* It so happens that *Agrobacterium* can transfer genetic material into plants, a process that occurs when genes are inserted that cause the plant to produce a tumorous growth called Crown Gall disease. Making use of this natural infection (and DNA transfer) process, the Calgene team mixed *Agrobacterium* carrying the antisense PG gene with tomato leaf pieces in a petri dish. Because the *Agrobacterium* infection process is not 100% efficient (only approximately 30% of tomato cells actually become infected), a screening test must be performed to determine which tomato leaf cells contain the new gene. This is impossible to determine just by looking at the plants, so a "marker" gene is piggybacked along with the antisense PG gene. The marker gene used is a bacterial gene that codes for a protein that gives the bacterium resistance to the antibiotic *kanamycin.* The gene is called the kanamycin resistance gene (K_r). By growing the infected leaf pieces in a solution containing kanamycin, only those cells that have incorporated the K_r gene into their chromosome structure will survive this screening test. The uninfected cells will be killed because they do not carry the kanamycin resistance gene. The infected cells carrying both the antisense PG gene and the K_r gene can now be grown into mature plants and the seeds harvested.

These fundamental genetic engineering techniques can be applied to any number of genes that produce in a plant desirable characteristics such as improved flavor and nutritional value, resistance to insects and viruses, and so forth. Although the Flavr Savr™ tomato was the pioneer in the agricultural biotechnology revolution, other products like Monsanto's Roundup Ready soybeans (which withstand herbicides that kill a wide variety of weeds) have followed and many more are in the pipeline awaiting regulatory approval.

FIGURE 14.8   Leaf cuttings. *Left.* Disks of *Peperomia* leaves that have been placed on moist filter paper. *Right.* A *Peperomia* leaf with its petiole in water.

of water with foil or plastic film, punch small holes in the covering, and insert the petioles so that they are immersed at the base. The water should be changed occasionally. After a few weeks, adventitious roots develop at the base of the petiole. The leaves may then be planted, leaving the blade exposed; new plants gradually appear from beneath the soil.

Rex begonias are sometimes propagated by punching out disks from the leaf blades with an instrument that resembles a cookie cutter; each disk should be about 1.9 centimeters (0.75 inch) in diameter. The disks are placed on damp filter paper in petri dishes, where they develop both roots and shoots. Rex begonia leaves may also be induced to form new plants by making cuts about 3 millimeters (0.16 inch) deep across the main veins on the lower surface. Leaves are then placed lower side down on moist potting medium, and the edges of the leaf are pinned down with toothpicks. Eventually, new plants develop at the vein cuts, while the original leaf withers.

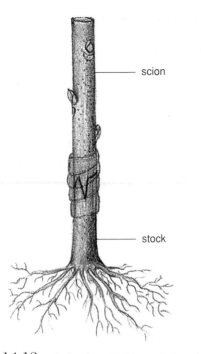

FIGURE 14.12 A simple graft. The rooted portion (*stock*) and portion to be grafted on to the stock (*scion*) are cut so that the two parts will fit together with the cambium of both portions in close contact.

Bulbs of hyacinth, grape hyacinth, squills, and similar plants can be induced to form bulblets by inverting them and cutting down about a third of the way with a sterilized knife, as though sectioning an apple, or by scooping out the center with a sterilized melon baller. After they are cut, the bulbs should be placed in dry sand in a warm place for about two weeks. During this time, callus tissue forms along the cuts. After the callus has formed, the bulbs should be removed from the sand and stored in trays at room temperature under humid but well-aerated conditions. Small bulblets form along the cuts during the ensuing two or three months. These are left attached, and the whole original bulb should be planted. After a year's growth, the bulblets can be separated and replanted independently. In addition to using sterile instruments, it may be necessary to use fungicides to prevent fungal diseases from damaging the materials.

Some bulbs, such as those of daffodils, tulips, and amaryllis, have dry scales on the outside and fleshy scales on the inside. Unlike those of the lilies, these bulbs do not respond to being separated. These plants and plants with corms (e.g., gladioli) naturally produce bulblets or cormlets, however, which may be separated and grown independently. The number of bulblets produced appears to be related to soil depth. Most develop when the original bulb is about 7.5 centimeters (3 inches) deep. Few, if any, are produced when the bulb is more than 17 centimeters (8 inches) deep.

Plants with fleshy storage roots, such as sweet potatoes or dahlias, can be propagated by dividing the root so that each section has a shoot bud. This is done in early spring, just before planting, when shoot buds have appeared.

Sweet potatoes can also be cut in pieces and induced to form new plants by suspending each segment with toothpicks from the top of a jar partially filled with water. The segments form adventitious shoots and additional roots in a relatively short time.

GRAFTING

Simple grafting basically involves the insertion of a short portion of stem, called a scion, into another stem with a root system, the **stock** (Fig. 14.12). A segment of stem, called an *interstock,* is sometimes grafted between the stock and the scion, or a single bud with a little surrounding tissue may be grafted onto a stock. In all grafts, *the cambia of the stock and scion must be placed tightly in contact with each other, and all parts being grafted must be related to one another.* For example, different varieties of apple may be grafted together, but it is not possible to graft elm or maple wood to an apple tree.

After the graft is made, the fit may be sufficiently tight without additional support, but it is customary to tie flat rubber strips, raffia, plastic tape, or waxed string around the union and cover it with grafting wax. The grafting wax is needed to prevent the tissues from drying and also to reduce the possibility of disease organisms gaining access to the interior of the stem. If the grafts are buried beneath the surface of the ground, they should be checked after growth begins to see if the wrapping is breaking down. If they are not below ground, the wrapping should be removed when it threatens to hinder the increase in girth that occurs when the cambium adds new tissue.

Discussions and illustrations of the major types of grafting are given in Appendix 4.

Summary

1. Genetic engineering involves the alteration of the DNA of a cell by artificially mixing large numbers of enzymatically broken DNA strands from two sources. Fragments from one source may associate with those from another source, creating recombinant DNA with parts of both original strands. The recombinant plasmids are inserted into live bacterial cells where they replicate. Since bacteria are easy to propagate, large amounts of recombinant DNA may be obtained in this fashion. Such recombinant DNA may be used in the future for improving crops and possibly eliminating diseases of both plants and animals.

2. Most economically important plants have been developed from wild ancestors through controlled plant breeding from available gene pools. Hybridization (which often produces hybrid vigor),

inbreeding, outcrossing, polyploidy, and mutations have all been employed in the development of new varieties. Norman Borlaug, the "Father of the Green Revolution," successfully increased productivity of crops through hybridization, but his methods are controversial because of growth requirements and adverse long-range effects on the ecology of agricultural lands.

3. Tissue culture, which involves the stimulation of isolated cells and tissues to grow on artificial media under sterile conditions, offers potential for crop improvement with considerably less space than is needed with field techniques. Tissue culture also lends itself to genetic engineering, protoplast fusion, artificial seed production, and clonal variant techniques.

4. Shoot meristem culture (mericloning) is employed in the mass production of certain plants, such as orchids; normal plant development from seed to flower is significantly reduced, and variations inherent in sexual reproduction are eliminated.

5. Vegetative propagation, which involves the growing of new plants from portions of other plants, began in antiquity. Cuttings of stems, leaves, and roots can be induced to grow into plants that are identical to the parent material. Such propagation is particularly useful in the perpetuation of favorable mutations, such as those of the navel orange. Layering may be used to propagate plants from the tips of flexible stems or from aerial stems. Almost any specialized plant stem or root can be propagated from portions of the parent material.

6. Grafting involves the union of an isolated stem portion (scion) with a rooted stem (stock). The cambium of both the stock and scion must be in close contact for the graft to be successful.

Review Questions

1. What is genetic engineering, and how is it accomplished?

2. How do restriction and repair enzymes perform their functions?

3. Give some current uses of genetic engineering.

4. Are there any potentially negative aspects to genetic engineering? Explain.

5. Explain the roles of hybridization, polyploidy, and mutations in plant breeding.

6. What is tissue culture, and how is it used commercially?

7. What is meant by the term *mericloning*?

8. Explain the difference between artificial and true seeds.

9. Give the major ways in which vegetative propagation is accomplished. What advantages are there to vegetative propagation as opposed to using seeds to propagate?

10. What is the difference between a stock and a scion?

Discussion Questions

1. Genetic engineering has already greatly improved the production of certain medicines, but research has been heavily restricted, despite the availability of funds for the work. Do you think restrictions should be lifted? Why?

2. Tissue culture and mericloning techniques have dramatically facilitated the production of desirable plants. What might be any long-range disadvantages to this and other forms of vegetative propagation?

3. What advantages might there be to having several varieties of apple or plum grafted on the same tree?

Additional Reading

Bajaj, Y. P. 1992. *High tech and micropropagation*, Nos. 1 & 2. New York: Springer-Verlag.

Bajaj, Y. P. (Ed.). 1989. *Plant protoplasts and genetic engineering*. New York: Springer-Verlag.

Bhojwani, S. S. (Ed.). 1991. *Plant tissue culture: Applications and limitations*. New York: Elsevier Science, Inc.

Blick, B. R., and J. J. Pasternak. 1994. *Molecular biotechnology: Fundamentals and applications of recombinant DNA*. Washington, DC: American Society of Microbiologists.

Endress, R. 1994. *Biotechnology of secondary products of plant cells*. New York: Springer-Verlag.

Hartmann, H. T., and D. E. Kester. 1990. *Plant propagation: Principles and practices*, 5th ed. Englewood Cliffs, NJ: Prentice-Hall.

Kyte, L. 1988. *Plants from test tubes: An introduction to micropropagation*. Portland, OR: Timber Press.

Lewis, R. 1992. *Life*. Dubuque, IA: Wm. C. Brown Publishers.

Stoner, C. H. 1992. *Biotechnology for hazardous waste*. Chelsea, MI: Lewis Publications, Inc.

Torrey, J. G. 1985. The development of plant biotechnology. *American Scientist* 73: 354–63.

Kahili ginger (Hedychium gardnerianum) *and Hawaiian tree ferns* (Cibotium *sp.*).

15 Evolution

Overview

The discussion of evolution begins with an abbreviated history of Charles Darwin and the principles or tenets involved in the theory of evolution through natural selection. Mechanisms of evolution, including mutations, hybridization and introgression, polyploidy, apomixis, and reproductive isolation are discussed. The chapter concludes with an examination of the role of isolation and mutation in modifying Darwinism and with an examination of the evidence for evolution.

Some Learning Goals

1. Understand how introgressive hybridization may lead to the development of new species.
2. Know the contributions of Charles Darwin to theories of organic evolution and the tenets of natural selection as he understood them.
3. Explain the significance of mutation and reproductive isolation to evolution.
4. Give reasons, past and present, for the controversy over evolutionary theory.

FIGURE 15.1 Charles Darwin.
(Courtesy National Library of Medicine)

Few historical events since 1859 have had a greater impact on society in general and the biological sciences in particular than Charles Darwin's book, *On the Origin of Species by Means of Natural Selection, or the Preservation of Favoured Races in the Struggle for Life.* Darwin's theory of evolution through natural selection has stimulated an enormous amount of thinking and research and has provided an explanation based on natural laws for the diversity of life around us. It also initially caused a great deal of controversy because, even though Darwin believed to his death in a divine creator, he also believed that the creator had used natural laws to bring all living things into being gradually over long periods of time, which ran against a literal interpretation of the Bible. Most of his contemporaries, guided by this literal interpretation of the biblical account of creation were convinced that all living things had been created in six days and had existed unchanged since the beginning.

Although the controversy has subsided, some disagreement still exists today. When the popular *Scofield Reference Bible* was first published in 1909, it included in the margin opposite the account of creation a date, "4004 B.C.," arrived at by the 17th century Irish archbishop James Ussher, who based his calculation on faulty interpretation of biblical genealogies. Ussher's date has been deleted from the margins of editions of the *Scofield Reference Bible* published since 1967, and the editors have observed that little evidence exists for fixing dates of biblical events prior to 2100 B.C.

Many *scientific creationists,* who since the 1970s increasingly have sought to have a nonevolutionary interpretation of the living world included in public school biology textbooks, do not necessarily believe the earth was created in 4004 B.C. or later, nor do they refuse to recognize the existence of minor variations in living organisms. The majority, however, believe the earth is not more than 30,000 years old and reject the foundations of evolution as incompatible with a literal interpretation of the biblical account of creation. In doing so, scientific creationists reject most of the evidence for evolution, including that which has accumulated since Darwin.

Part of the remaining disagreement also stems from a failure of people in diverse fields to define terms and to distinguish between fact and theory in evolutionary matters. Evolution itself, for example, has been broadly defined by some as simply being synonymous with change. We are told that anything from cars to computers to cultures is evolving and that even our thought processes evolve. To distinguish, therefore, between change in inanimate or intangible entities and progressive change in living organisms over time, we need to refer to the latter as *organic evolution*. Even this way we do not pin down the subject entirely, for there are variations in organic evolutionary theory. That understanding that pervades and unifies most biological thought today, however, finds its origin in the observations of Charles Darwin (Fig. 15.1) and of a contemporary of his, Alfred Wallace, who independently arrived at the same conclusions as Darwin.

CHARLES DARWIN

Charles Darwin was born in Shrewsbury, England, in 1809. His father, Dr. Robert Darwin, was a successful country physician and relatively wealthy. In 1825, at the age of 16, young Darwin was sent to the University of Edinburgh Medical School to follow in his father's footsteps. He did not do well in his studies, however, and dropped out after two years. The following year, he went to Cambridge University to study for the ministry but did not excel academically in this field either. Part of the reason for his poor scholastic showing apparently was due to his spending much time in the countryside collecting beetles, rocks, and other specimens or talking at length with his Cambridge biology professors, who held him in high regard. He barely attained what today would be called a C average, but he managed to graduate with a degree in theology in 1831. He was then 22 years old but was still not really sure what he wanted to do with his life.

While Darwin was pondering such matters, King William IV of England commissioned a sailing vessel, the HMS *Beagle*, to undertake a voyage around the world for the purpose of charting coastlines, particularly those of South America. Young Darwin's Cambridge biology professors recommended him for the (unpaid) post of assistant naturalist and captain's companion on the voyage. His father was opposed to the idea, but his uncle sided with him, and Darwin accepted the position.

The voyage, which began December 27, 1831, and took five years, gave Darwin an opportunity to collect specimens on both sides of South America, as well as in the Galápagos Islands and along the coasts of Australia and New Zealand. He kept a daily journal and spent countless hours alone on horseback collecting and observing the living world around him. This gave him ample opportunity to think about the forms and distribution of the myriad new organisms he encountered. His thoughts slowly led to the development of ideas that later blossomed into his theory of evolution through natural selection.

Upon his return to England in 1836, Darwin obtained the financial support of his father and his cousin Emma, whom he married. Although still young, he retired to the country and began working on his collections and journal. He also carried on a voluminous correspondence with other biologists and made extensive investigations into pollinating mechanisms, earthworm ecology, geographical distributions of plants and animals, and several other areas of natural history.

Throughout all of his activities, he was guided by a concept that he had adapted from an essay written by Malthus in 1798 on human populations and food supplies. Darwin realized that, although human beings might artificially improve or increase their food supply through selective breeding and cultivation, plants and animals could not do so and were therefore vulnerable to a process of selection in nature, which would explain changes in natural populations. He was reluctant to publish his ideas, however, and did not begin putting together his book on the origin of species until 1856.

Meanwhile, an English naturalist by the name of Alfred R. Wallace (1823–1913), who made major contributions to our knowledge of animal geography, independently concluded that natural selection contributed to the origin of new species and sent Darwin a brief essay on the topic in 1858. At the urging of friends, Darwin published a statement of his views jointly with Wallace, and, in 1859, Darwin's classic book *On the Origin of Species by Means of Natural Selection* was published.

Our understanding of the term *species* has been modified and refined since Darwin's time, but, as discussed in the next chapter, most biologists today think of a species in terms of a "population of individuals capable of interbreeding freely with one another but not freely with those of another species."

TENETS OF NATURAL SELECTION

In essence, Darwin's theory of evolution through natural selection is based on four principles or tenets:

1. *Overproduction.* Many living organisms produce enormous quantities of reproductive cells or offspring. For example, a single maple tree produces thousands of seeds each year, most being capable of becoming a new tree. Some fungi produce trillions of spores, each with the potential to become a new fungus.

2. *Struggle for Existence.* All the germinating seeds, spores, and other reproductive cells of living organisms compete for available moisture, light, nutrients, and space. The amounts of these elements available in nature are not sufficient to support all of these organisms, and many die as a result.

3. *Inheritance and Accumulation of Favorable Variations.* All living organisms vary. Those hereditary variations with survival value are inherited from generation to generation and accumulate with time, while other variations not important to the survival of the species are gradually eliminated.

4. *Survival and Reproduction of the Fittest.* Those forms of organisms best adapted to the environment have the best chance to survive and reproduce, while others less well adapted may succumb. A tree with thicker bark, for example, has a variation favorable to surviving cold temperatures; hence it may survive to reproductive age and leave more offspring, which will bear the inherited features of the thicker-barked tree.

One of the criticisms of Darwin's theory, published in 1859, was that it did not explain how hereditary variations originated and developed. It should be remembered, however, that Mendel's findings were not published until 1866, and the details of mitosis and meiosis did not become known until 1900 to 1906. Today, with our far greater knowledge of how variations occur and are inherited, we have come to understand the mechanisms of evolution in populations much better than was possible in Darwin's time.

All organisms, from aardvarks to zinnias, occur in populations that are composed of a few to billions of individuals.

Yet even within very large populations, it is virtually impossible to find two that are identical down to the last molecule. As humans, we are well aware of this within our own species, but variation exists in all living organisms, even though in simpler one-celled forms the differences may be much less obvious. Whether the differences are obvious or subtle, however, the general characteristics of a population will eventually change, if the environment or other factors favor certain hereditary variations and bring about the elimination of others.

In some instances, the environment may alter the phenotype slightly without affecting the genetic constitution of an organism. Plants that grow relatively tall at sea level may become dwarfed when they are transplanted to cooler areas in the mountains or drier areas near deserts, yet they are capable of breeding freely with plants at the original location if they are returned to that area.

If we apply fertilizers to plants, stimulate their growth with hormones, or prune them, the changes are not passed on to the offspring, for no permanent change occurs in a given population unless there is *inheritable variation*. The changes in transplanted, fertilized, or pruned plants are not transmitted to the offspring because the gametes of those plants will carry the same genetic information they would have carried if the transplanting, fertilizing, or pruning had not occurred.

Before the details of genetics became known, many prominent biologists believed that hereditary changes in populations over long periods of time (evolution) occurred as a result of the inheritance of acquired characteristics. One of the more prominent supporters of this widespread idea was Jean Baptiste Lamarck (1744–1829). He believed that giraffes acquired their long necks over many generations as a result of the gradual increase in neck length as shorter-necked animals stretched to reach leaves on the branches of trees. Slight stretching of the neck was supposed to have been passed on to the offspring as it occurred, and eventually numerous tiny increases due to individual stretching added up to the present great neck length of giraffes (Fig. 15.2).

If acquired characteristics could be inherited, we should be able to demonstrate it experimentally, and, indeed, many workers have attempted to do so, but all have failed. For example, one biologist surgically removed the tails of mice for many successive generations, but the average length of the tails of the last generation was exactly the same as that of the first generation. The experiment demonstrated that repeatedly removing tails in no way affects the hereditary characteristics carried in the genes within the cells. This is also the reason fruit trees that are pruned annually never produce seeds that grow into dwarfed trees, even after many generations.

Despite the experiments just discussed, short-tailed mice and dwarf fruit trees do occasionally occur but for reasons quite different from those proposed by Lamarck and his contemporaries. They come about as a result of a sudden change in a gene or chromosome. Such a change is called a **mutation,** a term introduced in 1901 by the Dutch botanist Hugo de Vries.

FIGURE 15.2   Lamarck and his contemporaries believed that the long neck of a giraffe developed over time as the animals stretched to reach higher leaves, and that the little length increases were inherited and became cumulative. This theory of inheritance of acquired characteristics was experimentally disproved and discarded when the true mechanisms of gene inheritance became known.

Changes within chromosomes may occur in several ways. A part of a chromosome may break off and be lost (*deletion*), or a piece of a chromosome may become attached to another (*translocation*). In some instances, a part of a chromosome may break off and then become reattached in an inverted position (*inversion*) (Fig. 15.3). A mutation of a gene itself may involve a change in one or more nucleotide pairs (nucleotides are discussed in Chapter 13).

Mutation rates vary considerably from gene to gene, but mutations occur constantly in all living organisms at an average estimated to be roughly one mutant gene for every 200,000 produced. *Mutator genes* that increase the rate of mutation in other genes have been discovered, but generally the mutation rate for a specific gene remains relatively constant unless changes in the environment (e.g., increases in cosmic radiation) occur. Most mutations are harmful, many times to the point of killing the cell. About 1% or fewer are either harmless or produce a characteristic that may help the organism survive changes in its environment.

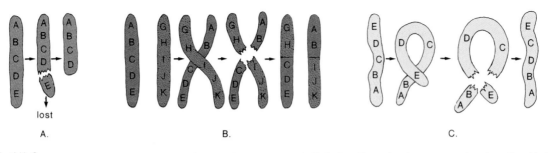

FIGURE 15.3 Some types of chromosomal changes that can occur. *A.* **Deletion.** Part of a chromosome breaks off and is lost. *B.* **Translocation.** Part of one chromosome becomes attached to another. *C.* **Inversion.** Part of a chromosome breaks off, becomes inverted from its original position, and then is reattached.

MODERN APPROACHES TO EVOLUTION

Modern theories of evolution combine Darwin's ideas with those of Mendel and include the Hardy-Weinberg law, discussed on page 220. They recognize today that evolution is based primarily on changes in gene frequencies resulting from mutation, *reproductive isolation,* and recombination of genes through hybridization.

THE ROLE OF HYBRIDIZATION IN EVOLUTION

Hybridization (the production of offspring by two parents that differ in one or more characteristics) seldom occurs naturally, but when it does take place, the hybrids may be significant or important in evolutionary change, depending on how the characteristics of the parents were combined. If, for example, the environment changes (e.g., average temperatures drop or annual precipitation increases), hybrids may have gene combinations that are better suited to the new environment than those of either parent.

If two species hybridize occasionally, *introgression* (backcrossing between the hybrids and the parents) may occur. If the backcrossing occurs repeatedly, some characteristics of the parents may eventually disappear from the population, if the new combinations of genes in the offspring happen to be better suited to the environment than those of the parents and as natural selection favors the offspring. Both parents and hybrids may, however, also evolve in other ways.

Polyploidy occurs occasionally in nature when, during mitosis, a new cell wall fails to develop between two daughter cells, even though the chromosomes have divided. As observed during our discussion of plant breeding in Chapter 14, this situation can result in a cell with twice the original number of chromosomes. If mitosis were to occur normally after such a cell was formed and the cell divided repeatedly until a complete organism resulted, that organism would have double the original number of chromosomes in all its cells.

The hybrids resulting from a cross between two species are often sterile because the chromosomes do not pair up properly in meiosis. If polyploidy does occur in such a hybrid, however, the extra set of chromosomes present from each parent provides an opportunity for any one chromosome to pair with its homologue in meiosis, and thus possibly the problem of sterility may be overcome. This type of polyploidy apparently occurred frequently in the past (in terms of geological time), and it is believed that more than 100,000 species of flowering plants that exist today originated in this fashion (Fig. 15.4).

Sterile hybrids also may reproduce asexually. Asexual propagation, called **apomixis,** may include the development and production of seeds without fertilization, as well as other forms of vegetative reproduction discussed in chapters 4 through 7 and 14, and in Appendix 4.

When species reproduce mainly by apomixis but sometimes also hybridize so that new combinations of genes are occasionally produced, they can be highly successful in nature. Dandelions and wild blackberries, for example, reproduce apomictically, as well as by other means, and are among the most successful plants known (Fig. 15.5).

REPRODUCTIVE ISOLATION

If new genes are produced in a freely interbreeding population, they will gradually be spread throughout the population, and the nature of the whole population will change in time. If some barrier divides the population, however, the two new populations eventually may become distinct from each other, sometimes in a relatively short period of time.

The log of a Portuguese sailing vessel of more than 600 years ago indicates that, for unknown reasons, some rabbits native to Portugal were released on one of the Canary Islands during a visit. When 20th century biologists examined the island descendants of those rabbits, which in Portugal forage during the day, they found them to be smaller than their continental ancestors, to be nocturnal in foraging habits, and to have larger eyes. In addition, attempts at breeding them back to their European ancestors failed because, in the short space of 600 years, a new species of rabbit had evolved.

FIGURE 15.4 Fireweed (*Epilobium angustifolium*)—a polyploid found primarily in mountainous areas of North America below 3,050 meters (10,000 feet). One subspecies has four sets of chromosomes, with a diploid chromosome number of $2n = 36$. A second subspecies has 8 sets of chromosomes, with a diploid chromosome number of $2n = 72$.

FIGURE 15.5 A common dandelion. In addition to reproducing by ordinary vegetative and sexual means, dandelions reproduce apomictically. Apomixis is a form of sexual reproduction through which seeds are produced without fertilization.

How do two populations of organisms that initially have the same gene pool come to have gene pools different enough to prevent their interbreeding? Geographic or other isolation of the two populations from each other prevents the flow of genes between the two populations, and random mutations, which are rarely identical, then spread only throughout the population in which they arise. In time, the genetic changes may become so great that, even when the isolation is removed, gene flow between the two populations no longer can occur.

In the United States, there are two closely related species of small trees or shrubs called *redbuds* that look very much alike. The eastern redbud occurs on the borders of streams, mostly east of the Mississippi River between the Canadian border and Florida; the western redbud is native to stream areas in California, Utah, Nevada, and Arizona. The two species can be artificially hybridized, but each is so adapted to its own wild habitat and associated climate that specimens of either species succumb when transplanted to the other's wild habitat (Fig. 15.6).

Several other factors contribute to the development of new species from geographically isolated populations with a common ancestry. When separation first occurs it is most unlikely that both populations will have genes that are identical in all respects, and a small population will have only a small percentage of the genetic variation present throughout the original population. In addition, geographically isolated populations normally will be subjected to selection pressures from numerous subtle to conspicuous differences in environment. The Canary Island rabbits, for example, initially probably found it difficult to compete with other animals for food during the day, and as mutations for improved night vision occurred, those acquiring the genes were able to survive while those without them perished.

Isolation leading to the development of new species is not limited to physical barriers, such as mountains or oceans, or to climate. Soils may play a role, or a mutant form within

A.

B.

FIGURE 15.6   Two species of redbud, both native to North America. *A*. Eastern redbud (*Cercis canadensis*). *B*. Western redbud (*Cercis occidentalis*).

A.

B.

FIGURE 15.7   *A*. Dutchman's breeches (*Dicentra cucullaria*). *B*. Squirrel corn (*Dicentra canadensis*).

a population may flower at a different time, preventing exchange of genes between it and nonmutant forms.

In the temperate deciduous woods of eastern North America and in the Columbia River basin in the Pacific Northwest, there are many populations of early spring-flowering herbs called *Dutchman's breeches,* which are discussed further in Chapter 16. These have delicate, highly dissected leaves, which are nearly indistinguishable from those of related but slightly later-flowering squirrel corn plants often found growing right among the Dutchman's breeches in eastern North America (Fig. 15.7). So closely do the plants resemble each other vegetatively that early botanists and many lay persons assumed they were the same species.

It is believed, however, that at some point in geological history a mutation or mutations occurred that caused some plants to begin flowering after other plants had already set seed. As a result, one group became reproductively isolated from the other group. In due course, other mutations af-

fecting the form and fragrance of the flowers and the shape and pigmentation of food-storage bulblets beneath the surface also occurred, but the plants continued to occupy the same habitats. In short, we now have two closely related but distinct species growing together without interbreeding, simply because it is no longer possible for them to do so.

Other isolating mechanisms may be mechanical. In orchids, for example, the pollen is usually produced in little sacs called *pollinia* that stick to the heads or bodies of visiting insects. If pollination is to occur, the pollinia must be inserted within a concave stigma. Each species of orchid is constructed so that it is highly unlikely that a pollinium of one species will be inserted within the stigma of another species. As a result, many species of related orchids can be *sympatric* (occupy the same territorial range) without genes being exchanged.

Four closely related sympatric species of Peruvian *Catasetum* orchids, which can easily be artificially hybridized, have no known natural hybrids, despite their being pollinated by a single species of bee. Microscopic examination of the pollinators has shown that the pollinium of one species is attached to the insect's head, that of another is attached to the insect's back, that of a third to the abdomen, and that of the fourth only to the left front leg. Even after visits to hundreds of flowers, none of the pollinia is misplaced (Fig. 15.8).

Even if pollen from one species is placed on or within the stigma of another species, however, fertilization frequently does not follow because the sperm is chemically or mechanically prevented from reaching the egg. Other isolating mechanisms include the failure of hybrids to survive or breed and the failure of embryos to develop.

RATES OF EVOLUTION

Darwin believed that evolution by natural selection was a slow and gradual process. A number of contemporary biologists, however, favor variations of *punctuated equilibrium* theories that hold that major changes have taken place in spurts of maybe 100,000 years or less, followed by periods of millions of years during which changes, if any, have been minor. They base their theories on fossil records, which have large "chains" of missing organisms. Although missing-link fossils are occasionally discovered, the record does little to support Darwin's concept of gradual, long-term change, even though it is widely acknowledged that possibly as few as one organism in a million ever became a fossil. Others opposed to theories of evolution through sudden change argue that because such a tiny percentage of organisms becomes fossilized and usually only the harder parts of organisms (e.g., bones, teeth) are preserved, drawing definite conclusions from fossil evidence about evolution through either gradual or sudden change is not warranted.

As indicated in Chapter 21, the conditions necessary for an organism to become a fossil are very specialized and limited in occurrence and probably also were in the past. This makes it quite improbable that large numbers of missing-link fossils will ever be found. Proponents of evolution through periods of rapid change argue that, under conditions of changing climates or other situations exerting strong selection pressures on forms with adaptive mutations, new species of organisms could arise in less than 100 generations, making 100,000 years ample time for considerable evolution to occur. Since it is not possible to prove or disprove the various theories experimentally, the debate on evolution rates will undoubtedly continue indefinitely, until or unless new evidence convincingly supports one theory more than another.

FIGURE 15.8 *Catasetum* orchids.

DISCUSSION

If we now know and understand some of the mechanisms of evolution, why does there yet remain any controversy over the broad subject itself? Obviously, ignorance is a factor, but it is not the sole reason. Science deals with tangible facts and evidence that can be experimentally tested; beliefs stemming from metaphysics and religion are outside the realm of science to prove or disprove.

Evidence in support of evolution is drawn from several additional areas, including the form and ecology of living organisms and the way they are related to each other today. Other evidence comes from the structure and relationships of proteins and other molecules and from fossils, which are remnants of previously living forms (Fig. 15.9). The simplest fossils are generally found in the oldest geological strata, while more complex forms tend to be found in younger strata.

Still further evidence is drawn from the geographical distribution of organisms. Many groups are confined to a single continent or island. In some instances, where similar

FIGURE 15.9   A fossil fern.

organisms occur on more than one land mass, there is evidence that the land masses concerned were once linked together, which would have permitted terrestrial migration. Still other conclusions are drawn from the physiology and chemistry of the organisms and, in animals, particularly, from the presence of vestigial organs (structures that are apparently now functionless but are presumed to represent previously functional organs in predecessor species). Among the best known of such structures in humans are the appendix and wisdom teeth.

Most scientists who have studied the evidence feel that some form of evolutionary process is the only plausible explanation for the unity of life at the molecular and cellular level and the extraordinary diversity of the organisms now around us. There is little unanimity of thought, however, as to precisely how evolution proceeded in the past. One authority is convinced that a certain group evolved from another, while other equally eminent authorities maintain that the exact reverse occurred. Part of the reason for such paradoxes is that the historical record is quite incomplete.

Although the fossil evidence for the evolution of a few organisms, such as certain mollusks and the horse, is fairly substantial, other fossil evidence is very fragmentary. As mentioned in the previous section, possibly fewer than one in each million organisms that once existed ever became a fossil, and there are probably no fossils at all of many herbaceous and soft-bodied organisms. With such evidence, scientists can deal only in probabilities or possibilities, and it is inevitable that various, sometimes conflicting, interpretations result.

Objective scientists freely acknowledge that numerous problems concerning the interpretation of the geological past and the pathways of organic evolution exist, but they ask their detractors to suggest more plausible alternatives. At this point, some people apply the tenets of religious faith, which like history, are not subject to scientific experimentation. Some see no conflict between science and religion, and others are convinced that the two are mutually exclusive. Depending on individual points of view, an impasse may result, with persons of different persuasions—including, unfortunately,

some scientists—becoming dogmatic on the topic. Virtually no objective thinkers will deny, however, the extraordinary impact theories of evolution have had on modern peoples and on their concepts of the living world around them.

Summary

1. The theory of organic evolution received its greatest impetus from Charles Darwin's *On the Origin of Species.*

2. Darwin's theory is based on four principles or tenets: (1) overproduction; (2) struggle for existence; (3) variation and inheritance; and (4) survival and reproduction of the fittest.

3. Before 1900, many biologists, notably Lamarck, believed that evolution occurred as a result of the inheritance of acquired characteristics. This theory was discredited experimentally.

4. Mechanisms of evolution include mutations in chromosomes or genes, hybridization (particularly introgression), polyploidy, apomixis, and reproductive isolation.

5. If new genes are produced in a freely interbreeding population, they will gradually be spread throughout the population, and the nature of the whole population will change in time. If a population is divided by a barrier, genes occurring in the one population will not spread throughout the isolated population as before. In time, because of the isolation, each new population may develop into separate species incapable of breeding with one another.

6. Darwin believed that evolution through natural selection was a gradual process over great periods of time. Some contemporary biologists believe that evolution has taken place in spurts between long periods of little change, based on evidence from the fossil record. Interpretation of the fossil record is, however, controversial and will be debated indefinitely.

7. Additional evidence in support of evolution is drawn from the form, the ecology, and relationships of living organisms; molecular structure; geographical distribution; fossils; and vestigial organs.

8. Most scientists feel some form of evolution is the only plausible explanation for the unity of life at the molecular and cellular level and the great diversity of life, but there is little agreement among them as to precisely how evolution proceeded in the past.

9. There is a variety of opinions and convictions on origins, but few advocates can deny the major impact that theories of evolution have had on modern peoples and on their concepts of life.

Review Questions

1. What is the difference between organic evolution and other forms of evolution?

2. How did Darwin's theory of evolution differ from Lamarck's?

3. What basic modifications have been made in evolutionary theory since Darwin's time?

4. Why is a hybrid often sterile?

5. What is meant by a *mechanism of evolution*?

6. Why is reproductive isolation necessary for evolution to occur?

7. What evidence is there to support modern concepts of evolution? Are there any problems with the evidence?

Discussion Questions

1. One of Darwin's tenets was that there is a struggle for existence among living organisms. Do plants struggle with one another to survive? If so, how do they do it?

2. Some populations change noticeably in form within a hundred years. If only one gene in every 200,000 per cell mutates and if most mutations are harmful, how is such change possible?

3. Do you think there might be viable alternatives to organic evolutionary theory to account for the diversity of life around us? Explain.

Additional Reading

Avers, C. J. 1989. *Process and pattern in evolution*. New York: Oxford University Press.

Ayala, F. J. 1991. *Origin of species*, 2d ed. In J. J. Head (Ed.), *Carolina Biology Readers*. Burlington, NC: Carolina Biological Supply Company.

Babb, N. 1994. *Evidence for evolution*. New York: Vantage Press, Inc.

Baltscheffsky, H., et al. (Eds.). 1987. *Molecular evolution of life*. New York: Cambridge University Press.

Darwin, C. 1975. *On the origin of species,* facsimile 1st ed. of 1859. New York: Cambridge University Press.

Grant, V. 1991. *The evolutionary process: A critical review of evolutionary theory*, rev. ed. New York: Columbia University Press.

Kerkut, G. A. 1960. *Implications of evolution*. Oxford and New York: Pergamon Press.

Stebbins, G. L. 1977. *Processes of organic evolution*, 3d ed. Englewood Cliffs, NJ: Prentice-Hall.

Stone, E. M., and R. J. Schwartz. 1990. *Intervening sequences in evolution and development*. New York: Oxford University Press.

A flower of Chinese hibiscus (Hibiscus rosa-sinensis), *the most widely cultivated of some 200 hibiscus species. In India the flowers are used for shining shoes. Other species are grown commercially as a source of fiber for canvas, rope, and sails.*

16 Plant Names and Classification

Overview

This chapter begins with a discussion of the problems involved in the use of common names for plants, as illustrated by a survey of such names for two related species of American spring-flowering perennials. It continues with a brief historical account of the events that led to the development and acceptance of Linnaeus's Binomial System of Nomenclature. The history of the development of a five-kingdom classification is presented, along with a list of the divisions and classes included in the kingdoms covered in this text. A dichotomous key to the kingdoms and divisions of organisms is provided. The chapter concludes with a brief discussion of cladistics.

Some Learning Goals

1. Understand several problems associated with the application of common names to organisms.
2. Know what the Binomial System of Nomenclature is, how it developed, and how it is currently used.
3. Learn several reasons for recognizing more than two kingdoms of living organisms.
4. Understand the bases for Whittaker's five-kingdom system.
5. Construct a dichotomous key to 10 different objects (e.g., pencils, golf balls, crayons, balloons, clocks).

FIGURE 16.1 Plantain (*Plantago major*). These plants have at least 300 different common names.

The general public thinks biologists are slightly pompous or weird when they refer to peanuts as *Arachis hypogaea* or an African hoopoe bird as *Upupa epops*. The biologists aren't, however, merely showing off or being difficult. Rather, they're identifying organisms by their *scientific names* through a system that has evolved over the last few centuries. It has now become vital to us, regardless of the location or language, to be able to distinguish among the 10 million existing types of organisms, as well as those that have become extinct, and to do so in a way that will identify them anywhere.

At present, all living organisms are given a single two-part Latin scientific name, and most also have *common names*. Only one correct scientific name applies to all individuals of a specific kind of organism, no matter where they're found, but many common names may be given to the same organism, and one common name may be applied to a number of different organisms.

The scientific name *Dicentra cucullaria*, for example, has been given to a pretty spring-flowering plant native to eastern North America and the Columbia River basin in the West. Its unique flower shape, which recalls the baggy pants of a traditional Dutch costume, has resulted in its being given the common name *Dutchman's breeches*. It also, however, has the common names *little-boys' breeches, kitten's breeches, breeches-flower, boys-and-girls, Indian boys-and-girls, monkshood, white eardrops, soldier's cap, colicweed, little blue stagger, white hearts, butterfly banners, rice roots,* and *meadow bells*. In addition to these English common names, the plant has Indian names, and in the Canadian province of Quebec it has French names.

Often growing with *Dicentra cucullaria* is a related plant, with similar leaves but with slightly differently shaped flowers, having the scientific name *Dicentra canadensis*. Because of the similarities between the two plants and their close association in the woods where they occur, some people in the past assumed that they were merely two different forms of the same plant. This is reflected in the fact that some of the common names of *Dicentra cucullaria* (e.g., colicweed, Indian boys-and-girls, little blue stagger) have also been applied to *Dicentra canadensis*, in addition to the names *squirrel corn, turkey corn, turkey pea, wild hyacinth, fumitory, staggerweed,* and *trembling stagger*. But the problem of the diversity and the overlapping of common names for these two plants doesn't stop there, for the names *monkshood, soldier's cap, rice roots, meadow bells,* and *turkey corn* have also been applied to completely unrelated plants. Both species of *Dicentra* are shown in Figure 15.7.

In Europe, with its many languages, common names can become very numerous indeed. The widespread weed with the scientific name *Plantago major*, for example, is often called *broad-leaved plantain* in English, but it also has no fewer than 45 other English names, 11 French names, 75 Dutch names, 106 German names, and possibly as many as several hundred more names in other languages, with literally dozens of these common names also applying to quite different plants (Fig. 16.1). If it weren't for the early

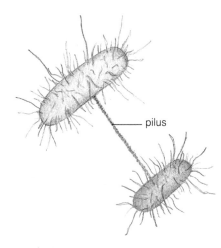

FIGURE 17.3 Conjugation in bacteria. Part of the DNA strand of the donor cell migrates through the hollow, tubelike *pilus* to the recipient cell, where it is incorporated into the DNA of its new cell. Shorter pili cover the surfaces of both cells.

Size, Form, and Classification of Bacteria

Except for the blue-green bacteria, which are discussed separately, and a few other giant bacteria, which attain lengths of up to 60 micrometers, most are less than 2 or 3 micrometers in diameter, and a few of the smaller species approach 0.15 micrometer. The latter are so small that 6,500 of them arranged in a row would not quite extend across the head of a pin. Because of their small size, bacteria are not visible individually to the unaided eye and are studied in the laboratory with electron microscopes or with the highest magnifications of light microscopes.

The thousands of different kinds of bacteria occur primarily in three forms: *cocci* (singular: *coccus*), which are spherical or sometimes elliptical; *bacilli* (singular: *bacillus*), which are rod-shaped or cylindrical; and *spirilli* (singular: *spirillum*), which are in the form of a helix, or spiral (Fig. 17.4). Further classification is based on numerous visible features including, for example, presence of pigments, development of slimy or gummy capsulelike sheaths around cells, presence of hairlike or budlike appendages, development of internal thicker-walled *endospores,* and gliding movements by which threadlike groups of bacteria appear to slide back and forth.

Some bacteria have slender flagella, usually about 5 to 10 micrometers in length, which propel them through fluid media. Others have somewhat shorter tubelike pili, which resemble flagella but do not function in locomotion. The pili apparently enable bacteria to attach themselves to surfaces or to each other.

Bacteria are also grouped into two large categories based on their reaction to a dye. After a heat-fixed smear of cells has been stained blue-black with a violet dye a dilute iodine solution, alcohol, or acetone is added. When so

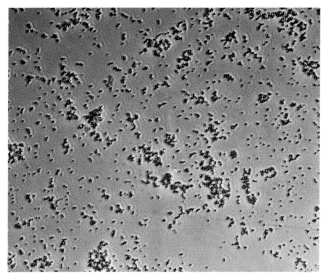

A.

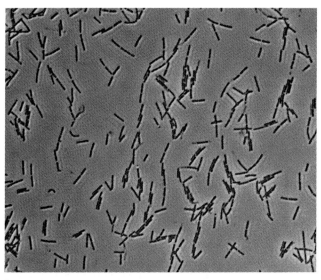

B.

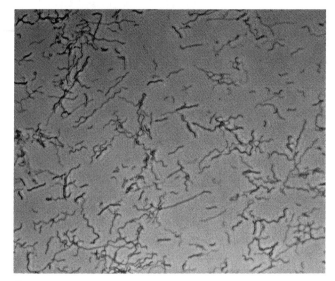

C.

FIGURE 17.4 Three basic forms of bacteria. *A*. Cocci. *B*. Bacilli. *C*. Spirilli, ×1,500.

treated, some species rapidly lose their color (but absorb pink safranin dye), and are called *gram-negative*. Others, called *gram-positive*, retain most of the blue-black color. Gram's stain, named after Christian Gram who discovered it in 1884, has many variations, but all the variations produce similar results. Other characteristics used in classifying bacteria are indicated in the discussions that follow.

SUBKINGDOM ARCHAEBACTERIOBIONTA— THE ARCHAEBACTERIA

The *archaebacteria* represent one of two quite distinct lines of the most primitive known living organisms. They are fundamentally different in their metabolism from the other line of bacteria, the *eubacteria,* and also differ in the unique sequence of bases in their RNA molecules, the lack of muramic acid in their walls, and in the production of distinctive lipids.

In the early 1980s, these basic differences led University of Illinois microbiologist Carl Woese and his colleagues, who have conducted considerable research on the archaebacteria, to suggest that the organisms belong in a kingdom of their own. At present, three distinct groups of bacteria are included in the archaebacteria.

The Methane Bacteria

The *methane bacteria,* which comprise the lion's share of the archaebacteria, are killed by oxygen and so are active only under anaerobic conditions found in swamps, ocean and lake sediments, hot springs, animal intestinal tracts, sewage treatment plants, and other areas not open to the air. Their energy is derived from the generation of methane gas from carbon dioxide and hydrogen.

Methane, or "marsh gas," is a principal component of natural gas and may have been a major part of the earth's atmosphere in early geologic times. It still is present in the atmospheres of the planets Jupiter, Saturn, Uranus, and Neptune and is the main ingredient of firedamp, which causes serious explosions in mines. Methane will burn when it constitutes only 5% to 6% of the air, and a flitting, dancing light called *ignis fatuus,* or "will-o'-the-wisp," which is occasionally seen at night over swamps and marshy places, is said to be due to the spontaneous combustion of the gas.

The Salt Bacteria

Commercial salt evaporation ponds and other shallow areas in bodies of water with high salt content often have a unique appearance from above. They can be strikingly red due to the presence of a distinctive group of archaebacteria. These are the *salt bacteria,* whose metabolism enables them to thrive under conditions of extreme salinity that instantly kills other living cells. The bacteria carry on a simple form of photosynthesis with the aid of a membrane-bound red pigment

FIGURE 17.5 The north end of Utah's Lake Bonneville, which has a very high salt content. The reddish areas are due to a red pigment produced by salt bacteria. The pigment, called *bacterial rhodopsin,* is involved in a form of photosynthesis.

called *bacterial rhodopsin*. The concentration of salt inside the cells is much lower than in their surroundings, but their metabolism is so closely tied to their environment that the bacteria die if placed in waters with lower salt concentrations (Fig. 17.5).

The Sulpholobus Bacteria

The *sulpholobus bacteria* constitute a third group of archaebacteria whose members occur in sulphur hot springs. The extraordinary metabolism of these bacteria permits them to thrive at very high temperatures—mostly in the vicinity of 80°C (176°F), with some doing very well at only 10°C below the boiling point of water. Their environment is also exceptionally acidic, the hot springs often having a pH of less than 2 (the neutral point on the 14-point pH scale is 7). One genus of bacteria (*Thermoplasma*) in this group is bounded by only a plasma membrane and has no cell wall. It is found only in the embers of coal tailings. Another genus (*Thermoproteus*) appears to be confined to the geothermal areas of Iceland.

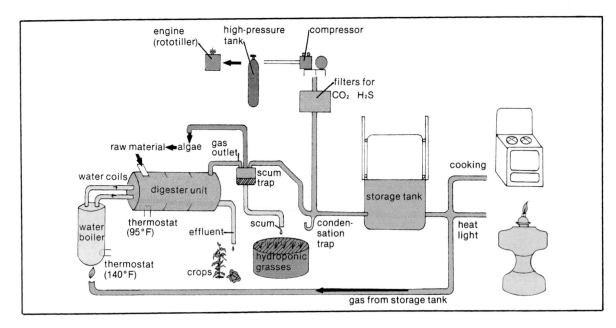

FIGURE 17.6 Integrated organic digester operation for the production of methane gas.
(Reprinted from *Producing Your Own Power* © 1974 by Rodale Press, Inc. Permission granted by Rodale Press, Inc., Emmaus, PA, 18049.)

Recently, James Lake and his colleagues discovered that the shape of the ribosomes of sulpholobus bacteria is significantly different from those of other archaebacteria, eubacteria, and eukaryotes and that the chemistry of sulphur-dependent bacteria also distinguishes them from other archaebacteria. Lake and his colleagues have proposed a third kingdom of prokaryotes named *Eocytes* (dawn cells). Other microbiologists are hesitant to base kingdom status on ribosome shape, but the controversy over the significance of these discoveries will undoubtedly continue as further research into the matter is pursued.

Human Relevance of the Archaebacteria

Archaebacteria are significant to humans in several ways. In the future, methane bacteria may be used on a large scale to furnish energy for engine fuels and for heating, cooking, and light, since the methane gas they produce can be substituted for natural gas. Methane has an octane number of 130 and has been used as a motor fuel in Italy for over 40 years. It is clean, nonpolluting, safe, and nontoxic, and it prolongs the life of automobile engines, also making them easier to start. The methane is given off by the bacteria as they "digest" organic wastes in the absence of oxygen. Nine kilograms (20 pounds) of horse manure or 4.5 kilograms (10 pounds) of pig manure fed daily into a methane digester will produce all the gas needed for the average American adult's cooking needs. Considerably less green plant material is required to produce the same amount.

The digester basically consists of an airtight drum connected by a pipe to a storage tank with a means of drawing off the sludge left after the gas has been produced (Fig. 17.6). The sludge itself makes an excellent fertilizer, although many sludges that originate from municipal and large-scale agricultural plants carry with them toxic levels of metals.

France had over 1,000 methane plants in operation by the mid-1950s, and in India, where cows produce over 812 million metric tons (800 million tons) of manure per year, many villages satisfy their fuel needs with manure-fed methane gas digesters. They are now being employed on a small scale in rural areas in the United States. Methane is the primary source of hydrogen in the commercial production of ammonia.

SUBKINGDOM EUBACTERIOBIONTA— THE TRUE BACTERIA

The Unpigmented, Purple, and Green Sulphur Bacteria

The *eubacteria* (true bacteria) have *muramic acid* in their cell walls and are also fundamentally different from archaebacteria in their RNA bases, metabolism, and lipids. They constitute quite a heterogeneous assemblage, with most being **heterotrophic** (organisms that cannot synthesize their own food and therefore depend on other organisms for it).

The majority of heterotrophic bacteria are **saprobes** (living organisms that obtain their food from nonliving organic matter). Saprobic bacteria, along with fungi, are primarily responsible for decay and for recycling all types of organic matter in the soil. Some of their recycling activities are discussed in the section on the nitrogen cycle in Chapter 25. Other heterotrophic bacteria are *parasites* (living organisms that depend on other living organisms for their food).

Several parasites that cause important human diseases are discussed later in the chapter.

Autotrophic Bacteria

A few groups of true bacteria are similar to green plants in being **autotrophic;** that is, they are capable of synthesizing organic compounds from simple inorganic substances. Some carry on photosynthesis without, however, producing oxygen as a by-product. Included in the photosynthetic bacteria are the *purple sulphur bacteria,* the *purple nonsulphur bacteria,* and the *green sulphur bacteria.* The *blue-green bacteria* and the *chloroxybacteria,* which do produce oxygen through photosynthesis, are discussed separately later in this chapter. Cells of the first two groups appear purplish, or occasionally red to brown, because of the presence of a mixture of greenish, yellow, and red pigments. Their greenish *bacteriochlorophyll* pigments (there are several closely related ones) are very similar to the chlorophyll *a* of higher plants.

No plastids are present in bacteria and their pigments are, instead, located in thylakoids or small spherical bodies. In their photosynthesis, the purple bacteria substitute hydrogen sulphide, the bad-smelling gas given off by rotten eggs, for the water used by higher plants. The generalized equation for the process is as follows:

$$CO_2 + 2 H_2S \xrightarrow[\text{light}]{\text{bacterio-}\atop\text{chlorophyll}} (CH_2O)n + H_2O + 2S$$

carbon hydrogen   carbo-   water sulphur
dioxide sulphide   hydrate

The equation for the purple nonsulphur bacteria is similar, but hydrogen from organic molecules is used instead of hydrogen sulphide. Green sulphur bacteria do use hydrogen sulphide, but their chlorophyll, called *chlorobium chlorophyll,* differs significantly in its chemistry from the chlorophylls of higher plants.

Other groups of bacteria are *chemoautotrophic;* that is, they are capable of obtaining the energy they require from various compounds or elements through chemical reactions involving the oxidation of reduced inorganic groups, such as NH_3, H_2S, and Fe^{++}, or the oxidation of hydrogen gas (as discussed in the section on oxidation-reduction reactions in Chapter 10).

Chemoautotrophic bacteria include *iron bacteria,* which transform soluble compounds of iron into insoluble substances that accumulate as deposits (e.g., in water pipes); *sulphur bacteria,* which can convert hydrogen sulphide gas to elemental sulphur and sulphur to sulphate; and *hydrogen bacteria,* which flourish in soils where they use molecular hydrogen derived from the activities of anaerobic and nitrogen-fixing bacteria. Nitrifying and nitrogen-fixing bacteria are a group of ecologically important bacteria that are discussed, along with their roles in the nitrogen cycle, in Chapter 25.

HUMAN RELEVANCE OF THE UNPIGMENTED, PURPLE, AND GREEN SULPHUR BACTERIA

Compost and Composting

Long before the existence of bacteria was known or even suspected, primitive agriculturists made rough piles of weeds, garbage, manure, and other wastes. They watched as the piles became reduced to a fraction of their original volume and changed into **compost,** a dark, fluffy material that conditions and enriches soil when mixed with it. Today, the compost pile is at the heart of the activities of numerous organic gardeners and farmers. With many communities instituting ordinances forbidding the burning of leaves and refuse and with space for waste disposal becoming scarce, a significant number of cities and towns have turned to the composting of street leaves and other materials. They have found that they not only save space but also produce a useful, ecologically compatible product (Fig. 17.7).

Because bacteria are ubiquitous, a single leaf or nonliving tissues in a pile of any size will eventually be decomposed. Ideally, however, a compost pile is 2 meters wide, 1.5 meters deep, and at a minimum, 2 meters long (6 feet wide, 4.5 feet deep, and at least 6 feet long). If the pile is of lesser length or breadth or greater depth, the conditions that favor the breakdown of materials (created by the bacteria themselves) develop at a slower pace.

Any accumulation of household garbage, leaves, weeds, grass clippings, and/or manure may be heaped together, and it has been demonstrated that no so-called starter culture of microorganisms is needed to initiate decomposition. Once the everpresent decay organisms have material to decompose, their numbers increase rapidly, and much heat is generated. In fact, the temperature in the center of the pile often rises to 70°C (158°F). As it rises, populations of organisms adapted to higher temperatures replace those not as well adapted. The remains of the latter are then added to the

FIGURE 17.7 Compost. *Left.* Garbage and leaves before composting. *Right.* After composting.

Table 17.2

SOME DISEASES OF HUMANS CAUSED BY BACTERIA OR VIRUSES

BACTERIAL DISEASES	VIRAL DISEASES
Anthrax	AIDS
Botulism	Chicken pox/shingles
Brucellosis	Common colds
Bubonic plague	Diabetes (some)
Cholera	Encephalitis
Diphtheria	Genital herpes
Dysentery (some)	Infectious hepatitis
Gonorrhea	Influenza
Leprosy	Measles
Mastitis	Mononucleosis
Meningitis	Mumps
Pneumonia (some)	Pneumonia (some)
Syphilis	Poliomyelitis
Tetanus	Rabies
Tuberculosis	Rubella
Typhoid fever	Smallpox
Whooping cough	Yellow fever

compost. The high temperatures also kill many weed seeds and most disease organisms.

The proportion of carbon to nitrogen present in a compost pile largely determines the pace at which the microbial activities proceed, a ratio of 30 carbon to 1 nitrogen being optimal. Microbial growth stops if the proportion of carbon gets much greater, and nitrogen is lost in the form of ammonia if the ratio drops lower.

If the materials are kept moist and turned occasionally to aerate the pile, composting may be completed in as little as two weeks. Shredding the materials exposes much more surface area for the microorganisms to work on and speeds up the process. Composters generally avoid or keep to a minimum a few materials, such as *Eucalyptus* and walnut leaves, which contain substances that inhibit the growth of other plants, when the compost is intended for use as a soil conditioner or crop fertilizer. Bamboo and some types of fern fronds are particularly resistant to decay bacteria and will decompose much more slowly than other materials. Domestic cat manure may contain stages of a parasite that can infect humans, although the organism should be killed if proper composting techniques are employed. Generally, however, almost any organic materials are suitable.

While the value of compost to the soil is indisputable, its value as a fertilizer has certain limitations. The nitrogen content of good compost is about 2% to 3%, phosphorus 0.5% to 1.0%, and potash 1% to 2%, as compared with five to ten times those amounts in chemical fertilizers, which also require less labor to produce. Nevertheless, with the spiraling costs of chemical fertilizers, increasing awareness of the problems accompanying their use, and greater public en-

lightenment concerning the accumulation and disposal of solid wastes, compost is undoubtedly destined to play a major role in pertinent agricultural practices of the future.

True Bacteria and Disease

It has been calculated that plant diseases alone cause American farmers losses of more than $4 billion per year. Although many plant diseases are caused by fungi, a number involve bacteria. Bacteria are involved in diseases of pears, potatoes, tomatoes, squash, melons, carrots, citrus, cabbage, and cotton. Bacteria also cause huge losses of foodstuffs after they have been harvested or processed. They are even better known, however, as the culprits in many serious diseases of animals and humans (Table 17.2).

Since they are so tiny and often incapable of independent movement, how are they transmitted, and what is it about their activities that causes them to fell creatures billions of times their size? We have learned much about their activities in the past century and know now that they can gain access to human tissues in a variety of ways.

Modes of Access of Disease Bacteria

Access from the Air   Every time a person coughs, sneezes, or just speaks loudly, an invisible cloud of tiny saliva droplets is produced (Fig. 17.8). Each droplet contains bacteria or other microorganisms and a tiny amount of protein. The moisture usually evaporates almost immediately, leaving minute protein flakes to which live bacteria adhere. As normal breathing takes place, these may soon find their way into the respiratory tracts of other humans or animals, particularly

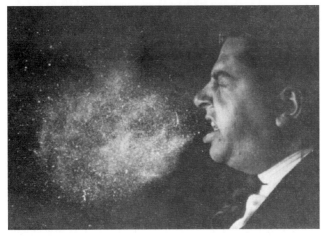

FIGURE 17.8   Stop-action photograph of a sneeze.
(Reprinted with the permission of Marshall W. Jennison, Syracuse University)

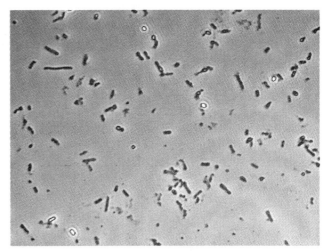

FIGURE 17.9   Botulism bacteria, ×1,500.
(Courtesy Robert McNulty)

if they have been confined within the air of a room. Fortunately, the natural resistance of most of those who acquire the bacteria in this way prevents the bacteria from multiplying to the point of causing a disease. When the resistance is not there, however, a number of diseases, including diphtheria, whooping cough, some forms of meningitis and pneumonia, and strep throat can develop.

Psittacosis, a disease carried by birds, is caused by the inhalation of *chlamydias,* which are exceptionally minute organisms unable to manufacture their own ATP molecules. They are apparently energy parasites, depending on their host cells for the energy needed to carry on their own functions. Some chlamydias are transmitted sexually and in recent years have become a widespread human problem.

Access Through Contamination of Food and Drink

Food Poisoning and Diseases Associated with Natural Disasters   In the past, open sewers and unsanitary conditions for food preparation caused a number of bacterial diseases, including cholera, dysentery, *Staphylococcus* and *Salmonella* food poisoning, and typhoid and paratyphoid fevers to reach epidemic proportions. Although *Staphylococcus* food poisoning, which is seldom fatal, is still fairly common today, civilized countries now rarely see epidemics caused by other bacteria unless a natural disaster, such as a flood or a typhoon, disrupts normal sewage disposal. The diseases are more often spread by carriers who handle food, or by houseflies.

The United States and other countries with ocean shorelines ban the harvesting of shellfish at certain times because *Salmonella* bacteria may multiply enough in clams and mussels to cause illness in humans who eat them—particularly if they are not well cooked. The *Salmonella* bacteria multiply in the intestinal tract or spread from there to other parts of the body.

Legionnaire's Disease   Some bacteria are very common on algae in freshwater streams, lakes, and reservoirs. One such bacterium, *Legionella pneumophila,* causes Legionnaire's disease, which killed 34 members of the American Legion attending a convention in a Philadelphia hotel in 1976. An estimated 50,000 Americans are infected annually by Legionnaire's disease bacteria, which nearly always pass through the human digestive tract harmlessly without multiplying. On rare occasions, however, something unknown triggers their reproduction, mostly in older males who are heavy smokers or alcoholics. The results are often fatal.

Botulism   The most deadly of all known biological toxins is produced by a bacterium with the scientific name of *Clostridium botulinum* (Fig. 17.9). The name comes from *botulus,* the Latin word for *sausage,* since the bacterium was first discovered after people had died from eating some contaminated sausage at a picnic.

Unlike *Salmonella* food poisoning, **botulism** is not an infection but is poisoning from a substance produced by bacteria that can grow and multiply anaerobically (in the absence of oxygen) in improperly processed or stored foods. Home-canned beans, beets, corn, and asparagus in particular have been known to permit the development of botulism bacteria. Just 1 gram (0.035 ounce) of the toxin is enough to kill 14 million adults, and a little more than half a kilogram (1.1 pound) could eliminate the entire human race.

The bacteria, which are present in most soils, produce unusually heat-resistant spores and are likely to be present on any soil-contaminated foods. They may not be destroyed during canning unless the food is heated for 30 minutes at 80°C (176°F) or boiled for 10 to 15 minutes. They ordinarily will not grow in foods that have been preserved in brines containing at least 10% salt (sodium chloride) or in fairly acid media, such as the juices produced by most stone fruits. The toxin is absorbed directly from the stomach and the intestines, affecting nerves and muscles and causing paralysis. While some antidotes are available, they are ineffective after symptoms have become advanced.

Deaths reported from botulism in the United States reached a peak of about 25 per year in the 1930s but have declined to five or six per year since then. There is evidence that some deaths of infants due to unknown causes are actually due to botulism. For reasons that are not clear, botulism bacteria, which pass harmlessly through the digestive tracts of humans over the age of 1 year, may germinate in the intestines of infants less than a year old. Between 1976 and 1980, 170 cases of sudden infant death reported worldwide appear to have been due to botulism, and the evidence now suggests that botulism is responsible for about 5% of the 8,000 cases of sudden infant death that occur each year in the United States alone.

Access Through Direct Contact

Syphilis and Gonorrhea   Bacteria responsible for diseases, such as syphilis, gonorrhea, anthrax, and brucellosis, enter the body through the skin or mucous membranes (i.e., those membranes lining tracts with openings to the exterior of the body). Both syphilis and gonorrhea are transmitted through sexual intercourse or other forms of direct contact and rarely through the use of public washroom towels or toilets. In 1976, gonorrhea accounted for the largest number of reported cases of any communicable disease in the United States, and currently, despite increasing precautions taken by the public since the escalation of AIDS, more than 1 million Americans become infected every year. The symptoms of both syphilis and gonorrhea, which include persistent sores or discharges from the genitalia, sometimes disappear after a few weeks, leading victims to believe the body has healed itself. More often than not, however, symptoms reappear in different parts of the body at a later date. Both diseases are curable when treated promptly, but it is very important that such treatment be sought, since failure to do so can lead to sterility, blindness, and even death.

Anthrax   Anthrax, which is primarily a disease of cattle and other farm animals in addition to wild animals, is sometimes transmitted to humans, particularly workers in the wool and hide industries. Like syphilis and gonorrhea, it can be effectively eliminated if treated early enough but may be fatal if allowed to progress. Brucellosis, another disease of farm animals, is occasionally transmitted to humans through direct contact or through the consumption of contaminated milk. It is sometimes called *undulant fever* because of a daily rise and fall of temperature apparently associated with the release of toxins by the bacteria.

Access Through Wounds

Tetanus and Gas Gangrene   When one steps on a dirty nail or is wounded in such a way that dirt is forced into body tissues, tetanus (lockjaw) bacteria, which are common soil organisms, may gain access to dead or damaged cells. There they can multiply and produce a deadly toxin so powerful that 0.00025 gram (0.00000088175 ounce) is enough to kill an adult. In contrast, about 150 times that amount of strychnine is needed to achieve the same result. Control of tetanus through immunization is very effective and has become widespread. Only 21 cases of tetanus were reported in the United States in 1989. Several related bacteria that gain access to the body in the same way are responsible for poten-

tially fatal gas gangrene, which used to be feared on the battlefields in the past but is now controlled through the use of antibiotics and aseptic techniques.

Access Through Bites of Insects and Other Organisms

Bubonic Plague   Bubonic plague (Black Death) and tularemia are two bacterial diseases transmitted by fleas, deerflies, ticks, or lice that have been parasitizing infected animals, particularly rodents.

Rat fleas, found on infected rats that inhabit dumps, sewers, barnyards, and ships (Fig. 17.10), acquire the bacteria for plague and then pass them on to humans through their bites. The disease has been found in ground squirrels and other rodents in the United States, particularly in the West, since 1900.

In the past, bubonic plague has spread with great speed and reached devastating epidemic proportions. In 1665 in London, hundreds of thousands of humans perished from the disease, and between 1347 and 1349, it is believed to have killed one-fourth of the entire population of Europe (about 25 million persons). Today, plague is rare in North America, but it still occasionally manifests itself in port cities of Asia, Europe, and South America. Control depends on control of rats and fleas, which are virtually impossible to eradicate entirely, and the use of vaccines, which produce immunity for about 6 to 12 months.

Tularemia, Rickettsias, and PPLOs   Tularemia is primarily a disease of animals, but infected ticks or deerflies may transmit the disease to humans through bites. It is an occupational disease of meat handlers and is fatal in 5% to 8% of the cases. Ticks, lice, and fleas may also transmit rickettsias, which cause typhus and spotted fevers. Rickettsias are extremely tiny bacteria that live within eukaryotic cells.

Pleuropneumonialike organisms (PPLOs) are also minute bacteria that may be transmitted by various means. These have no cell walls and therefore are quite plastic. They're found in many plants, in hot springs, and in the moist surfaces of the respiratory and intestinal tracts of animals and humans. They're responsible for a form of human pneumonia and for numerous plant diseases. They're the only group of prokaryotes known to be resistant to penicillin.

Lyme Disease   Lyme disease, which was known in Europe before 1900, has spread rapidly throughout the United States since 1975 when an outbreak occurred at Lyme, Connecticut. Its arthritislike effects are caused by a bacterium (*Borrelia burgdorferi*) hosted by deer and field mice. It is injected into the human bloodstream by deer ticks.

Koch's Postulates

Since there are so many bacteria present everywhere, how can one be certain that a given bacterium obtained from an infected person is actually the organism responsible for the observed disease? Robert Koch, a German physician, became known during the latter half of the 19th century for his investigations of anthrax and tuberculosis. As a result of his work, Koch formulated rules for proving that a particular microorganism is the cause of a particular disease. His rules, with minor modifications, are still followed today. They

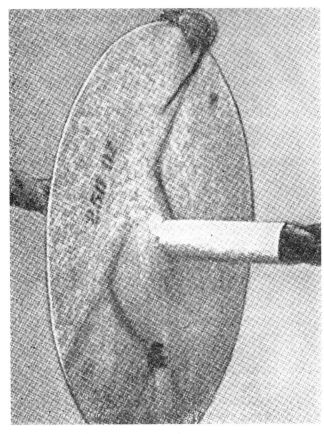

FIGURE 17.10   A rat climbing over a barrier on a ship's mooring line.
(Courtesy U.S. Public Health Service)

A.

B.

FIGURE 17.11   A tomato hornworm (*A*) before, and (*B*) 3 days after spraying with *Bacillus thuringiensis*.

have come to be known as *Koch's Postulates*, and their essence is as follows:

1. The microorganism must be present in all cases of the disease.

2. The microorganism must be isolated from the victim in pure culture (i.e., in a culture containing only that single kind of organism).

3. When the microorganism from the pure culture is injected into a susceptible host organism, it must produce the disease in the host.

4. The microorganism must be isolated from the experimentally infected host and grown in pure culture for comparison with that of the original culture.

True Bacteria Useful to Humans

For many years, we've controlled insect pests of food plants mostly through the use of toxic sprays. Residues of the sprays remaining on the fruits and vegetables have accumulated in human tissues, often with adverse effects, while at the same time many organisms have become immune or resistant to the toxins. In addition, the sprays have killed useful organisms, and precipitation runoff has washed the toxins into streams, lakes, and oceans, harming or killing aquatic organisms.

As we've become aware of the undesirable effects of the use of toxic sprays, we've looked for alternative means of controlling crop pests. Today, many harmful pests and even weeds can be significantly limited through the use of "biological controls," which are discussed in Appendix 2.

Bacillus Thuringiensis *and* B. Popillae

Three biological control bacteria have been registered for use by the U.S. Department of Agriculture. One, *Bacillus thuringiensis* (often referred to as *BT*), has been remarkably effective against a wide range of caterpillars and worms, including peach tree borers, European corn borers, bollworms, cabbage worms and loopers, tomato and fruit hornworms (Fig. 17.11), tent caterpillars, fall webworms, leaf miners, alfalfa caterpillars, leaf rollers, gypsy moth larvae, and cankerworms.

The bacteria, which are easily mass-produced by commercial companies, are sold in the form of a stable wettable dust containing millions of spores. When the spores are sprayed on food plants, they are harmless to humans, birds, animals, earthworms, or any living creatures other than moth or butterfly larvae. When a caterpillar ingests any tissue with BT spores on it, the bacteria quickly become active and multiply within the digestive tract, soon paralyzing the gut. The caterpillar stops feeding within two or three hours and slowly turns black, dropping off the plant in two to four days.

Plant Viruses

The book *Hot Zone* and the movie *Outbreak* have created an awareness of emerging viruses and their dangers to the human population. The Ebola, Hanta, and HIV viruses are now everyday words that have become synonymous with death. There is another group of viruses that also has a significant impact—plant viruses—which cause an estimated $15 billion worth of crop loss per year worldwide. They infect plants and cause hundreds of diseases such as tomato spotted wilt disease, tobacco mosaic disease, maize stripe disease, and apple chlorotic leaf spot disease.

In total, nearly 400 plant viruses have been identified and classified by the International Committee on Taxonomy of Viruses (ICTV). A further 320 have been identified, but are awaiting final classification.

Surprisingly, the first viruses ever identified were in plants. In 1898, a Dutch professor of microbiology, Dr. Martinus Beijerinck, was working to identify the disease that caused tobacco leaves to become mottled with light green and yellow spots. He demonstrated that the condition was not caused by a bacterium as was commonly thought at the time, but rather by some other unknown pathogen in the sap of the tobacco plant. He proved this by collecting sap from a diseased plant that was then passed through a filter capable of straining out any bacteria. When the filtered solution was reinjected into the leaf veins of healthy plants and the disease was transmitted, he had made his point. He called this filtered sap a *contagium vivum fluidium* (a contagious living fluid), and introduced the term **virus** to describe its property of being able to reproduce itself within living plants. Dr. Beijerinck's virus was later named tobacco mosaic virus (TMV), consistent with the now-established practice of naming plant viruses both by the plant it infects and by describing the major disease symptom (e.g., mosaic, wilting, spotted, etc.).

Not only were the first viruses discovered in plants, but the understanding of their biochemical nature was first recognized through research on tobacco mosaic virus. Today we know that viruses are submicroscopic, infectious particles that are composed of a protein coat and a nucleic acid center. They can be seen only with an electron microscope. As obligate parasites, they can reproduce themselves only within a living cell. The biochemical nature of viruses remained unknown until 1935 when Dr. Wendell Stanley, an organic chemist in the United States, succeeded in crystallizing the protein coat of tobacco mosaic virus (TMV). Stanley, however, did not recognize the nucleic acid content of the virus that was later shown to be RNA. The fact that RNA could exist **separately** from DNA was a discovery that has had great influence on the development of molecular biology thought.

Today we know that tobacco mosaic virus is a rigid rod, 300 nm $\times$ 15 nm, composed of a protein coat of approximately 2,100 helically arranged protein subunits surrounding an axial canal that contains a single-stranded RNA molecule consisting of 6,400 nucleotides. It, like all plant viruses, is classified according to the type of nucleic acid that it contains, either DNA or RNA but never both, whether the nucleic acid is single- or double-stranded, and the shape of the virus particle (spheres, stiff rods, flexible rods).

---

another serious disease of both animals and humans. Agents of disease that could pass through filters became known as *filterable viruses,* although the word *filterable* is no longer used. Today, we know that not only smallpox and rabies are caused by these viruses but also measles, mumps, chicken pox, polio, yellow fever, influenza, fever blisters, warts, and the common cold.

Only those organisms that have certain unique features are now called *viruses*. These features, which include a complete lack of cellular structure, make viruses quite different from anything else in the five kingdoms of living organisms now recognized. In fact, some question whether viruses are even living organisms. In 1946, Wendell Stanley, an American chemist, received a Nobel Prize for demonstrating that a virus causing tobacco mosaic, a common plant disease, could be isolated, purified, and crystalized and that the crystals could be stored indefinitely and would always produce the disease in healthy plants any time they were placed in contact with them. We also know that viruses do not grow by increasing in size or dividing, nor do they respond to external stimuli. They cannot move on their own, and they cannot carry on independent metabolism.

Viruses are incredibly numerous. In 1989, for example, marine biologists at the University of Bergen in Norway discovered that a teaspoon of sea water typically contains more than 1 billion viruses. They are about the size of large molecules, varying in diameter from about 15 to 300 nanometers (Fig. 17.16). Thousands of the smallest ones could fit inside a single bacterium of average size.

Viruses consist of a nucleic acid core surrounded by a protein coat. The architecture of the protein coats varies considerably, but many have 20 sides and resemble tiny geodesic domes, while others have distinguishable head and tail regions. The nucleic acid core consists of either DNA or

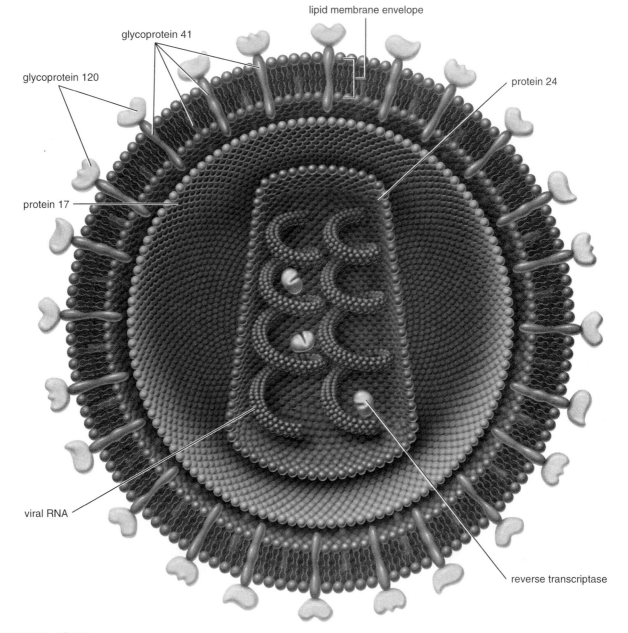

glycoprotein 41

glycoprotein 120

lipid membrane envelope

protein 24

protein 17

viral RNA

reverse transcriptase

FIGURE 17.19   A virus is a nucleic acid coated with protein. The human immunodeficiency virus (HIV), which causes AIDS, consists of RNA surrounded by several layers of proteins. Once inside a human cell (usually a T cell, part of the immune system), the virus uses an enzyme to convert its RNA to DNA, which then inserts into the host DNA. HIV damages the human body's protection against disease by killing T cells and by using these cells to make more of itself.

Human Relevance of Viruses

The economic impact of viruses in both developing and industrialized countries is enormous. The annual loss in work time due to common cold and influenza viruses alone amounts to millions of hours. While discomfort, adverse effects on employment, and even deaths due to viral diseases, such as chicken pox, measles, German measles, mumps, and yellow fever, have declined dramatically since immunizations against the diseases became widespread, they still take their toll. Another viral disease, infectious hepatitis, periodically still claims victims. Guillain-Barré Syndrome and Epstein-Barr infections are debilitating diseases caused by viruses that are apparently carried by nearly everybody, but what triggers them into action is as yet still unknown.

AIDS (acquired immune deficiency syndrome), a fatal disease caused by one or more viruses called *retroviruses* that are related to those that cause cancer, was unknown until its discovery in 1959 in a blood sample in what is now Zaire. Research has shown that two forms of the AIDS virus, HIV-1 and HIV-2, diverged from a common ancestor in the early 1950s (Fig. 17.19). Thirty years later, the AIDS virus had spread all over the world. Retroviruses mutate so rapidly that they are capable of evolving about a million times faster than cellular organisms. Since the initial appearance of AIDS, increasing research is being applied to the development of a vaccine with the use of a genetically engineered virus. If successful, the vaccine will cause the human body to develop a defense against AIDS viruses without the disease itself developing.

The production of vaccines for a number of other diseases is undertaken by a flourishing worldwide multimillion dollar industry. Other mass-produced viruses are used to infect ticks, insects, and other disease organisms of both animals and plants. Some viruses cause great losses when they infest creamery vats during the manufacture of dairy products or culture vats during the production of antibiotics. One phage attacks nitrogen-fixing bacteria in the roots of leguminous plants, while other phages attack diphtheria and tuberculosis bacteria. One natural virus called *Abby* (an abbreviation of Abington, which is one of 19 strains of nuclear polyhedrosis viruses) is found only in the caterpillars of gypsy moths. Gypsy moth caterpillars have been particularly destructive in forests of both North America and Europe, and it is hoped that the destruction may be greatly reduced by the dissemination of this virus in the forests under attack.

Summary

1. Kingdom Monera consists of prokaryotic organisms, which are grouped into archaebacteria and eubacteria. The bacteria occur as single cells, in colonies, or in the form of chains or filaments. Some cells may be motile or they may exhibit a gliding motion; most are nonmotile.

2. Bacterial nutrition occurs primarily by absorption of food in solution through the cell wall, but some bacteria are photosynthetic or chemosynthetic.

3. Reproduction is asexual by means of fission; some genetic recombination occurs by means of pili between cells.

4. Bacteria are mostly less than 2 or 3 micrometers in diameter. They occur as spheres (cocci), rods (bacilli), and spirals (spirilli) and are further classified on the basis of sheaths, appendages, and motion. They are also classified as gram-positive or gram-negative.

5. Prokaryotic cells have no nuclear envelopes or organelles. Each cell has one long DNA molecule and sometimes up to 30 or 40 small, circular DNA molecules called *plasmids,* which replicate independently of the large DNA molecule or chromosome.

6. Neither meiosis nor mitosis occurs, but fission takes place with the development of a transverse wall near the middle of the cell; gametes and zygotes are not produced. Conjugation facilitates genetic recombination. Transformation involves the incorporation of fragments of DNA released by dead cells; transduction involves the viral transfer of fragments of DNA from one cell to another.

7. Heterotrophic bacteria are saprobes or parasites. Autotrophic bacteria are photosynthetic but do not produce oxygen; chemoautotrophic bacteria obtain their energy through oxidation of reduced inorganic groups.

8. Any nonliving tissues will eventually be decomposed to compost by bacteria and fungi. Compost is definitely good for the soil but has limited value as a fertilizer.

9. Bacteria cause huge losses through plant diseases and food spoilage and many serious diseases in animals and humans. They gain access to their hosts by various means.

10. Koch formulated postulates (rules) for proving that a particular microorganism is the cause of a particular disease.

11. Bacteria useful to humans include *Bacillus thuringiensis, Bacillus thuringiensis* var. *israelensis,* and *Bacillus popilliae.* Bacteria also play a major role in the manufacture of dairy products, such as yogurt, sour cream, kefir, and cheese.

12. Bacteria are used in the manufacture of industrial chemicals, vitamins, flavorings, food stabilizers, and a blood plasma substitute; they play a role in the curing of vanilla, cocoa beans, coffee, and tea and in the production of vinegar and sauerkraut; and they aid in the extraction of linen fibers from flax stems, in the production of ensilage for cattle feed, and in the production of several important amino acids.

13. Blue-green bacteria are virtually ubiquitous in their occurrence.

14. The cells of blue-green bacteria occur in a variety of forms. They are distinguished from other bacteria in having chlorophyll *a,* in producing oxygen, and in having blue and red phycobilin pigments. They produce cyanophycin, a nitrogenous food reserve.

15. Blue-green bacteria have no flagella, but some species have gliding movements. Fragmentation may occur at heterocysts. Akinetes may also be produced.

16. Blue-green bacteria cells may have been the origin of chloroplasts, since they divide as chloroplasts do when their host cells divide.

17. Blue-green bacteria may become very abundant in bodies of polluted fresh water. Toxic substances are produced when the bacteria die and are decomposed. At least 40 species of blue-green bacteria are known to fix nitrogen.

18. Chloroxybacteria are similar in form to blue-green bacteria but have pigmentation similar to that of higher plants and lack phycobilins.

19. Vaccination was first performed by Jesty and Jenner in connection with smallpox. Smallpox is now believed to have been eradicated.

20. Viruses, which have no cellular structure, are about the size of large molecules. Some can be isolated, purified, and crystallized yet remain virulent. They cannot grow or increase in size and cannot be reproduced outside

of a living cell, upon whose DNA they depend for duplication.

21. Viruses consist of a core of nucleic acid surrounded by a protein coat. They are classified on the basis of the DNA or RNA in their core, their size and shape, the number of identical structural units in their cores, and the nature of their protein coats.

22. Bacteriophages are viruses that attack bacteria. In replicating, they become attached to a susceptible cell, which they penetrate, with their DNA or RNA directing the synthesis of new virus molecules from host material; the assembled new viruses are released when the host cell dies.

23. Cells of higher animals being invaded by viruses produce interferon, a protein that causes cells to produce a protective substance that inhibits duplication of viruses and also inhibits the capacity of viruses to transform normal cells into tumor cells.

24. Viral diseases, such as chicken pox, measles, mumps, and yellow fever, have declined since immunizations against the diseases have become widespread. Mass-produced viruses are used to infect insects and other pests of both plants and animals. Some viruses cause losses in creamery vats.

Review Questions

1. What is symbiosis? Give examples other than those mentioned in the text.

2. Why are bacteria not classified on the basis of visible features alone?

3. How do bacteria exchange DNA?

4. How does fission differ from mitosis?

5. Is photosynthesis the same in bacteria as it is in higher plants? Explain.

6. How do chemoautotrophic bacteria differ from photosynthetic bacteria?

7. What is the difference between nitrification and nitrogen fixation?

8. If decay bacteria use nitrogen, how does composting accumulate any nitrogen?

9. How are disease bacteria transmitted?

10. What are Koch's Postulates?

11. Why are many bacteria considered useful?

12. What do blue-green bacteria and other bacteria have in common? How do they differ?

13. How do blue-green bacteria survive freezing and desiccation?

14. What is an algal-bacterial bloom?

15. How do viruses differ from bacteria?

16. What is a vaccination?

17. What is a phage?

18. How do viruses multiply?

Discussion Questions

1. If a virulent phage were to eliminate all the bacteria in North America for one year, how would our lives be affected?

2. What would be the feasibility and the advantages or disadvantages of using only blue-green bacteria and other nitrogen-fixing bacteria for our agricultural nitrogen needs?

3. Methane gas produced by bacteria is proving to be sufficient to meet all the fuel needs of villages in India. Do you think we could produce and use methane in a similar fashion in the United States? Explain.

4. If blue-green bacteria are capable only of asexual reproduction, does this mean that species of these organisms can never change in form?

5. As long as viruses can multiply, why should there be any question as to whether they are living?

Additional Reading

Bos, L. 1992. *Introduction to plant virology*. New York: State Mutual Book and Periodical Service.

Brock, T. D., and M. Madigan. 1990. *Biology of microorganisms*, 6th ed. Englewood Cliffs, NJ: Prentice-Hall.

Cann, A. 1993. *Molecular virology*. San Diego, CA: Academic Press.

Carr, N. G., and B. A. Whitton (Eds.). 1982. *The biology of Cyanobacteria*. Berkeley, CA: University of California Press.

Dimmock, N. J., and S. B. Primrose. 1987. *Introduction to modern virology*. London: Blackwell Scientific Publications.

Dixon, B. 1994. *Power unseen: How microbes rule the world*. San Francisco: W. H. Freeman.

Gunter-Schlegel, H. 1994. *General microbiology*, 7th ed. New York: Cambridge University Press.

Holt, J. G., and N. R. Krieg (Eds.). 1984. *Bergey's manual of determinative bacteriology*, vol. 1. Baltimore: Williams and Wilkins.

Kim, C. W. 1991. *Microbiology*, 10th ed. East Norwalk, CT: Appleton and Lange.

Pelczar, M. J., Jr. 1993. *Microbiology: Concepts and applications*, 5th ed. New York: McGraw-Hill.

Prescott, L., et al. 1993. *Microbiology*, 2d ed. Madison, WI: Brown and Benchmark.

Stolp, H. 1988. *Microbial ecology*. New York: Cambridge University Press.

Stoner, C. H. 1993. *Biotechnology for hazardous waste treatment*. Boca Raton, FL: Lewis Publications.

Tortora, G. J. 1991. *Microbiology: An introduction*, 4th ed. Menlo Park, CA: Benjamin/Cummings Publishing Co., Inc.

Woese, C. R. 1981. Archaebacteria. *Scientific American* 244(6): 98–122.

Giant kelp (Macrocystis pyrifera), *which grows to lengths of 30 meters (100 feet). One specimen from the Pacific Coast was 45.7 meters (148 feet) long. Giant kelps are harvested for their* algin, *a useful substance discussed in this chapter.*
(Photo by Digital Stock.)

18 Kingdom Protoctista

Overview

After giving an introduction and summary of the features of members of Kingdom Protoctista, this chapter discusses the divisions of algae, slime molds, chytrids, and water molds. A brief discussion of the life cycle of a representative of each division is included.

First, the Chrysophyta (yellow-green algae, golden-brown algae, diatoms, and cryptophytes) are discussed. Next, the Division Pyrrophyta (dinoflagellates) is briefly examined and the role of members of the division in red tides and bioluminescence is explored. Asexual and sexual reproduction in the Chlorophyta (green algae) is shown in Chlamydomonas, Ulothrix, Spirogyra, *and* Oedogonium; *mention is made of* Chlorella, *desmids,* Acetabularia, Volvox, *and* Ulva. *After giving a brief overview of the Euglenophyta (euglenoids), the chapter takes up the Phaeophyta (brown algae) and Rhodophyta (red algae). Differences between the divisions of algae are shown in a table that lists food reserves, special pigments, and flagella. This is followed with a digest of the human and ecological relevance of the algae. The chapter concludes with a brief examination of the slime molds, chytrids, and water molds.*

Some Learning Goals

1. Know features that the members of Kingdom Protoctista share with one another, and note the basic ways in which they differ.
2. Understand how a diatom differs in structure and form from other members of the Division Chrysophyta.
3. Diagram the life cycles of *Chlamydomonas, Ulothrix, Spirogyra,* and *Oedogonium;* indicate where meiosis and fertilization occur in each.
4. Learn at least two features that distinguish the Chrysophyta, Pyrrophyta, Euglenophyta, Chlorophyta, Phaeophyta, and Rhodophyta from one another.
5. Know what *holdfasts, stipes, blades, bladders,* and *thalli* are.
6. Learn the function of each of the three thallus forms of a red alga, such as *Polysiphonia,* and know at least 20 economically important uses of algae, in addition to the numerous uses of algin that are given.
7. Understand why the slime molds, chytrids, and water molds are not true fungi, and what they have in common with other members of Kingdom Protoctista.

I was once the sole passenger on a ship carrying cargo from an Indian Ocean port to Boston. I had turned in early one night as we were nearing the equator, but I awoke around 2:00 A.M. and went to the bathroom to get a glass of water. I had not turned on the light and was startled to notice that the water in the toilet bowl was "winking" at me. Numerous little lights in the water were flashing on and off! I remembered that ships out of the harbor take in water from their surroundings for their sewage lines, and I hurried out on deck to see where we were headed. I was at once treated to a beautiful display of *bioluminescence* (discussed in Chapter 11) caused by millions of

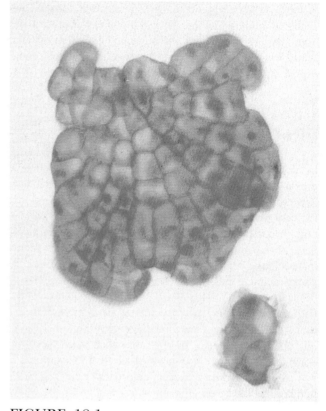

FIGURE 18.1 *Coleochaete,* a green alga that has several features of higher plants and is believed to be an indirect ancestor of land plants, ×500.

microscopic algae called *dinoflagellates.* As the waves broke, both at the bow and in the wake, there was a sparkling, shimmering glow as the tiny organisms, through their respiration, produced light. The dinoflagellates constitute but one of nine divisions of algae and other relatively simple eukaryotic organisms discussed in this chapter.

The fossil record indicates that less than 1 billion years ago all living organisms were confined to the oceans, where they were protected from drying out, ultraviolet radiation, and large fluctuations in temperatures. They also absorbed the nutrients they needed directly from the water in which they were immersed. The fossil record also suggests many times that, beginning about 400 million years ago, green algae, which are important members of Kingdom Protoctista, began making the transition from water to land, eventually giving rise to green land plants.

Coleochaete (Fig. 18.1), a tiny, freshwater green alga that grows as an **epiphyte** (an alga or plant that attaches itself in a nonparasitic manner to another living organism) on the stems and leaves of submerged plants, shares several features with higher plants and probably was an indirect ancestor of today's land plants. The features include somewhat parenchymalike cells, the development of a cell plate and phragmoplast during mitosis, a protective covering for the zygote, and the production of a ligninlike compound. Lignin adds mechanical strength to cell walls, and the discoverers of the substance suggest that it originated as a protection against microbes. The ligninlike compound was present 100 million years earlier than the presumed evolution of plant life from water to land, and its presence in algal cells could explain how such organisms adapted to land habitats.

FEATURES OF KINGDOM PROTOCTISTA

In contrast with members of Kingdom Monera, which have prokaryotic cells, all members of Kingdom Protoctista have *eukaryotic* cells. The organisms comprising this kingdom are very diverse and heterogeneous, but none have the distinctive combinations of characteristics possessed by members of Kingdoms Plantae, Fungi, or Animalia. Many, including the euglenoids, protozoans, and some algae, consist of a single cell, while other algae are multicellular or occur as colonies or filaments. Nutrition is equally varied, with the algae being photosynthetic, the slime molds and protozoans ingesting their food, the euglenoids either carrying on photosynthesis or ingesting their food, and the oomycetes and chytrids absorbing their food in solution.

Individual life cycles vary considerably, but reproduction is generally by cell division and sexual processes. Many *protists* (mostly single-celled members of Kingdom Protoctista) are motile, usually by means of flagella, but other members of the kingdom, especially those that are multicellular, are nonmotile, although most of the multicellular members produce some motile cells.

SUBKINGDOM PHYCOBIONTA—ALGAE

Children fortunate enough to have lived near ocean beaches where seaweeds are cast ashore by the surf have often enjoyed stamping on the bladders ("floats") of kelps to hear the distinct popping sound as they break. Some have collected and pressed beautiful, feathery red seaweeds, and others who have waded around the shores of freshwater lakes or in slow-moving streams have encountered slimy feeling pond scums. Anyone who has kept tropical fish in glass tanks have sooner or later had to scrape a brownish or greenish film from the inner surfaces of the tank, and those who have lived in homes with their own swimming pools have learned that the plaster of the pool soon acquires colored patches on its surface if chemicals are not regularly added to prevent them from appearing.

Although some of the seaweeds have flattened, leaflike blades, algae have no true leaves or flowers. They are involved in our everyday lives in more ways than most people realize (see the discussion on the human and ecological relevance of the algae, which begins on page 308). Seaweeds and some pond scums, fish tank films, and colored patches in swimming pools include but a few of the numerous kinds of algae all assigned to Kingdom Protoctista. The algae are grouped into several major divisions based on the form of their reproductive cells and combinations of pigments and food reserves.

DIVISION CHRYSOPHYTA— THE GOLDEN-BROWN ALGAE

If the roughly 6,000 members of this division were not primarily microscopic, many surely would become collectors' items in the art and antique shops of the world because of their exquisite form and ornamentation (Fig. 18.2; see also figs. 18.3 and 18.4).

The algae in this division can be grouped into four classes: *yellow-green algae* (Xanthophyceae), *golden-brown algae* (Chrysophyceae), *diatoms* (Bacillariophyceae), and *cryptophytes* (Cryptophyceae). Superficially, the organisms of each class may appear unrelated to each other, but they do share several features, including food reserves, specialized pigments, and other cell characteristics. Some members of each class produce a unique "resting" cell called a *statospore* (Fig. 18.3). These cells resemble miniature glass apothecary bottles, complete with plugs that dissolve or "uncork," releasing the protoplast inside. Many statospores are striking in form, with finely sculptured ornamentations on the surface.

Diatoms

Diatoms (Fig. 18.4) are the best-known and the most economically important members of the division. These mostly unicellular algae occur in astronomical numbers in both fresh and salt water but are particularly abundant in colder marine habitats. In fact, as a rule, the colder the water, the greater the number of diatoms present, and huge populations of diatoms are found on and within ice in both Antarctica and the Arctic. A major constituent of the foam that accumulates at the wave line on beaches is an oil produced by diatoms.

Diatoms also usually dominate the algal flora on damp cliffs, the bark of trees, bare soil, and the sides of buildings. More than 5,600 living species are recognized, with almost as many more known only as fossils. Some can withstand extreme drought, and one species is known to have become active after lying dormant in dry soil for 48 years.

Diatoms look like ornate, little, glass pillboxes, with half of the rigid, crystal-clear wall fitting inside the other overlapping half. Many marine diatoms are circular in outline when viewed from the top, while freshwater species viewed the same way tend to resemble the outline of a kayak. As much as 95% of the wall content is silica, an ingredient of glass, deposited in an organic framework of pectin or other substances.

The diatom walls usually have exquisitely fine grooves and pores that are exceptionally minute passageways connecting the protoplasm with the watery environment outside the shell. These fine grooves and pores are so uniformly spaced they have been used to test the resolution of

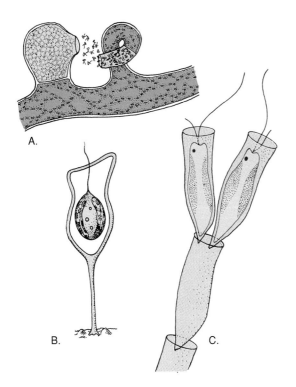

FIGURE 18.2   Some representatives of the Division Chrysophyta. *A. Vaucheria. B. Stipitococcus. C. Dinobryon.*

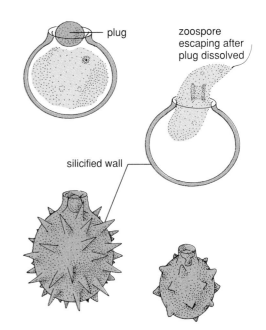

FIGURE 18.3   Statospores, which often resemble apothecary bottles, are formed by many of the golden-brown algae.

(After G.M. Smith. 1950. *The Freshwater Algae of the United States,* 2d ed. Redrawn by permission of the McGraw-Hill Book Co.)

microscope lenses. When *phycologists* (scientists who specialize in the study of algae) want to examine the markings on diatom walls, they often obtain a better view by dissolving the protoplasm with hot acid.

Each diatom may have one, two, or many chloroplasts per cell. In addition to chlorophyll *a,* the accessory pigments chlorophyll c_1 and chlorophyll c_2 are typically present. The chloroplasts usually are golden-brown in color because of the dominance of **fucoxanthin,** a brownish pigment also found in the brown algae. Food reserves are oils, fats, or the carbohydrate *chrysolaminarin.*

Freshwater diatoms that have a lengthwise groove called a *raphe* glide backward and forward with somewhat jerky motions, at a rate of up to three times their length in 5 seconds. It is believed that the movements occur in response to external stimuli such as light. Extremely tiny fibrils apparently take up water as they are discharged into the raphe or pores through which moving cytoplasm protrudes. When they come in contact with a surface, they stick and contract, moving the cell as the caterpillar treads on a tractor do and leaving a trail something like that of a snail.

Reproduction in diatoms is unique. Before any cell division can occur, an adequate source of silicon must be present in the surrounding medium. In culture, the number of cells produced is directly proportional to the amount of silicon added to an otherwise nutrient-balanced medium. Division in culture is also rhythmic, with all the cells dividing at the same time.

After a protoplast, which is diploid, has undergone mitosis and division, the two halves of the cell separate, with a

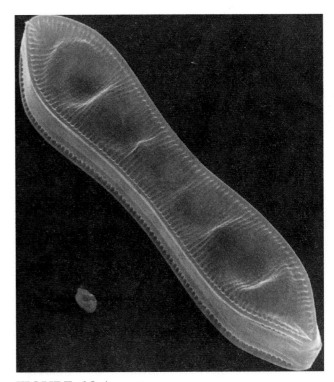

FIGURE 18.4   A diatom.

(Scanning electron micrograph courtesy J.D. Pickett-Heaps)

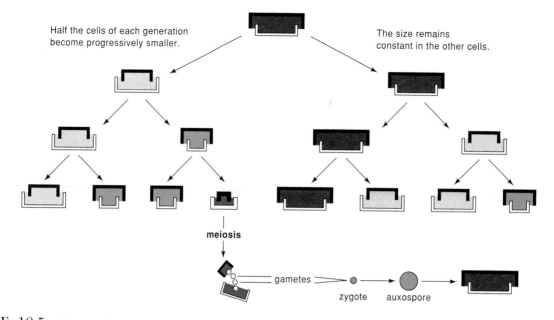

Half the cells of each generation become progressively smaller.

The size remains constant in the other cells.

meiosis

gametes → zygote → auxospore

FIGURE 18.5 Reproduction in diatoms. The two rigid halves of each pillboxlike cell separate and a new rigid half shell forms *inside* each original half. This results in half the new cells becoming smaller with each generation. At one point, however, the diploid nucleus of a reduced-size cell undergoes meiosis, and four gametes are formed. The zygote produced when the two gametes unite becomes considerably enlarged. The enlarged cell, called an *auxospore,* develops into a diatom of the same size as the original diatom.

daughter protoplast remaining in each half. Then a new half wall fitting *inside* the old half is formed. This occurs for a number of generations, with the result that half of the cells become progressively smaller. Eventually, however, a protoplast undergoes meiosis, producing four gametes, which then escape. These fuse with other gametes, becoming zygotes called *auxospores*. Auxospores are like any other zygotes except that they restore the original size of the diatom by rapidly ballooning before forming rigid pillboxlike walls (Fig. 18.5).

DIVISION PYRROPHYTA— THE DINOFLAGELLATES

Occasionally, visitors to an ocean beach in midsummer may notice a distinctly reddish tint to the water, usually as a result of a phenomenon known as a *red tide*. Red tides are caused by the sudden and not fully understood multiplication of unicellular organisms called *dinoflagellates* (Fig. 18.6). There are over 3,000 species of dinoflagellates, 300 of them capable of producing red tides. When a red tide appears, some biologists collect cups of seawater for examination with a microscope. (The material can be preserved indefinitely with the addition of a few drops of formaldehyde, vinegar, or other weak acid.)

Each cup of seawater collected usually contains a large number of dinoflagellate "shells," which are the best

known representatives of the Division Pyrrophyta. Some resemble armor-plated spaceships, while others may be smooth or have fine lengthwise ribs. The "armor" plates are located just inside the plasma membrane and are composed mostly of cellulose of varying thickness.

Dinoflagellates have two flagella that are distinctively arranged, usually attached near each other in two adjacent and often intersecting grooves. One flagellum, which acts as a rudder, trails behind the cell. The other, which encircles the cell at right angles to the first groove, gives the cell a spinning motion as it undulates in its groove like a tiny snake.

Most dinoflagellates have two or more disc-shaped chloroplasts, which contain distinctive brown pigments in addition to various other pigments, including chlorophylls a and c_2. About 45% of the species are, however, nonphotosynthetic, and some ingest food particles, whether or not chlorophyll is present. Some have an *eyespot* (a pigmented organelle that is sensitive to light), and all have a unique nucleus in which the chromosomes remain condensed and clearly visible throughout the life of the cell. The chromosomes contain a disproportionately large amount of DNA— as much as 40 times that found in human cells. The food reserve is starch, which in dinoflagellates is stored outside the chloroplasts.

Dinoflagellates occur in most types of fresh and salt water, but those that cause red tides have received the most publicity because about 40 of the species also produce powerful neurotoxins that accumulate in shellfish such as oysters,

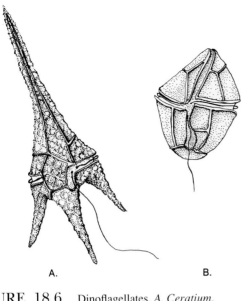

FIGURE 18.6 Dinoflagellates. *A. Ceratium.* *B. Gonyaulax.*

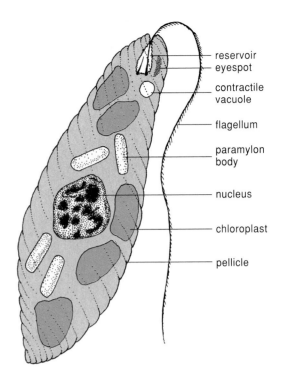

FIGURE 18.7 A single *Euglena.*

mussels, scallops, and clams. The poisons apparently don't harm the shellfish, but about 2,000 persons a year become ill—15% fatally—from eating contaminated shellfish. Although fish in open waters can swim away from affected waters, large numbers still die and wash ashore after a red tide, and the devastation to caged fish can be enormous. Even pelicans, dolphins, whales, and manatees have been poisoned by dinoflagellate toxins in their food chains.

The toxins, which are of three major types, have been studied for possible use in chemical warfare. They are so potent that as little as half a gram (enough to barely cover two of the periods printed on this page) can be fatal. In humans, they can cause nausea, abdominal cramps, muscular paralysis, amnesia, hallucinations, diarrhea, and respiratory failure resulting in death.

The havoc to the fishing industry caused by major red tides is so great that several laboratories have been conducting research with dinoflagellates and their marine habitats to try to determine the causes of their blooms and to find a way of preventing such destruction in the future.

Until 1970, blooms of toxin-producing dinoflagellates were known only from temperate waters of the Northern Hemisphere. By 1990, however, the blooms were also occurring with increasing frequency throughout both temperate and tropical waters of the Southern Hemisphere.

Reproduction is by cell division. Sexual reproduction appears to be rare.

Division Euglenophyta— THE EUGLENOIDS

Barnyard pools and sewage treatment ponds often develop a rich green bloom of algae. A superficial examination of water from such a pool with the aid of a microscope usually reveals large numbers of active green cells, and a closer inspection probably will reveal one to several of the more than 750 species of *euglenoids,* of which *Euglena* is a common example (Fig. 18.7).

A *Euglena* cell, which is spindle-shaped and has no rigid wall, can be seen to change shape even as the organism moves along. Just beneath the plasma membrane are fine strips that spiral around the cell parallel to one another. The strips and the plasma membrane are devoid of cellulose and together are called a *pellicle.* A single functional flagellum, which has numerous tiny hairs along one side, pulls the cell through the water. A second very short flagellum is present within a *reservoir* at the base of the functional flagellum.

Other features of *Euglena* include the presence of a *gullet,* or groove, through which food can be ingested, and in about a third of the 500 species there are several to many mostly disc-shaped chloroplasts present. A red **eyespot** is located in the cytoplasm near the base of the flagella, and a carbohydrate food reserve called *paramylon* normally is present in the form of small whitish bodies of various shapes.

Reproduction is by cell division. The cell starts to divide at the flagellar end and eventually splits lengthwise, forming two complete cells. Sexual reproduction has been suspected but has never been confirmed.

Some species of *Euglena* can live in the dark if appropriate food and vitamins are present. Others are known to reproduce faster than their chloroplasts under certain circumstances, so some chloroplast-free cells are formed. As long as a suitable environment is provided, these cells also can survive indefinitely. In the past, when only two

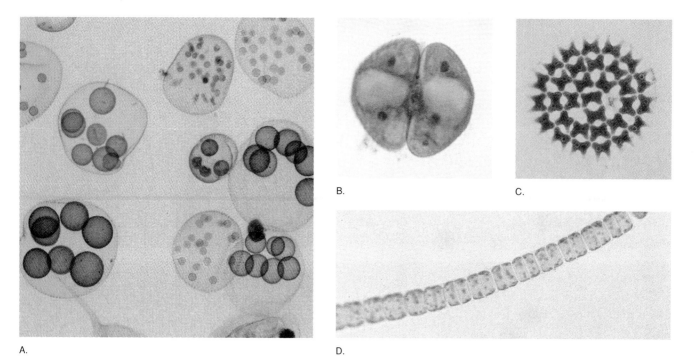

FIGURE 18.8 Representative green algae. *A. Volvox.* The cells form hollow, spherical colonies that spin on their axes as the flagella of each cell beat in such a way that the motion is coordinated. New colonies are formed within the old ones, ×1,000. *B. Cosmarium,* a desmid. These algae consist of single cells that often have a constriction in the center, ×1,200. *C. Pediastrum.* A colonial alga that forms flat plates, ×1,200. *D.* A filament of *Ulothrix,* ×800.

kingdoms were recognized, *Euglena's* capacity to satisfy its energy needs through either photosynthesis or ingestion of food resulted in its being treated as a plant in botany texts and as an animal in zoology texts.

DIVISION CHLOROPHYTA— THE GREEN ALGAE

Division Chlorophyta includes about 7,500 species of organisms commonly known as the *green algae.* They occur in a rich variety of forms and are very widespread, with some of the most beautiful chloroplasts of all photosynthetic organisms. Some are unicellular and microscopic. Others form threadlike filaments, platelike colonies, netlike tubes, or hollow balls (Fig. 18.8).

Some green algae are seaweeds, resembling lettuce leaves or long green ropes. Several unicellular species grow in greenish patches or streaks on the bark of trees, while others grow in large numbers on the fur of sloths and other jungle animals, providing them with a form of camouflage. Still others thrive in snowbanks, live in flatworms and sponges, or are found on the backs of turtles. They are the most common member in lichen "partnerships" (discussed in Chapter 19). The greatest variety, however, is found in freshwater ponds, lakes, and streams. Ocean forms are also varied; there they are an important part of the **plankton**

(free-floating, mostly microscopic organisms) and thus of food chains.

The cells of green algae resemble those of higher plants in having the same kinds of chlorophylls (*a* and *b*) and other pigments in their chloroplasts. The green algae also are believed to have been ancestral to the higher plants, and like the higher plants, they store their food within the chloroplasts in the form of starch. Although most green algae have a single nucleus in their cells, one group, the *bryopsids,* has multinucleate cells. Most green algae can undergo both asexual and sexual reproduction. The manner in which they do so is illustrative of the forms of reproduction found in most of the organisms discussed in the chapters to follow, and so several different representative green algae will be examined in some detail here.

Chlamydomonas

A lively little alga, *Chlamydomonas* (Fig. 18.9) is a common inhabitant of quiet freshwater pools. It has an ancient history among eukaryotic organisms, with fossil relatives occurring in rock formations reported to be nearly 1 billion years old. *Chlamydomonas* is unicellular, with a somewhat oval cell surrounded by a cellulose wall. A pair of whiplike flagella at one end propels the cell very rapidly. The flagella are, however, difficult to see with ordinary light microscopes, and the cell itself is usually less than 25 micrometers (one 10,000th of an inch) long. Near the base of the flagella are two or

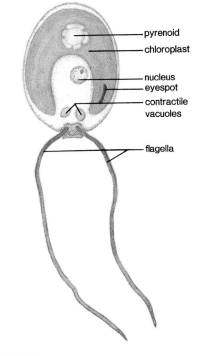

FIGURE 18.9   *Chlamydomonas.*

more vacuoles, which can contract and expand. They apparently regulate the water content of the cell.

A dominant feature of each *Chlamydomonas* is a single, usually cup-shaped chloroplast, which at least partially hides the centrally located nucleus. One or two roundish glistening **pyrenoids** are located in each chloroplast. Pyrenoids are proteinaceous structures thought to contain enzymes associated with the synthesis of starch. Most species also have a red eyespot on the chloroplast near the base of the flagella. The eyespot is sensitive to light, but it is

merely part of an organelle within a single cell and nothing like an eye in structure.

Asexual Reproduction

Before a *Chlamydomonas* reproduces asexually, the cell's flagella degenerate and drop off or are reabsorbed. Then the nucleus divides by mitosis, and the entire protoplasm becomes two cells within the cellulose wall. The two daughter cells, which develop flagella, escape and swim away as the parent cell wall breaks down. Once they have grown to their full size, they may repeat the process. Sometimes mitosis occurs more than once, producing 4, 8, or up to 32 little cells with flagella inside the parent cell. Occasionally, flagella do not develop, and the cells remain together in a colony. When growth conditions change, however, each cell of the colony may develop flagella and swim away. This type of reproduction brings about no changes in the number of chromosomes present in the nucleus, and all the cells remain *haploid.*

Sexual Reproduction

Under certain combinations of light, temperature, and additional unknown environmental forces, many cells in a population of *Chlamydomonas* may congregate together. Careful study of such events has revealed that pairs of cells appear to be attracted to each other by their flagella and function as gametes, which sometimes are of two types. The cell walls break down as the protoplasts slowly emerge and mate, fusing together and forming zygotes. A new wall, often relatively thick and ornamented with little bumps, forms around each zygote. This may remain dormant for several days, weeks, or even months, but under favorable conditions a dramatic change occurs. The protoplast, which is now *diploid,* undergoes meiosis, producing four haploid **zoospores** (motile cells that do not unite with other cells; many different kinds of algae produce zoospores). When the

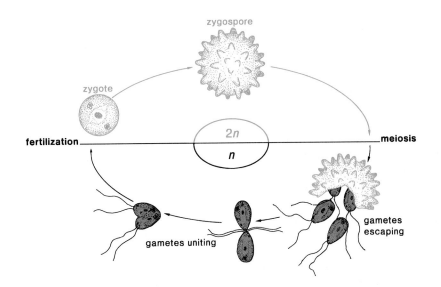

FIGURE 18.10   Sexual life cycle of *Chlamydomonas.*

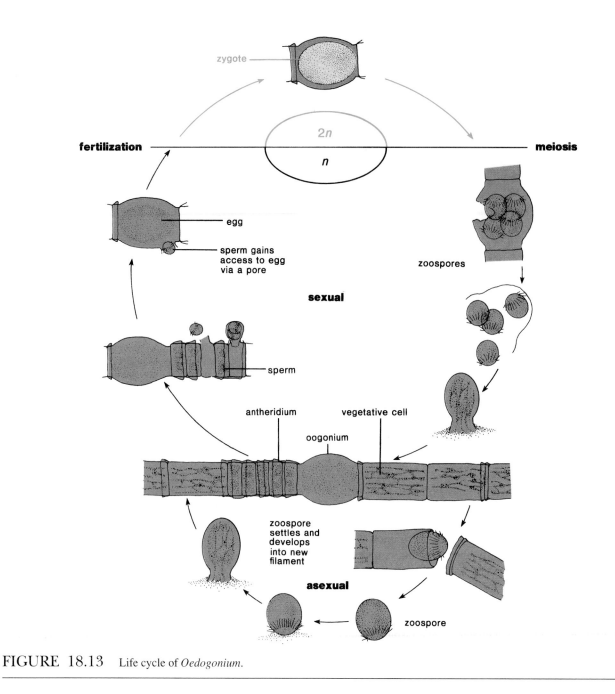

FIGURE 18.13 Life cycle of *Oedogonium.*

The zygotes usually develop thick walls and remain dormant for some time, often over the winter. Thick-walled zygotes are characteristic of most freshwater green algae. Eventually, their protoplasts undergo meiosis, producing four haploid cells. Three of these disintegrate, and a single new *Spirogyra* filament grows from the interior of the old zygote shell. The type of sexual reproduction shown by *Spirogyra* is called **conjugation.**

Oedogonium

Aquatic flowering plants and other algae often provide surfaces to which *Oedogonium* (pronounced ee-doh-goh-nee-um), a filamentous green alga, may attach itself. It is,

however, strictly epiphytic and in no way parasitic. The basal cells of the unbranched filaments form holdfasts, and the terminal cell of each filament is rounded. The remaining cells are cylindrical and attached end to end. The name *Oedogonium* comes from words meaning "swollen reproductive cell" and is quite apt, as the female reproductive cells do indeed bulge noticeably. Each cell contains a large netlike chloroplast, which rolls and forms a tube something like a loosely woven basket around and toward the periphery of each protoplast. There are pyrenoids at a number of the intersections of the net.

Asexual Reproduction

Akinetes (thick-walled overwintering cells) may occasionally be formed, but more commonly zoospores are produced singly in cells at the tips of the filaments (Fig. 18.13). Unlike the zoospores of most other algae, those of *Oedogonium* have about 120 small flagella forming a fringe around the cell toward one end. The zoospores look like tiny, balding, faceless heads. Their rapid rotating movements can be quite entertaining when viewed through a microscope. After they escape from their filaments, they eventually settle and form new filaments in the same manner as *Ulothrix*.

Sexual Reproduction

Oedogonium shows more sexual specialization than any of the other three green algae previously discussed. Short box-like cells called **antheridia** (singular: **antheridium**) are formed in the filaments alongside the ordinary vegetative cells. A pair of male gametes, or **sperms,** is produced in each antheridium. The sperms resemble the zoospores but are smaller. Certain cells become swollen and round to elliptical in outline. These cells, called **oogonia** (singular: **oogonium**), each contain a single female gamete or **egg,** which occupies nearly all of the cell. As the egg matures, a pore develops on the side of the oogonium. When sperms escape from the antheridia, they are attracted to the oogonia by a substance released by the eggs. One sperm eventually enters the oogonium through the pore and unites with the egg, forming a zygote.

In some species of *Oedogonium,* oogonia are produced only on female filaments and antheridia only on male filaments. Sometimes, the male filaments are dwarf. They attach themselves to the female filaments, and then both produce hormones that influence the other's development.

Zygotes may remain dormant for a year or more, but they eventually undergo meiosis, producing four zoospores, each of which is capable of developing into a new filament. The sexual reproduction exhibited by *Oedogonium,* in which one gamete is *motile* (capable of spontaneous movement) while the other gamete is larger and stationary, is called **oogamy.**

Other Green Algae

As indicated earlier, the green algae constitute a very diverse group, with an extraordinary variety of forms and chloroplast shapes. Each species obviously has to reproduce in order to perpetuate itself. Most undergo both sexual and asexual reproduction, but a few do not. For example, the worldwide algae that make parts of some tree trunks appear as though they had received a light brushing or spattering of green paint usually are unicellular or colonial forms that reproduce only asexually.

These and *Chlorella,* another widespread green alga composed of tiny spherical cells, reproduce by forming either daughter cells or aplanospores through mitosis. The daughter cells often remain together in packets, while the aplanospores of *Chlorella,* which are formed inside the parent cells, grow to full size as the parent cell wall breaks down.

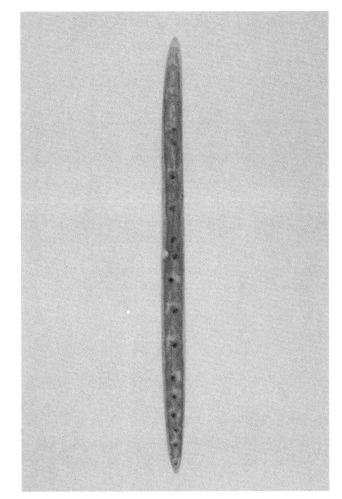

FIGURE 18.14 A single cell of the common desmid, *Closterium.*

Chlorella is very easy to culture and is a favorite organism of research scientists. It has been used in many major investigations of photosynthesis and respiration, and in the future it may become important in human nutrition. (See the section on human and ecological relevance of the algae, which begins on page 308.)

Chlorella could also play a key role in long-range space exploration. Present exploration is severely limited by the weight of oxygen tanks and food supplies needed on a spacecraft, and so scientists have turned to *Chlorella* and similar algae as portable oxygen generators and food sources. Future spacecraft may be equipped with tanks of such algae. These would carry on photosynthesis, using available light and carbon dioxide given off by the astronauts, furnishing the astronauts with oxygen. As the algae multiplied, the excess could either be eaten or fed to freshwater shrimp, which could, in turn, become food for the astronauts. Still other algae and bacteria could recycle other human wastes. Such a self-perpetuating closed system, as it is called, has already been successfully tested with mice and

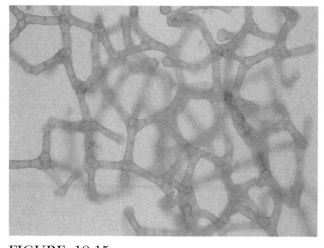

FIGURE 18.15 A portion of a water net (*Hydrodictyon*), a green alga whose cells form a network in the shape of a tube.

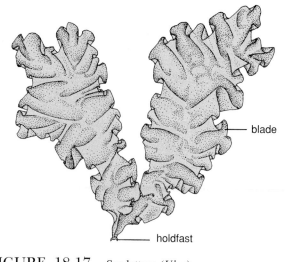

FIGURE 18.17 Sea lettuce (*Ulva*).

FIGURE 18.16 Mermaid's wineglass (*Acetabularia*).

other animals. Many algae, however, are known to produce traces of deadly carbon monoxide gas, and until this problem is resolved, humans will not be subjected to such research.

Desmids (Fig. 18.14), whose 2,500 species of crescent-shaped, elliptical, and star-shaped cells are mostly free-floating and unicellular, reproduce by conjugation. In the beautiful *water nets* (*Hydrodictyon*) (Fig. 18.15), sexual reproduction is isogamous, with up to 100,000 flagellated gametes being produced in a single cell.

Sexual reproduction is also isogamous in the *mermaid's wineglass* (*Acetabularia*), a marine alga consisting of a single huge cell shaped like a delicate mushroom (Fig. 18.16). Each cell is up to 5 centimeters (2 inches) long. This alga has been used in classic experiments demonstrating the influence of the nucleus on the form of the cell. If the cap of an alga is removed and the nucleus is replaced with a nu-

cleus taken from another species, the base regenerates a cap identical to the previous one. If this new cap is also removed, however, the next cap that develops shows form characteristics of both species. If the intermediate cap is then removed, the next cap that develops is identical to that of the species from which the nucleus originally came. Clearly, the original nucleus directs development of cytoplasmic substances regulating cap form, and when these are gone, the replacement nucleus exerts its own influence.

Volvox (see Fig. 18.8 A) is representative of a line of green algae that forms colonies, apparently by means of single cells similar to those of *Chlamydomonas*, held together in a secretion of gelatinous material. In some colonies, the cells are actually connected to one another by cytoplasmic strands. The flagella of individual cells beat separately but pull the whole colony along. A *Volvox* colony may consist of several hundred to many thousands of cells arranged so they resemble a hollow ball, which appears to spin on its axis as it moves. Reproduction may be either asexual or sexual, with smaller daughter colonies being formed inside the parent colony. The daughter colonies are released when the parent colony breaks apart.

Sea lettuce (*Ulva*) is a multicellular seaweed with flattened, crinkly edged green blades that may attain lengths of 1 meter (3 feet) or more (Fig. 18.17). The blades, which may be either haploid or diploid, are anchored to rocks by means of a holdfast at the base. Diploid blades produce spores that develop into haploid blades bearing gametangia. The gametes from the haploid blades fuse in pairs, forming zygotes, each of which can potentially grow into a new diploid blade. Except for the reproductive structures, the haploid and diploid blades of sea lettuce are indistinguishable from one another.

Cladophora is a branched, filamentous green alga whose species are represented in both fresh and marine

FIGURE 18.18 *Chara*, a stonewort.
(Courtesy John Z. Kiss)

waters. Unlike other green algae, the cells of *Cladophora* and its relatives are mostly multinucleate.

Stoneworts (Fig. 18.18), which are aquatic and often calcium-encrusted, loosely resemble small horsetail plants (discussed in Chapter 21). Most botanists now place them in their own division (Charophyta), but some include them with the green algae because of their pigmentation and a few other green algal features. Their sexual reproduction is oogamous, and the antheridia in which the sperms are produced are multicellular. Other features of both their vegetative growth and reproduction are more complex than those of any of the other green algae. Some botanists have considered them more closely related to mosses than to algae.

DIVISION PHAEOPHYTA— THE BROWN ALGAE

Most seaweeds that are brown to olive-green in color are assigned to this division, which includes between 1,500 and 2,000 species of *brown algae* of various sizes. Many are relatively large, and none are unicellular or colonial. Only four of the 260 known genera occur in fresh water, the vast majority being found in colder ocean waters, usually in shallower areas, although the giant kelp (Fig. 18.19) may be found in water 30 meters (100 feet) or more deep. One giant kelp measured 274 meters (710 feet) in length, which is believed to be a record for any single living organism.

Many have a body that is differentiated into a **holdfast,** a **stipe,** and flattened leaflike **blades** (Fig. 18.20). The holdfast is a tough, sinewy structure resembling a mass of intertwined roots. It holds the seaweed to rocks so tenaciously that even the heaviest pounding of surf will not readily dislodge it.

The stalk that constitutes the stipe is often hollow, with a meristem either at its base or at the blade junctions. Since the meristem produces new tissue at the base, the oldest parts of the blades are at the tips.

The blades, which like most of the rest of the body are photosynthetic, may have gas-filled floats called *bladders* toward their bases. The bladder gases may include as much as 10% by volume of deadly carbon monoxide. The function of the carbon monoxide is not known.

A.

B.

FIGURE 18.19 Kelps. *A*. The tip of a giant kelp. *B*. Sea palms (*Postelsia*). The heaviest pounding surf seldom dislodges these brown algae from the rocks.

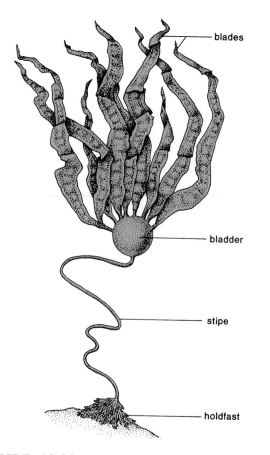

FIGURE 18.20 Parts of the brown alga *Nereocystis*, a kelp.

FIGURE 18.21 *Sargassum,* a floating brown alga from which the Sargasso Sea got its name. It is also found in other marine waters.

The color of the brown algae, which can vary from light yellow-brown to almost black, reflects the presence of varying amounts of the brown pigment *fucoxanthin,* in addition to chlorophylls *a* and *c* and several other pigments in the chloroplasts.

The main food reserve is a carbohydrate called *laminarin.* **Algin** or *alginic acid* (see Table 18.2 and the discussion of human and ecological relevance of the algae, page 308) occurs on or in the cell walls and can represent as much as 40% of the dry weight of some kelps. Reproductive cells are unusual in that the two flagella are inserted laterally (i.e., on the side) instead of at the ends. The only motile cells in the brown algae are the reproductive cells.

In some localities off the coast of British Columbia and Washington, herring deposit spawn (eggs) in layers up to 2.5 centimeters (1 inch) thick on both surfaces of giant kelp blades in late spring. In the past, and to a limited extent at present, native North Americans have harvested these spawn-covered blades, sun dried them, and used them for winter or feast food. Even today, some schoolchildren are given small pieces of the dried or preserved material for lunchbox snacks.

The Sargasso Sea gets its name from a brown rockweed, *Sargassum,* which is washed up in large quantities on the shores along the Gulf of Mexico after tropical storms (Fig. 18.21). A species occurring in the Pacific Ocean has been used, in chopped form, as a poultice on cuts received from coral by native Hawaiians. This and several other brown algae reproduce asexually by fragmentation, while some produce aplanospores. Sexual reproduction takes several different forms, depending on the species. The conspicuous phases of the life cycles are usually diploid.

In the common rockweed, *Fucus* (Fig. 18.22), separate male and female **thalli** (singular: **thallus;** the term for a body that is usually flattened, multicellular, and not organized into leaves, stems, and roots) are produced. Somewhat puffy fertile areas called *receptacles* develop at the tips of the branches of the thallus. The surface of each receptacle is dotted with pores (visible to the naked eye) that open into special, spherical hollow chambers called *conceptacles.*

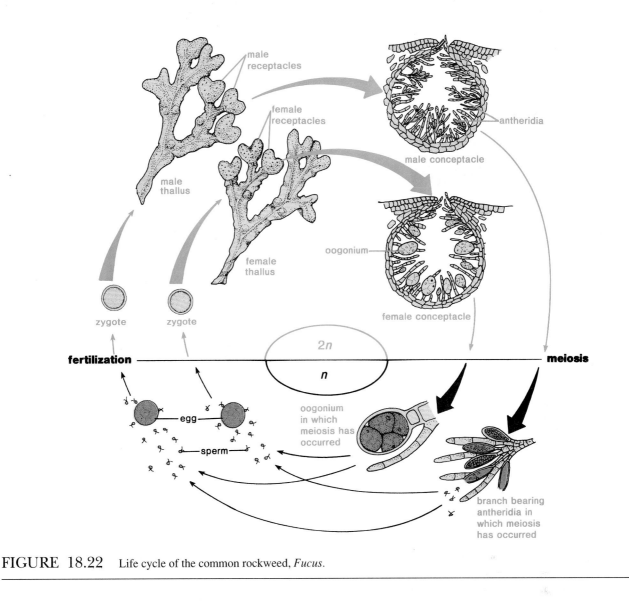

FIGURE 18.22   Life cycle of the common rockweed, *Fucus*.

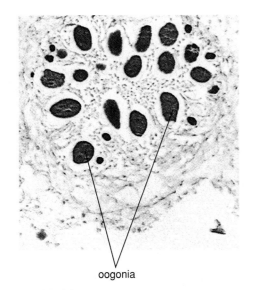

oogonia

FIGURE 18.23   A sectioned female conceptacle of *Fucus*, ×100.

Within the conceptacles, **gametangia** (cells or structures in which gametes are produced) are formed. Eight eggs are produced in each *oogonium* (female gametangium) as a result of a single diploid nucleus undergoing meiosis followed by mitosis (Fig. 18.23). Meiosis also occurs in each *antheridium* (male gametangium), but three mitotic divisions follow meiosis, so 64 sperms are produced. Eventually, both eggs and sperms are released into the water, where fertilization takes place, and the zygotes develop into mature thalli, completing the life cycle.

DIVISION RHODOPHYTA— THE RED ALGAE

Like the brown algae, most of the more than 5,000 species of *red algae* are seaweeds (Fig. 18.24). However, they tend to occur in warmer and deeper waters than their brown

FIGURE 18.24 Representative red algae. *A. Botryocladia. B. Stenogramme. C. Gigartina. D. Gelidium.*

counterparts. Some grow attached to rocks in intertidal zones, where they may be exposed at low tide. Others grow at depths of up to 200 meters (656 feet) where light barely reaches them, and in 1984 a new species of red algae was discovered at a depth of 269 meters (884 feet), where the light is only 0.0005% of peak surface sunlight. A few are unicellular, but most are filamentous. The filaments frequently are so tightly packed that the plants appear to have flattened blades or to form branching segments. Some develop as beautiful feathery structures that have the appearance

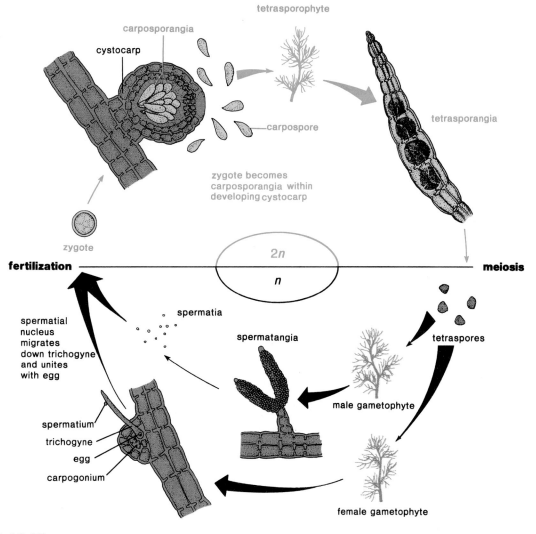

FIGURE 18.25   Life cycle of the red alga *Polysiphonia.*

of delicate works of art.[1] None match the large kelps in size, the largest species seldom exceeding 1 meter (3 feet) in height.

The red algae have relatively complex life cycles, often involving three different types of body structures.

1. Most seaweeds produce their own "glue" and are easy to mount on paper for display. Fresh specimens can be laid directly on clean, high rag-content paper, covered with a layer or two of cheesecloth, and pressed between sheets of blotting paper for a day or two until they are dry. Feathery types will make better specimens if they are placed in a shallow pan of water so that their delicate structures float out, as they do in the ocean. Paper should be slid under them and then carefully lifted so that the seaweed spreads out naturally on the surface. Once the specimens are dry, the cheesecloth (which keeps them from sticking to the blotting paper) is removed and they remain glued to the paper. They can then be displayed or stored indefinitely. Green and brown seaweeds can be treated in similar fashion.

Certain marine algae, which form crusty growths or jointed-appearing upright structures on rocks, accumulate calcium salts as they grow and often contribute to the development of coral reefs. These coralline algae need no special treatment to be displayed, although some may lose their natural pinkish or purplish color when they die.

Meiosis usually occurs on a special body called a *tetrasporophyte,* while gametes are produced on separate male and female bodies. All of the reproductive cells are nonmotile and are carried passively by water currents. Zygotes may migrate from one cell to another through special tubes, which form bizarre loops in some species.

In *Polysiphonia* (Fig. 18.25), a feathery red alga that is widespread in marine waters, the three types of thalli (male gametophyte, female gametophyte, and tetrasporophyte) all outwardly resemble one another. They are about 2 to 15 centimeters (1 to 6 inches) tall and are branched into many fine, threadlike segments. The male sex structures, called *spermatangia,* slightly resemble dense clusters of tiny grapes on slender branches of the male gametophyte thallus. Each spermatangium, as it is released from its branch, functions as a nonmotile male gamete, or *spermatium.*

The female sex structures, called *carpogonia,* are produced on the female gametophyte thallus. Each carpogonium consists of a single cell that looks something like a microscopic bottle with a long neck called the *trichogyne.*

$\mathcal{T}$able 18.1

COMPARISON OF THE DIVISIONS OF ALGAE

DIVISION	FOOD RESERVES	SPECIAL PIGMENTS[1]	FLAGELLA
Chrysophyta			
6,000 species of yellow-green algae, golden-brown algae, diatoms, and cryptophytes	oils, fats, chryso-laminarin	chlorophyll c_1 chlorophyll c_2	2, 1, or none, usually unequal, at or near apex
Pyrrophyta			
3,000 species of dinoflagellates	starch	chlorophyll c_2	2; 1 trailing, 1 girdling; few with none
Euglenophyta			
750 species of euglenoids	paramylon	chlorophyll b	2, 1, or 3, at or near apex
Chlorophyta			
7,500 species of green algae	starch	chlorophyll b	2, 4, or more at apex, or none
Phaeophyta			
1,500 to 2,000 species of brown algae	laminarin, mannitol	chlorophyll c	2, lateral and unequal
Rhodophyta			
More than 5,000 species of red algae	floridean starch, mannitol	chlorophyll d (some species), phycobilins	none

1. All divisions of algae have chlorophyll a and yellow to orange pigments called carotenes and xanthophylls, although the combinations and specific types of xanthophylls vary with the divisions. The fucoxanthin of the golden-brown algae and the brown algae is a xanthophyll.

A single nucleus at the base of the carpogonium functions as the female gamete, or *egg*. Since the spermatia have no flagella, they cannot move of their own accord, but currents may carry them considerable distances. If a spermatium should brush against a trichogyne, it may become attached. The walls between the spermatium and the trichogyne then break down, the nucleus of the spermatium migrates to the egg nucleus, and the two nuclei unite, forming a zygote.

Next, the zygote develops into a cluster of clublike *carposporangia* toward the base of a cystocarp (an urn-shaped body that the female gametophyte thallus forms around the carposporangia). Diploid asexual spores called *carpospores* are produced in the carposporangia and released, to be carried away by ocean currents.

When a carpospore lodges in a suitable location (e.g., a rock crevice or the hull of a ship), it usually germinates and grows into a *tetrasporophyte,* which closely resembles a gametophyte thallus. Tetrasporangia are formed along the branches of the tetrasporophytes. Each tetrasporangium undergoes meiosis, giving rise to four haploid *tetraspores.* When tetraspores germinate, they develop into male or female gametophytes, thereby completing the life cycle.

The red to purplish colors of most red algae are due to the presence of varying amounts of red and blue accessory pigments called *phycobilins,* which are closely related to those found in the blue-green bacteria. The similarity has led to the belief that the red algae may have been derived from the blue-green bacteria. Several other pigments, including chlorophyll a and sometimes chlorophyll d, are also present in the chloroplasts. The principal reserve food is a carbohydrate called *floridean starch.* A number of red algae also produce **agar** and other important gelatinous substances discussed in the next section.

A comparative summary of the food reserves, special pigments, and flagella of the divisions of eukaryotic algae is given in Table 18.1.

HUMAN AND ECOLOGICAL RELEVANCE OF THE ALGAE

Diatoms

We have already noted that numbers of blue-green bacteria are at the bottom of aquatic food chains, but members of all the protoctistan algal divisions play similar roles. Diatoms, for example, are consumed by fish that feed on plankton. Up to 40% of a diatom's mass consists of oils that are converted into cod and other liver oils, which are rich sources of vitamins for humans. The oils also may in the past have contributed to petroleum oil deposits. The National Renewable Energy Laboratory in Golden, Colorado, is presently working on a project to convert oil from cultured diatoms into a "clean" diesel fuel substitute.

Diatoms also have other extensive and more direct industrial uses. In the past, they have apparently been at least as numerous as they are now. As billions upon billions of them have reproduced and died, their microscopic, glassy shells have accumulated on the ocean floor, forming deposits of *diatomaceous earth*. These deposits have accumulated to depths of hundreds of meters in some parts of the world and are quarried in several areas where past geological activity has raised them above sea level. At Lompoc, California, beds of diatomaceous earth are more than 200 meters (650 feet) deep, while in the Santa Maria oil fields of California, deposits reach a depth of 1,000 meters (3,280 feet).

Diatomaceous earth is a light, porous, and powdery looking material that contains about 6 billion diatom shells per liter (1.057 quarts), yet a liter weighs only eight 10ths of a kilogram (1.76 pounds). It also has an exceptionally high melting point of 1,750°C (3,182°F) and is insoluble in most acids and other liquids. These properties make it ideal for a variety of industrial and domestic uses, including many types of filtration. The sugar industry uses diatomaceous earth in sugar refining, and its use for swimming pool filters is widespread. It is also used in silver and other metal polishes, in toothpaste, and in the manufacture of paint that reflects light, which is used on highway markers and signs and on the automobile license plates of some states. It is packed as insulation around blast furnaces and boilers. It is also manufactured into the construction panels used in prefabricated housing.

Pacific Coast fish farmers have experienced losses of salmon and cod when dense concentrations of *Chaetoceros* diatoms have developed in the aquaculture pens. The diatoms have long hollow spines that break off and penetrate the fish gills, disrupting gas exchange and causing bleeding. This damage, in turn, may permit secondary infections and excessive mucus production to occur.

Uses of Green Algae

A few green algae have occasionally been used for food, but members of this division have generally been used less by humans than have those of other divisions. With dwindling world food supplies, this could change. Sea lettuce has been used for food on a limited scale in Asian countries for some time, and several countries are experimenting with the suitability of plankton for human consumption. Except for vitamin C, *Chlorella* contains most of the vitamins needed in human nutrition, and since it is so easy to culture, it may become an important protein source in many parts of the globe. *Chlorella* has also been investigated as a potential oxygen source for atomic submarines, in addition to its possible use in space exploration.

Algin

Commercially produced ice cream, salad dressing, beer, jelly beans, latex paint, penicillin suspensions, paper, textiles, toothpaste, ceramics, and floor polish today all share a common ingredient, *algin*, produced by the giant kelps and other brown algae. It is now used in so many products (Table 18.2) that one might wonder how the world used to get along without it.

Algin has the unique ability to regulate water "behavior" in a wide variety of products. It can, for example, control the development of ice crystals in frozen foods, regulate the penetration of water into a porous surface, and generally stabilize any kind of suspension, such as an ordinary milkshake or other thick fluid containing water. It is produced by several kinds of seaweeds, but a major source is the giant kelp found in the cooler ocean waters of the world, usually just offshore where there are strong currents (see Fig. 18.19 A). This large seaweed, which has been known to attain a length of 92 meters (300 feet), sometimes grows at the rate of 3 to 6 decimeters (1 to 2 feet) per day. Specially equipped oceangoing vessels (Fig. 18.26) harvest the kelp by mowing off the top meter (3 feet) of growth, taking the chopped material aboard and then transferring it to processing centers onshore where it is extracted and refined.

Minerals and Food

Brown algae also produce a number of other useful substances. Many seaweeds, but particularly kelps, build up concentrations of iodine to as much as 20,000 times that of the surrounding seawater. Although it is cheaper to obtain iodine from other sources in the United States, dried kelp has been used in other parts of the world in the treatment of goiter, which results from iodine deficiency. Kelps are relatively high in nitrogen and potassium and have been used as fertilizer for many years. Before such use, the seaweeds need to be rinsed to rid them of salt. They also have been used as livestock feed in northern Europe and elsewhere. In the Orient, many marine algae are used for food—in soups, confections, meat dishes, vegetable dishes, and beverages. In Japan, acetic acid is produced through fermentation of seaweeds.

During the Irish famine of 1845 to 1846, *dulse,* a red seaweed, became an important substitute for the potato crop that had been destroyed by blight. Dulse also occurs along both the Atlantic and Pacific coasts of North America. In Maine and eastern Canada, where it is a popular snack food, dulse is referred to as "Nova Scotia Popcorn." Another red seaweed, *purple laver (nori)*, which occurs in both American and Asian waters, is used extensively for food, particularly in the Orient. In Japan, it is cultured on nets or bamboo stakes set out in shallow marine bays (Fig. 18.27). It is harvested when the thin, crinkly, gelatinous blades are several centimeters (2 to 3 inches) in diameter and is used in meat and macaroni dishes, soups, and dry, spiced delicacies.

Irish moss is another important edible red alga. It is also used in bulking laxatives, cosmetics, and pharmaceutical

Table 18.2

SOME USES OF ALGIN

Food

1. As a thickening agent in toppings, pastry fillings, meringues, potato salad, canned foods, gravies, dry mixes, bakery jellies, icings, dietetic foods, flavored syrups, candies, puddings
2. As an emulsifier and suspension agent in soft drinks and concentrates, salad dressings, barbecue sauces, frozen food batters
3. As a stabilizer in chocolate drinks, eggnog, ice cream, sherbets, sour cream, coffee creamers, party dips, buttermilk, dairy toppings, milkshakes, marshmallows

Paper

1. Provides better ink and varnish holdout on paper surfaces; provides uniformity of ink acceptance, reduction in coating weight and improved holdout of oil, wax, and solvents in paperboard products; makes improved coating for frozen food cartons
2. Used for coating greaseproof papers

Textiles

1. Thickens print paste and improves dye dispersal. Reduces weaving time and eliminates damage to printing rolls or screens

Pharmaceuticals and Cosmetics

1. As a thickening agent in weight-control products, cough syrups, suppositories, ointments, toothpastes, shampoos, eye makeup
2. As a smoothing agent in lotions, creams, lubricating jellies; as a binder in manufacture of pills; as a blood anticoagulant
3. As a suspension agent for liquid vitamins, mineral oil emulsions, antibiotics; as a gelling agent for facial beauty masks, dental impression compounds

Industrial Uses

1. Used in manufacture of acidic cleaners, films, seed coverings, welding rod flux, ceramic glazes, boiler compounds (prevents minerals from precipitating on tubes), leather finishes, sizing, various rubber compounds (e.g., automobile tires, electric insulation, foam cushions, baby pants), cleaners, polishes, latex paints, adhesives, tapes, patching plaster, crack fillers, wall joint cement, fiberglass battery plates, insecticides, resins, tungsten filaments for light bulbs, digestible surgical gut (disappears by time incision is healed), oil well-drilling mud
2. Used in clarification of beet sugar; mixed with alfalfa and grain meals in dairy and poultry feeds

Brewing

1. Helps create creamier beer foam with smaller, longer-lasting bubbles

preparations. Blancmange is a dessert made from Irish moss and milk. *Carrageenan* is a mucilaginous substance extracted from Irish moss and used as a thickening agent in chocolate milk and other dairy products. *Funori,* obtained from yet another red alga, is used as a laundry starch, as an adhesive in hair dressings, and in some water-based paints.

Agar

One of the most important of all algal substances is **agar,** produced most abundantly by the red alga *Gelidium.* This substance, which has the consistency of gelatin, is used (with nutrients added) around the world in laboratories and medical institutions as a culture medium for the growth of bacteria. When various nutrients are added to it, it can also be used as a culture medium for the growth of both plant and animal cells. Full-sized plants have been induced to develop from pollen grains sown on nutrient agar. Orchid tissues are cultured commercially on it and induced to grow into full-sized plants (as discussed in the section on mericloning in Chapter 14), and its use in making the capsules containing drugs and vitamins is now worldwide. It is also used as an agent in bakery products to retain moistness, as a base for cosmetics, and as an agent in gelatin desserts to promote rapid setting.

Current research involving red algae and other seaweeds indicates they contain a number of substances of

FIGURE 18.26 A specially equipped vessel harvesting kelp off the California coast. The ship moves backward through the kelp beds as the machinery at the stern mows the kelp and conveys it into the hold.

(Courtesy Kelco Company)

potential medicinal value. More than 20 seaweeds have been used in preparations designed for the expulsion of digestive tract worms, control of diarrhea, and the treatment of cancer. Some have shown considerable potential as antibiotics and insecticides. Chemical relatives of DDT have been found to be produced by certain red algae. The sea hare and other marine animals feed on such algae, and it is possible that such animals may degrade (break down) the DDT-like compounds to simpler substances—a feat unknown among land animals.

OTHER MEMBERS OF KINGDOM PROTOCTISTA

Protozoans (Phylum Protozoa) and *sponges* (Phylum Porifera) are included within Kingdom Protoctista; they have, however, traditionally been regarded as animals and are not covered in this book. The *slime molds,* which appear to be related to the protozoa, are discussed next.

SUBKINGDOM MYXOBIONTA— THE SLIME MOLDS

In the Ripley's Believe It or Not pavilion at the Chicago World's Fair in 1933 there was an exhibit of "hair growing on wood." The "hairs," while indeed superficially resembling short human hair, were actually the reproductive

FIGURE 18.27 Beds of purple laver ("nori") ready to harvest in shallow marine water at Sendai, Japan.

(Courtesy I.A. Abbott)

FIGURE 18.28   Reproductive bodies (sporangia) of the slime mold *Stemonitis.*

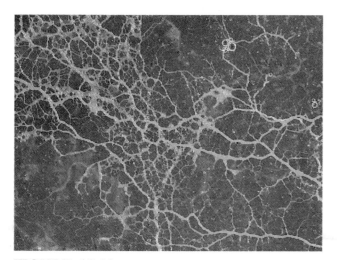

FIGURE 18.29   A plasmodium of the slime mold *Physarum.*

structures of a species of **slime mold** (Fig. 18.28). These curious organisms are totally without chlorophyll and thus are incapable of producing their own food. They are a bit of a puzzle to biologists because they are distinctly animal-like during much of their life cycle but also distinctly funguslike when they reproduce.

The tiny roundish *spores* of the more than 500 species of slime mold average only 10 to 12 micrometers in diameter and thus are individually invisible to the naked eye. Nevertheless, they are present nearly everywhere and are especially abundant in airborne dusts.

If you place almost any dead leaf or piece of bark in a dish and add a food source, such as a few dry oatmeal flakes to which a drop or two of water has been added, the odds are that slime mold spores will be present and will germinate. If the dish is covered, the spores will sometimes germinate in as little as 15 minutes, but germination usually takes several hours or longer. Within a few days, a curious glistening mass of active slime mold material looking something like the netted veins of a leaf may appear. This material, whose "veins" tend to merge into the shape of a fan at its leading edges, is a slime mold **plasmodium** (Fig. 18.29).

If the plasmodium is examined with a microscope, it becomes apparent that there are no cell walls present. The protoplasm in the veins, particularly toward the center, flows very rapidly and rhythmically. The protoplasmic movement may stop momentarily at regular intervals and then resume its flow, sometimes in the opposite direction.

Plasmodia are often white, but they also may be brilliantly colored in shades of yellow, orange, blue, violet, or black. A few are colorless and more or less transparent. They are found on damp forest debris, under logs, on old shelf or bracket fungi, sometimes on older mushrooms, and in other moist places where there is dead organic matter. They tend to flow forward at a rate of up to 2.5 centimeters (1 inch) or more per hour, often against slow moisture seepage, feeding on bacteria and other organic particles as they go. They contain many diploid nuclei, all of which divide often and simultaneously as growth occurs. With an adequate food supply, a plasmodium may increase to 25 times its original size in just one week.

A.

B.

FIGURE 18.30 Common slime mold sporangia. *A. Lamproderma. B. Lycogala.*
(*B.* courtesy L.L. Steimley)

Dramatic events take place when significant changes in food supplies, available moisture, light, and other environmental features occur. The plasmodium usually is converted, often quite rapidly, into many separate small **sporangia** (Fig. 18.30), each containing thousands of minute, one-celled **spores.** The sporangia often are globe-shaped, but in some species they develop as long or wide stationary bodies that may resemble a jumbled network of tubes, or they may resemble erect hairs or end up as a shapeless blob. The sporangia may have slender stalks, depending on the species. Others exhibit combinations of body forms. The spores are often distributed throughout a jumbled mass of threads called a *capillitium.*

When a spore is formed, a single nucleus and a little cytoplasm become surrounded by a wall. Meiosis takes place in the spore, and three of the four resulting nuclei degenerate. When the spore germinates, one or more amoeba-like cells called *myxamoebae* emerge. Sometimes, these have flagella, in which case they are called *swarm cells.* Either form may become like the other through the development or loss of flagella. At first, myxamoebae or swarm cells feed on bacteria and other food particles. Sooner or later, however, they function as gametes, fusing in pairs and forming zygotes. A new plasmodium usually develops from the zygote, although occasionally zygotes or small plasmodia may fuse and form larger plasmodia (Fig. 18.31).

Cellular Slime Molds

Two or more divisions of slime mold are recognized. Most of the species have typical plasmodia and follow the patterns just discussed. About two dozen species of *cellular slime molds* are, however, evidently not closely related to the other slime molds. Individual amoebalike cells of cellular slime molds feed independently, dividing and producing separate new cells from time to time. When the population reaches a certain size, they stop feeding and clump together, forming a mass called a *pseudoplasmodium.* The pseudoplasmodium resembles and crawls like a garden slug. It eventually becomes stationary and is transformed into a sporangiumlike mass of spores.

Human and Ecological Relevance of the Slime Molds

The slime molds, like the bacteria, contribute to ecological balance in forests and woodlands by breaking down organic particles to simpler substances; they also reduce bacterial populations in their habitats. One atypical species occasionally attacks cabbages and another causes powdery scab of potatoes. Yet another species causes a disease of watercress, but slime molds otherwise are of little economic significance.

SUBKINGDOM MASTIGOBIONTA, DIVISION CHYTRIDIOMYCOTA— THE CHYTRIDS

If you were to immerse dead leaves or flowers, old onion bulb scales, dead beetle wing covers, or other organic material in water that had been mixed with a little soil, within a day or two thousands of microscopic **chytrids** (pronounced kítt-ridds) probably would appear on the surfaces

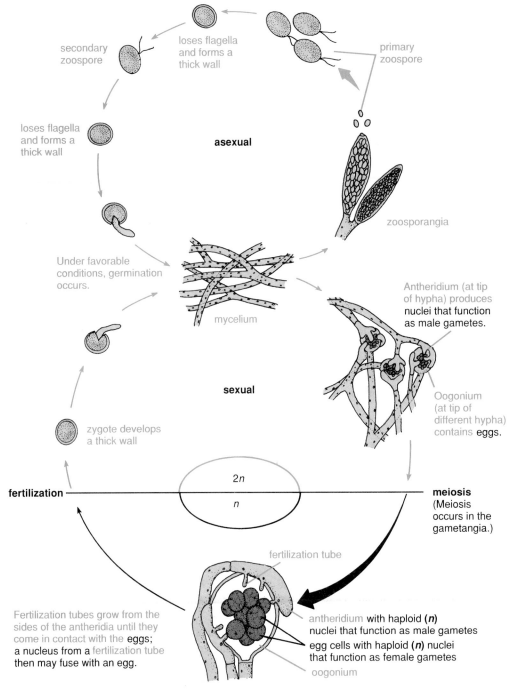

FIGURE 18.34 Life cycle of the water mold *Saprolegnia*.

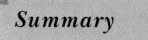

Summary

1. All living organisms were confined to the ocean less than 1 billion years ago. About 400 million years ago, green algae began making the transition from water to land and gave rise to land plants.

2. Kingdom Protoctista includes many diverse organisms that all have eukaryotic cells. Members may be unicellular or multicellular and occur as either colonies or filaments. Modes of nutrition include photosynthesis, ingestion of food, a combination of both, or absorption of food in solution. Some members are nonmotile but most are motile.

3. The four classes of Chrysophyta are the yellow-green algae, the true golden-brown algae, the diatoms, and

the cryptophytes. Some members of each class produce statospores.

4. Diatoms are very abundant; they have a glassy shell that consists of two halves that fit together like a pillbox.

5. Fucoxanthin gives a golden-brown color to many diatoms. Diatoms move in caterpillar fashion by contact of the cytoplasm with a surface as it protrudes through the pores.

6. Diatoms reproduce asexually by mitosis and sexually through the fusion of gametes that form an auxospore (zygote).

7. Dinoflagellates (Pyrrophyta) are unicellular organisms with two flagella. Some cause red tides that can kill fish and poison humans. Dinoflagellates in tropical waters exhibit bioluminescence (emission of light) when disturbed.

8. Euglenoids (Euglenophyta) have no rigid cell wall; only one functional flagellum, a gullet, and paramylon as a carbohydrate food reserve. Reproduction is by cell division. Sexual reproduction has not been confirmed.

9. Green algae (Chlorophyta) have cells with the same pigments and reserve food (starch) as those of higher plants. The chloroplasts of *Chlamydomonas* have pyrenoids; the cells have two or more vacuoles and often a red eyespot. Asexual reproduction is by mitosis; sexual reproduction is isogamous.

10. *Ulothrix* is a filamentous green alga that is attached to objects by means of a holdfast; each cell contains a bracelet-shaped chloroplast. Asexual reproduction is by zoospores; sexual reproduction is isogamous.

11. *Spirogyra* is a floating, filamentous green alga with spiral, ribbonlike chloroplasts. Asexual reproduction is by fragmentation. Sexual reproduction is by conjugation.

12. *Oedogonium* is an epiphytic, filamentous green alga; it has cylindrical, netted chloroplasts. In asexual reproduction, zoospores are produced. Sexual reproduction is oogamous.

13. Other green algae include *Chlorella*, desmids, *Acetabularia*, *Volvox*, *Ulva*, and *Cladophora*.

14. Stoneworts are either included with the green algae or placed in Division Charophyta.

15. Many brown algae (Phaeophyta) are large seaweeds; their thalli often may be differentiated into a stipe, flattened blades, and a holdfast.

16. The color of brown algae is largely due to fucoxanthin; the main carbohydrate food reserve is laminarin. Some brown algae produce algin, a gelatinous substance useful to humans. The reproductive cells have lateral flagella. The common rockweed, *Fucus*, produces eggs and sperms that form zygotes in the water.

17. Most red algae (Rhodophyta), represented by *Polysiphonia*, are seaweeds with life cycles that involve three different types of thalli and nonmotile gametes.

18. Red and blue phycobilins are partially responsible for the colors of red algae; chlorophyll *d* may be present in the chloroplasts. The main carbohydrate food reserve is floridean starch. Some red algae produce agar, an economically important gelatinous substance.

19. Algae are ecologically and economically important. Diatomaceous earth is used for filtering, polishes, insulation, and reflectorized paint. *Chlorella* is a potential food and oxygen source. Algin is used as a stabilizer and thickening agent in hundreds of products.

20. Some brown algae are a source of fertilizer, iodine, and food. Red algae are a source of agar and food and have potential medicinal value.

21. Subkingdom Myxobionta includes the slime molds, which are animal-like in their vegetative state; they consist of a multinucleate mass of protoplasm called a *plasmodium* that flows over damp surfaces, ingesting food particles.

22. Slime molds form stationary sporangia containing spores, from which myxamoebae or swarm spores emerge upon germination. Myxamoebae or swarm spores function as gametes, with new plasmodia developing from the zygotes.

23. Cellular slime molds produce a pseudoplasmodium, which crawls like a slug that converts to a stationary, sporangiumlike mass of spores.

24. Subkingdom Mastigobionta includes aquatic chytrids and water molds. Chytrids consist mostly of spherical cells with rhizoids at one end; they reproduce only asexually, with zoospores having a single flagellum.

25. Water molds, which have coenocytic mycelia, include organisms that cause diseases of fish and other aquatic organisms. Asexual reproduction involves zoospores; gametes are produced in oogonia and antheridia.

Review Questions

1. How do cells of diatoms differ from those of other organisms?

2. What forms of sexual and asexual reproduction occur in the green algae?

3. Which groups of algae produce the following important products: (a) agar, (b) algin, (c) nerve poisons, (d) abrasives for polishes?

4. How would you distinguish *Chlamydomonas* from *Euglena*?

5. *Spirogyra, Ulothrix,* and *Oedogonium* all form filaments. How can you tell them apart?

6. In the green algae studied, where in the life cycles does the chromosome number change from *n* to 2*n* and vice versa?

7. Where and how is algin obtained?

8. Is there any difference in structure between the holdfasts of microscopic green algae and those of brown algae?

9. Why are some green algae red and some red algae green?

10. Which divisions of algae have only unicellular representatives?

Discussion Questions

1. Some algae are attached to solid objects or other organisms, while others are free-floating. What are the advantages and disadvantages of each type of growth?

2. The variety and sizes of algae found in the oceans are considerably greater than those of freshwater forms. What reasons can you suggest?

3. Seaweeds that grow in intertidal zones, where they may be exposed between tides, are often more gelatinous than their continually submerged counterparts. Explain.

4. Why do some algae grow on one side of a tree and not all around the trunk?

5. Should the bladders (floats) of some of the kelps give these algae any advantage over those algae that do not possess such structures?

Additional Reading

Abbott, I. A. 1984. *Limu: An ethnobotanical study of some edible Hawaiian seaweeds,* 3d ed. Lawai, HI: Pacific Tropical Botanical Garden.

Abbott, I. A., and E. Y. Dawson. 1978. *How to know the seaweeds,* 2d ed. Pictured Key Nature Series. Dubuque, IA: Wm. C. Brown Publishers.

Abbott, I. A., and G. J. Hollenberg. 1976. *Marine algae of California.* Palo Alto, CA: Stanford University Press.

Becker, E.W. (Ed.). 1985. *Production and use of microalgae.* Forestburgh, NY: Lubrecht & Cramer.

Bold, H. C., and M. J. Wynne. 1985. *Introduction to the algae,* 2d ed. Englewood Cliffs, NJ: Prentice-Hall.

Buetow, D. E. (Ed.). 1968–1989. *The biology of Euglena: General biology and ultrastructure,* 4 vols. San Diego, CA: Academic Press.

Chapman, V. J., and D. J. Chapman. 1980. *Seaweeds and their uses,* 3d ed. New York: Chapman and Hall.

Fritsch, F. E. 1935, 1945. *Structure and reproduction of the algae,* 2 vols. Ann Arbor, MI: Books Demand.

Hallegraeff, G. M. 1993. A review of harmful algal blooms and their apparent global increase. *Phycologia* 32(2): 79–99.

Irvine, D., and D. M. John. (Eds.). 1985. *Systematics of the green algae.* San Diego, CA: Academic Press.

Lee, R. E. 1989. *Phycology,* 2d ed. New York: Cambridge University Press.

Palmer, C. M. 1962. *Algae in water supplies.* Washington, DC: Government Printing Office, U.S. Department of Health, Education and Welfare, Public Health Service.

Prescott, G. W. 1978. *How to know the freshwater algae,* 3d ed. Pictured Key Nature Series. Dubuque, IA: Wm. C. Brown Publishers.

Round, F. E. 1984. *The ecology of algae.* New York: Cambridge University Press.

Scagel, R. F., et al. 1984. *Plants: An evolutionary survey.* Belmont, CA: Wadsworth Publishing Co.

Smith, G. M. 1950. *Freshwater algae of the United States,* 2d ed. New York: McGraw-Hill.

Sze, P. 1992. *A biology of the algae,* 2d ed. Dubuque, IA: Wm. C. Brown Publishers.

Turner, N. J. 1974. *Food plants of British Columbia Indians. Part 1: Coastal peoples.* Victoria, Canada: British Columbia Provincial Museum.

*Fly agaric mushrooms (*Amanita muscaria*).*

Kingdom Fungi and Lichens

19

After a short introduction, the distinctions between Kingdoms Protoctista and Fungi are discussed. A summary of the features of Kingdom Fungi and a review of how the kingdom came to be recognized are given.

Kingdom Fungi includes two divisions and several classes of true fungi. Life cycles and discussions of the economic importance of representative members of each division are presented. Nematode-trapping fungi, Pilobolus, truffles, morels, ergot, yeasts, stinkhorns, puffballs, bracket fungi, bird's-nest fungi, smuts, rusts, poisonous and hallucinogenic fungi, shiitake mushrooms, mushroom culture, antibiotics, industrial products obtained from fungi, and fungi in nature are discussed. The chapter concludes with an examination of various forms of lichens. Natural dyeing is briefly explored, and the economic importance of lichens is reviewed.

Some Learning Goals

1. Know the general features that distinguish Kingdom Fungi from the other kingdoms.
2. Distinguish the divisions and classes of fungi from one another on the basis of their cells or hyphae and their reproduction.
3. Identify and describe sporangium, conidium, coenocytic, mycelium, dikaryotic, zygospore, ascus, and basidium.
4. Assign each of the following to its class of true fungi: athlete's foot, Dutch elm disease, *Pilobolus, Penicillium* mold, stinkhorn, yeast, ergot, chestnut blight fungus, puffball, smut.
5. Know five economically important fungi in each of four different classes of true fungi.
6. Understand how lichens are classified and identified.
7. Learn the basic structure of a lichen.

FIGURE 19.1   Typical fungal hyphae, as seen with the aid of a dissecting microscope.

W hen it comes to breaking down organic materials of all kinds and making them available for recycling, fungi and bacteria are the most important organisms known. The major oil spill that occurred in Alaska in 1989 intensified efforts to find better and faster ways of cleaning up such disasters than was possible in the past. Although there are bacteria that break down oil molecules, very few such organisms survive the saltwater of the ocean for any length of time. Attention now, however, is being given to a fungus that breaks down many types of oils to harmless simpler molecules, such as carbon dioxide.

This fungus, known as *Corollospora maritima*, occurs naturally on almost all the beaches of the world and flourishes in ocean surf, which, as mentioned in the previous chapter, contains oils produced by diatoms. Ways of culturing the fungus inexpensively and in large quantities are being explored, and also strains that produce the oil-degrading enzymes in significantly larger amounts are being sought. This chapter examines the nature, reproduction, and other features of this and the thousands of other known true fungi found all over the world.

The word *fungus* may evoke images of mushrooms or some sort of powdery or spongy, creeping growth. Although mushrooms are indeed fungi, and while many fungi do appear to be creeping along the ground, the forms of fungi seem almost infinite. There are about 100,000 known species of mushrooms, rusts, smuts, mildews, molds, stinkhorns, puffballs, truffles, and other organisms in Kingdom Fungi, and over 1,000 new species are described each year. There may be as many as 200,000 species yet undiscovered or undescribed, and our daily ravaging of rain forests and other fungal habitats undoubtedly is dooming many others to extinction before they can be found.

As they grow, most fungi produce an intertwined mass of delicate threads, which tend to branch freely and also often fuse together. The individual threads, which usually are more or less tubular, are called **hyphae** (singular: **hypha;** hyphae may be partitioned into cylindrical cells or they may become extensive, branched, multinucleate cells) (Fig. 19.1).

A mass of hyphae is collectively referred to as a **mycelium.** When appropriate food is available, fungi grow very rapidly. In fact, despite the microscopic size of individual hyphae, all those produced by a single fungus in one day when laid end to end could extend more than 1 kilometer (0.6 mile). Some fungi thrive in freezers if the temperature is not lower than −5°C (23°F), while others reproduce freely under temperature conditions of 55°C (131°F) or higher.

Besides being vital to the natural recycling of dead organic material, fungi, along with bacteria, also cause huge economic losses through food spoilage and disease; these topics are explored in the sections on human and ecological relevance of fungi later in the chapter. Scientists who study fungi are known as *mycologists,* and consumers of fungi are called *mycophagists* (from the Greek word *myketos* for "a fungus").

DISTINCTIONS BETWEEN KINGDOMS PROTOCTISTA AND FUNGI

In the past, the true fungi, slime molds, and bacteria were all placed in a single division of the Plant Kingdom. Once the fundamental differences between prokaryotic and eukaryotic cells became known, however, the bacteria were placed in the prokaryotic Kingdom Monera. Then it became increasingly apparent that the metabolism, reproduction, and general lines of diversity of fungi were different from those of members of the Plant Kingdom, with the fungi evidently having been independently derived from ancestral unicellular organisms. Accordingly, fungi were placed in their own kingdom.

All true fungi are filamentous or unicellular *saprobes* (decomposer organisms that live on dead organic matter) or are parasites. Both saprobes and parasites absorb their food in solution through their cell walls. The slime molds, which engulf their food and were discussed in Chapter 18, appear to be related to protozoa, which are single-celled protists included in Kingdom Protoctista. Chytrids and water molds are also funguslike, but like most protists their reproductive cells have flagella, and they, too, are presently included in Kingdom Protoctista.

The members of Kingdom Fungi are placed in two divisions and several classes. They are all filamentous and do not have motile cells. All true fungi produce hyphae, which grow at their tips. Structures such as mushrooms are formed from hyphae tightly interwoven and packed together. The cell walls of true fungi consist primarily of chitin, a material also found in the shells of arthropods (e.g., insects, crabs). Fungi exhibit a variety of forms of sexual reproduction. The food substances, which most fungi absorb through their cell walls, are often broken down with the aid of enzymes secreted to the outside by the cells. Because of the great variety of form and reproduction throughout Kingdom Fungi, a neat pigeonholing of all the members into distinct groups is difficult. Broad groups can, however, be recognized.

KINGDOM FUNGI— THE TRUE FUNGI

Division Zygomycota— The Coenocytic True Fungi

Although black bread molds are probably the best known members of this division, they are not the only fungi that grow on bread. So many organisms can, in fact, contribute to bread spoilage that nearly all commercially baked goods have, in the past, had chemicals, such as calcium propionate, added to the dough to prevent or retard the growth of such organisms. There is now a trend toward eliminating preservatives from bread, pies, and other bakery items since the chemicals that have made the goods a less suitable medium for the growth of fungi apparently are unhealthy for humans after prolonged use. Alternative ways of retarding fungal growth and spoilage are being sought.

Rhizopus (Fig. 19.2), a well-known representative black bread mold, has spores that are exceedingly common everywhere. They have been found in the air above the North Pole, over jungles, on the inside and outside of buildings, in soils, in clothing, in automobiles, and hundreds of kilometers out to sea, easily carried there by prevailing winds and breezes.

When a spore lands in a suitable growing area, it germinates and soon produces an extensive mycelium, which, like that of the water molds, is *coenocytic* (not partitioned into individual cells) and contains numerous haploid nuclei. After the mycelium has developed, certain hyphae called **sporangiophores** grow upright and produce globe-shaped **sporangia** at their tips (Fig. 19.3). Numerous black **spores** are formed within each sporangium. When these spores are released through the breakdown of the sporangium wall, they may blow away and repeat the cycle.

Such reproduction, involving no union of gametes, is asexual, but black bread molds also reproduce sexually by conjugation. Although there is no visible difference in form, black bread mold mycelia occur in two different mating strains. When a hypha of one strain encounters a hypha of the other, swellings called *progametangia* develop opposite each other on the hyphae. These protuberances grow toward each other until they touch. A crosswall is formed a short distance behind each tip, and the two *gametangia* merge, becoming a single large multinucleate cell in which the nuclei of the two strains fuse in pairs. A thick ornamented wall then develops around this cell with its numerous diploid nuclei. This structure, called a *zygospore,* is the characteristic sexual spore of members of this division. A zygospore may lie dormant for months, but eventually it may crack open, and one or more sporangiophores with sporangia at their tips grow out. Meiosis apparently takes place just before this occurs and thousands of black spores are produced in the sporangia. In some species, the spores are produced externally on hyphae instead of being formed in sporangia. Zoospores are unknown in this division.

One dung-inhabiting genus of fungi in this class has the scientific name of *Pilobolus,* derived from two Greek words meaning "cap thrower." The name is quite appropriate, as the mature sporangia are catapulted a distance of up to 8 meters (26 feet), where they adhere to grass or other vegetation (Fig. 19.4). When the vegetation is ingested by animals, the spores germinate in the digestive tract and are already growing in the dung when it is released.

Pilobolus fungi release their sporangia with force precisely in the direction of light, to which they are very

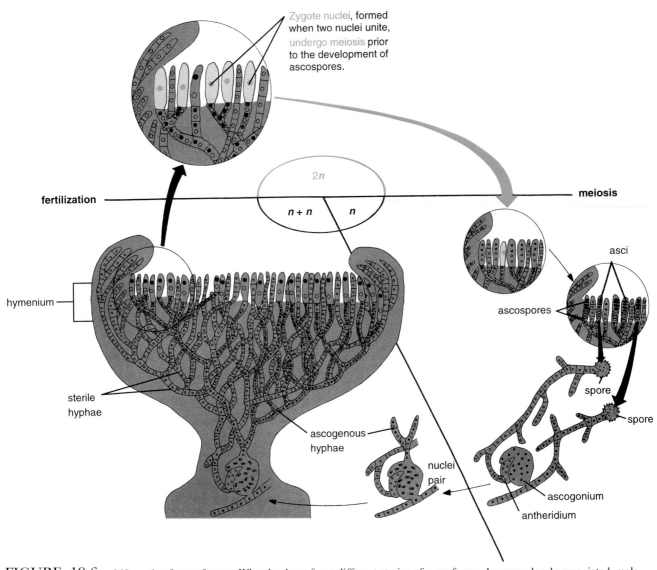

Zygote nuclei, formed when two nuclei unite, undergo meiosis prior to the development of ascospores.

fertilization

2n

n + n   n

meiosis

asci

ascospores

spore

spore

hymenium

sterile hyphae

ascogenous hyphae

nuclei pair

ascogonium

antheridium

FIGURE 19.6   Life cycle of a sac fungus. When hyphae of two different strains of a sac fungus become closely associated, male antheridia may be formed on one and female ascogonia on the other. Male nuclei migrate into an ascogonium and pair but do not fuse with the female nuclei present. Then new hyphae (ascogenous hyphae), whose cells each contain a pair of nuclei, grow from the ascogonium. In a process involving fusion of the pairs of nuclei (followed by meiosis), fingerlike asci, each containing four or eight haploid nuclei, are formed in a layer called a *hymenium*, which lines an ascocarp. The haploid nuclei become ascospores, which are discharged into the air. They are potentially capable of initiating new mycelia and repeating the process.

on one and female *ascogonia* on the other. Male nuclei migrate into an ascogonium and pair but do not fuse with the female nuclei present. Then new hyphae (*ascogenous hyphae*), whose cells each contain one male and one female nucleus, grow from the ascogonium. The cells divide in a unique way, producing new cells with one of each kind of nucleus. At maturity, the pairs of nuclei in the cells at the tips of each ascogenous hypha unite, and tubular asci develop in a layer (referred to as the *hymenium*) at the surface of a structure called an *ascocarp*. The diploid zygote nucleus undergoes meiosis, and the resulting four haploid nuclei usually divide once more by mitosis so there is a row of eight nuclei in each ascus. These nuclei become *ascospores* as they are walled off from one another with a little cytoplasm (Fig. 19.6).

Thousands of asci may be packed together in an ascocarp, which often is cup-shaped (Fig. 19.7) but also may be completely enclosed (Fig. 19.8) or flask-shaped with a little opening at the top. Truffles are really enclosed ascocarps. Cup-shaped ascocarps may be several centimeters (2 to 3 inches) in diameter and may be brilliantly colored on the inside. When ascospores are mature, they are often released with force from the asci. If an open ascocarp is jarred at maturity, it may appear to belch fine puffs of smoke consisting of thousands of ascospores.

When an ascospore lands in a suitable area, it may germinate, producing a new mycelium, and then repeat the process. In many instances, however, a number of asexual generations involving conidia are produced between the sexual cycles.

FIGURE 19.7   An ascocarp of a sac fungus.

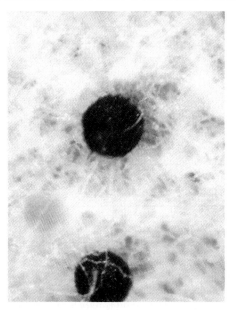

FIGURE 19.8   Closed ascocarps (*cleistothecia*) of a powdery mildew on an oak leaf, ×200.

Human and Ecological Relevance of the Sac Fungi

Morels, which have been called by some people "the world's most delicious mushrooms," and *truffles* have been prized as food for centuries. Neither is, however, a true mushroom. True mushrooms are included in the *club fungi* discussed in the next section. Wealthy Romans and Greeks used to insist on preparing morels personally according to various recipes

FIGURE 19.9   A morel.
(© Doug Sherman/Geofile)

they had concocted, and they are still prized as gourmet food today. Prior to 1982, numerous unsuccessful attempts were made to cultivate them under controlled conditions, but morels now can be mass-produced commercially and are beginning to be marketed. Morels (Fig. 19.9) are the size of a medium to smallish mushroom, are tan in color, and have a wrinkled, somewhat conelike top on a stalk that resembles a miniature tree trunk. The numerous depressions between the wrinkles each contain thousands of asci. Although morels are perfectly edible by themselves, there are unconfirmed reports that some persons have become ill after consuming them with alcohol. Caution in this regard is advised.

A related "mushroom" called a *false morel* or *beefsteak morel* is considered a delicacy by many but has caused death in others. This points to the peculiar unpredictability of the interaction between the human metabolism and certain fungi. For some people, the fungi are quite safe to eat; for others they are not—a literal example of "one man's meat being another man's poison."

When rye and, to a lesser extent, other grains come into flower in a field, they may become infected with *ergot* fungus (Fig. 19.10). This fungus seldom causes serious damage to the crop, but as it develops in the maturing grain it produces several powerful drugs. If the infected grain is harvested and milled, a disease called **ergotism** may occur in those who eat the contaminated bread. The disease can affect the central nervous system, often causing hysteria, convulsions, and sometimes death. Another form of ergotism causes gangrene of the limbs, which may result in loss of the affected limb. It can also be a serious problem for cattle grazing in infected fields. They frequently abort their fetal calves and may succumb themselves.

Ergotism was common in Europe in the Middle Ages. Known then as *St. Anthony's Fire,* it killed 40,000 people in an epidemic in A.D. 994. In 1722, the cavalry of Czar Peter the Great was felled by ergotism just as the czar was about to conquer Turkey, and the conquest never took place. Some historians have implicated ergotism in violent social upheavals such as the French Revolution and the Salem Witchcraft Trials. As recently as 1951, an outbreak hospitalized 150 victims in a French village. Five persons died and 30 became temporarily insane, imagining they were being chased by snakes and demons.

In small controlled doses, ergot drugs have proved medically useful. They stimulate contraction of the uterus to initiate childbirth and have been used in abortions and in the treatment of migraine headaches. Ergot is also an initial source for the manufacture of the hallucinogenic drug lysergic acid diethylamide, popularly known as *LSD.*

Sac fungi play a basic role in the preparation of baked goods and alcohol. Enzymes produced by yeast aid in fermentation, producing ethyl alcohol and carbon dioxide in the process. The carbon dioxide produced in bread dough forms bubbles of gas, which are trapped, causing the dough to rise and giving bread its porous texture. Part of the flavor of individual wines, beers, ciders, sake, and other alcoholic beverages is imparted by the species, or strain, of yeast used to ferment the fruits or grains.

Yeasts are at least indirectly involved in the manufacture of a number of other important products. Ephedrine, a drug also produced by mormon tea, a leafless western desert shrub, is obtained commercially from certain yeasts. The drug is used in nose drops and in the treatment of asthma. Yeasts are also a rich source of B vitamins and are used in the production of glycerol for explosives. Ethyl alcohol is used in industry as a solvent and in the manufacture of synthetic rubber, acetic acid, and vitamin D. Yeast contains about 50% protein and makes a nourishing feed for livestock. Its enzyme, invertase, catalyzes the conversion of sucrose into glucose and fructose and is used to soften the centers of chocolate candies after the chocolate coating has been applied.

Several very important plant diseases are found in this class of fungi. *Dutch elm disease,* a disease originally described in Holland, has devastated the once numerous and stately elm trees in many towns and cities in the midwestern and eastern United States and has now spread to the West

FIGURE 19.10 Ergotized rye. The prominent darker objects scattered throughout the ears of rye are kernels infected with ergot fungus.

(Copyright © Loran Anderson)

and South. When Dutch elm disease was first discovered, limited control of the disease was achieved by spraying the trees with DDT to kill the elm bark beetles that spread the disease from tree to tree. DDT killed many useful organisms as well, however, and its use for such purposes was eventually banned.

Other sprays have since been used, again with limited success, and biological controls are now being sought. One such control was reported in 1980 by Gary Strobel of Montana State University. He injected *Pseudomonas* bacteria into 20 diseased trees and saved seven of them when the bacteria multiplied and killed the Dutch elm disease fungus. In

FIGURE 19.11 A peach leaf infected with peach leaf curl.

FIGURE 19.12 A common stinkhorn. Note the slimy mass of spores on the cap.

(Courtesy Leland Shanor)

1987, Dr. Stroebel injected a genetically altered strain of *Pseudomonas* bacteria with more powerful toxins against Dutch elm disease into 15 elm trees but then had to cut the trees down because he had not previously obtained permission from the Environmental Protection Agency for his experiment. Whether this control technique will meet with widespread success remains to be seen, but it is the type of control that is much to be preferred to poisonous sprays that upset delicate ecological balances (see Appendix 2, which deals with biological controls).

Chestnut blight has virtually eliminated the once numerous American chestnut trees from the eastern deciduous forests. Attempts to control the disease have met with very limited success thus far. Much better success has been obtained in controlling *peach leaf curl,* a disease that attacks the leaves of some stone fruits, especially peach trees (Fig. 19.11). Sprays that contain copper or zinc salts apparently inhibit the germination of spores of peach leaf curl and have been effective in preventing serious damage when applied to trees before the dormant buds swell and open in the spring.

Class Basidiomycetes— The Club Fungi

At the end of my first year in college, I did odd jobs during the summer, including various types of yard work. On one occasion while cleaning dead leaves and debris from a shaded garden area, I came across two or three growths, about the width of a pencil and the length of a finger, rising above the surface of the ground. On closer inspection, I observed that these growths had the consistency and appearance of a sponge and the tips were partially covered with a slimy and evil-smelling substance. The growths turned out to be fungi called *stinkhorns* (Fig. 19.12), whose odor attracts flies, which disseminate the sticky spores that stick to their bodies.

A.

B.

C.

FIGURE 19.13 *A.* Common woods mushrooms. *B.* Earth stars. *C.* A bracket fungus.
(*B.* courtesy Perry J. Reynolds)

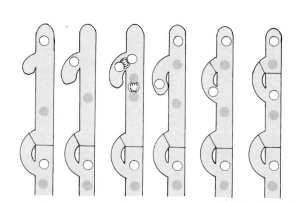

FIGURE 19.14 Development of a clamp connection. *Left* to *right.* Protrusion appears in wall of cell; one nucleus migrates into loop; both nuclei undergo mitosis; loop carrying daughter nucleus turns to cell wall; clamp connection established; daughter nuclei pair in two cells.

Stinkhorns are interesting but relatively unimportant representatives of another large class of true fungi, the *club fungi*. Included in this class are mushrooms (Fig. 19.13 A) or

toadstools (the only distinctions between mushrooms and toadstools are based on folklore or tradition—botanically there is no difference), puffballs, earth stars (Fig. 19.13 B), shelf or bracket fungi (Fig. 19.13 C), rusts, smuts, jelly fungi, and bird's-nest fungi. They are called *club fungi* because in sexual reproduction, they produce their spores at the tips of swollen hyphae that often resemble baseball bats or clubs. These swollen hyphal tips are called **basidia** (singular: **basidium**). The hyphae, like those of the cup or sac fungi, are divided up into individual cells. These cells, however, have either a single nucleus or, in some stages, two nuclei. The crosswalls have a central pore that is surrounded by a swelling, and both the pore and swelling are covered by a cap. This cap, with some exceptions, blocks passage of nuclei between cells but allows cytoplasm and small organelles to pass through.

Asexual reproduction is much less frequent in club fungi than in the other classes of fungi. When it does occur, it is mainly by means of conidia, although a few species produce buds similar to those of yeasts, and others have hyphae that fragment into individual cells, each functioning like a spore and forming a new mycelium after germination.

Sexual reproduction in many mushrooms begins in the same way as it does in members of the two classes previously discussed. When a spore lands in a suitable place—often an area with good organic material and humus in the soil—it germinates and produces a mycelium just beneath the surface. The hyphae of the mycelium are divided into cells that each contain a single haploid nucleus. Such a mycelium is said to be **monokaryotic**. Monokaryotic mycelia of club fungi often occur in four mating types, usually designated simply as types 1, 2, 3, and 4. Only types 1 and 3 or types 2 and 4 can mate with each other.

If the growth of the hyphae of compatible mating types happens to bring them close together, cells of each mycelium may unite, initiating a new mycelium in which each cell has two nuclei. Such a mycelium is said to be

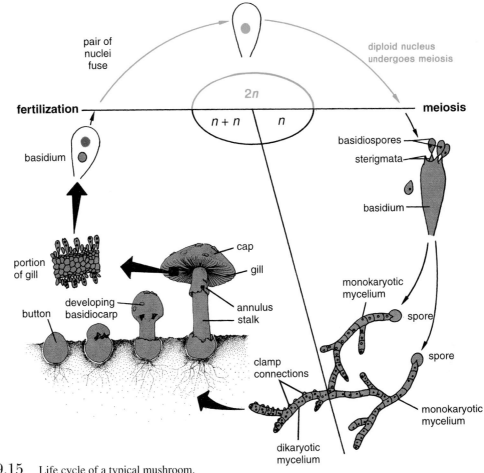

FIGURE 19.15   Life cycle of a typical mushroom.

dikaryotic. Dikaryotic mycelia usually have little walled-off bypass loops called *clamp connections* between cells on the surface of the hyphae (Fig. 19.14). The clamp connections develop as a result of a unique type of mitosis that ensures that each cell will have within it one nucleus of each original mating type.

After developing for a while, the dikaryotic mycelium may become very dense and form a compact, solid-looking mass called a *button.* This pushes above the surface and expands into a *basidiocarp,* commonly called a **mushroom** (Fig. 19.15). Most mushrooms have an expanded umbrella-like *cap* and a *stalk* (**stipe**). Some have a ring called an *annulus* on the stalk. It is the remnant of a membrane that extended from the cap to the stalk and tore as the cap expanded. Some mushrooms, such as the notorious *death angel mushroom,* also have a cup called a *volva* at the base (Fig. 19.16). Thin, fleshy-looking plates called **gills** radiate out from the stalk on the underside of the cap. Microscopic examination of a gill reveals the compacted hyphae of which it is composed and innumerable **basidia** oriented at right angles to the flat surfaces of the gill.

As each basidium matures, the two nuclei unite and then undergo meiosis. The four nuclei that result from meiosis become **basidiospores** when they migrate through four (in a few species, two) tiny pegs at the tip of the basidium, walls forming around the nuclei in the process. The tiny pegs, called *sterigmata* (singular: *sterigma*), serve as stalks for the basidiospores. One large mushroom may produce several billion basidiospores within a few days. These are forcibly discharged into the air between the gills, and they blow away with the slightest breeze.

If you remove a mushroom stalk and place the cap gill-side down on a piece of paper, covering it with a dish to eliminate air currents, the spores will fall and adhere to the paper in a pattern perfectly reflecting the arrangement of the gills. The dish and cap can be removed a day later, and the *spore print* (Fig. 19.17) can be made more or less permanent with the application of a little clear varnish or shellac. Professional mycologists use such spore prints as an aid to identification, employing white paper for dark-colored spores and black paper for white or light-colored spores.

FIGURE 19.16   A death angel mushroom (*Amanita*). Note the egglike volva at the base.
(Courtesy Dr. T. Duffy)

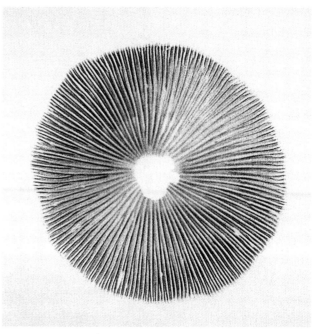

FIGURE 19.17   A spore print.

FIGURE 19.18   A bolete. These mushrooms have pores instead of gills beneath their caps.

In nature, some of the basidiospores eventually repeat the reproductive cycle. Often, a dikaryotic mycelium radiates out from its starting point, periodically producing basidiocarps in so-called *fairy rings*. If conditions are favorable, the mycelium continues to grow at the edges for many years while dying in the center as food resources are depleted. Some mycelia have been known to grow in this fashion for over 500 years.

Some mushrooms produce their spores on the surfaces of thousands of tiny pores instead of on gills (Fig. 19.18).

Shelf or *bracket fungi* (Fig. 19.19) grow out horizontally from the bark or dead wood to which they have become attached, some adding a new layer of growth each year. Perennial species can become large enough and so securely attached that they can support the weight of an adult.

Other members of this class produce spores within parchmentlike membranes, forming somewhat ball-like basidiocarps called *puffballs*. Puffballs, which generally are edible, range in diameter from a few millimeters to 1.2 meters (0.125 inch to 4 feet) (Fig. 19.20). They have no stalks and rest in contact with the ground. Literally trillions of spores may be produced by a large puffball. These are released through a pore at the top or from random locations when the outer membrane breaks down. *Earth stars* (see Fig. 19.13 B), which are similar to puffballs, differ from them in having a ring of appendages at the base that look like a set of woody petals around a flower.

FIGURE 19.19 A shelf, or bracket, fungus (*Phacolus*).
(Courtesy Richard Critchfield)

FIGURE 19.21 A bird's-nest fungus.

FIGURE 19.20 Giant puffballs. The largest weighed 6.12 kilograms (13.5 pounds).
(Courtesy Louise White, Redding, CA, *Record-Searchlight*)

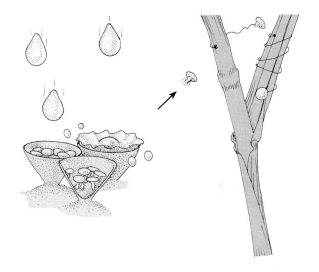

FIGURE 19.22 How the "eggs" in a bird's-nest fungus are dispersed.

(After Harold J. Brodie. 1951. "The splash-cup dispersal mechanism in plants." *Canadian Journal of Botany* 29: 224–34. Reproduced by permission of the National Research Council of Canada from the *Canadian Journal of Botany,* 29: 224–34)

Bird's-nest fungi (Fig. 19.21), which grow on wood or manure, form nestlike cavities in which "eggs" containing basidiospores are produced. In some species, each "egg" has a sticky thread attached to it. When raindrops fall in the nests, the eggs may be splashed out, and as they fly through the air, the sticky threads catch on nearby vegetation, whipping the eggs around it (Fig. 19.22). When animals graze on the vegetation, the spores pass unharmed through the intestinal tract.

Smuts are parasitic club fungi that do considerable damage to corn, wheat, and other grain crops. In corn smut (Fig. 19.23), the mycelium grows between the cells of the host. The hyphae absorb nourishment from these cells and also secrete substances that stimulate them to divide and enlarge, forming tumors on the surfaces of the corn kernels. These eventually break open, revealing millions of sooty black spores, which are blown away by the wind. Some smuts affect only the flowering heads or grains, while others infect the whole plant.

FIGURE 19.23   Corn smut fungus on an ear of corn.

Rusts, which also are parasites, attack a wide variety of plants. Some rusts grow and reproduce on only one species of flowering or cone-bearing plant. Others, however, require two or more different hosts to complete their life cycles. *Black stem rust of wheat,* which has reduced wheat yields by millions of bushels in a single year in the United States alone, has plagued farmers ever since wheat was first cultivated thousands of years ago. More than 300 races of black stem rust are now known. This rust requires both common barberry plants and wheat to complete its life cycle (Fig. 19.24).

Since two hosts are necessary for black stem rust of wheat to complete its life cycle, control of the disease could be accomplished through eradication of common barberry bushes. An estimated 600 million such plants have been destroyed in the United States since 1918, in an attempt to eradicate the disease, but it has proved impossible to eliminate the species altogether. Producing rust-resistant strains of wheat has also helped, but even as new strains are developed, the rusts themselves hybridize or mutate, producing new races capable of attacking previously immune varieties of cereals—a striking example of evolution in action.

Another serious rust with two hosts is the *white pine blister rust,* which has caused huge losses of valuable timber trees in both the eastern and western United States. Basidiospores infect the pine trees, and when the basidiospores germinate, other types of spores are produced. These different spores in turn infect currant and gooseberry bushes, and the spores formed on the currants and gooseberries eventually give rise to new basidiospores, completing the cycle. The U.S. Forest Service had a program of gooseberry bush eradication in operation for many years in an attempt to alleviate the problem, but the program was only partially successful. Spraying programs have more recently been implemented with some success, and rust-resistant trees also are being selected and bred as alternatives.

Other rusts with two hosts include *apple rust* (alternate host: cedar trees), *poplar leaf spot* (alternate host: larch or tamarack trees), and *corn rust* (alternate host: sorrel plants).

A relative of black stem rust was recently discovered in the Rocky Mountains of Colorado. This rust causes its rock cress host plants to produce fake flowers that look and smell so real many insects are fooled by them. When bees and butterflies visit the fake flowers, they find a sugary, nectarlike secretion. While gathering the sticky fluid, they inadvertently also pick up fungal sex cells, which are spread to other rock cress plants.

Human and Ecological Relevance of the Club Fungi

Of the approximately 25,000 described species of club fungi, fewer than 75 are known to be poisonous. Many of the latter are, however, common and not readily distinguishable by amateurs from edible species. Also, some edible forms, such as the inky cap and shaggy mane mushrooms, which cause no problems by themselves, may make one very ill if consumed with alcohol. Few of the poisonous forms normally are fatal, but unfortunately some—such as the death angel, which causes 90% of the fatalities attributed to mushroom poisoning—are relatively common.

Poisoning from death angels and similar species is due to alpha-amatin (an alkaloid). Symptoms of the poison, which completely blocks RNA synthesis, usually take from 6 to 24 hours to appear. Until a few years ago, successful treatment was impossible by the time the intense stomachache, blurred vision, violent vomiting, and other symptoms occurred.

At present, hope for survival of the victims of mushroom poisoning lies with the administration of a drug known as *thioctic acid.* Even thioctic acid does not always prevent a fatality, however, and either the drug or the mushroom poison usually leave the patient with hypoglycemia (a blood sugar deficiency). Some wild mushroom lovers have fed parts of their collections to dogs or cats, and when nothing happened to the animals after an hour or two, they have eaten the mushrooms themselves. Both they and the animals later succumbed. Others have mistakenly believed that the toxic substances are destroyed by cooking. Records show, however, that before the discovery of thioctic acid as an antidote, death ensued in 50% to 90% of those who had eaten just one or two of the deadly mushrooms, cooked or raw, and

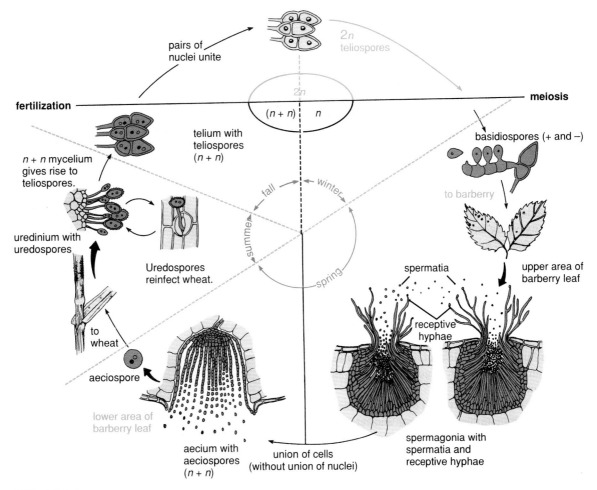

FIGURE 19.24   Life cycle of black stem rust of wheat.

that even as little as 1 cubic centimeter (less than 0.5 cubic inch) can be fatal.

A number of widespread but very unreliable beliefs exist concerning distinctions between edible and poisonous mushrooms. One holds that a silver coin placed in the cooking pan will turn black while the mushrooms are cooking if any poison is present. Some members of both edible and poisonous species will turn such a coin black, and some will not. Another superstition holds that edible species can be peeled, while poisonous ones cannot. Again this is a fallacy, for death angels peel quite easily. Still other erroneous beliefs are that poisonous mushrooms appear only in the fall or the early spring, that all mushrooms eaten by snails and beetles are edible, that all purplish-colored mushrooms are poisonous, and that all mushrooms growing in grassy areas are edible. Again, these notions simply are not supported by the facts. It is foolhardy for anyone to eat mushrooms that have not been correctly identified by a knowledgeable authority.

Some poisonous mushrooms cause hallucinations in those who eat them. During the Mayan civilization in Central America, *teonanacatl* ("God's flesh") sacred mushrooms (Fig. 19.25) were used in religious ceremonies. The consumption of these mushrooms, which has continued among native groups in Mexico and Central America to the present, results in sharply focused, vividly colored visions.

FIGURE 19.25   Teonanacatl mushrooms.
(Courtesy Drug Enforcement Administration)

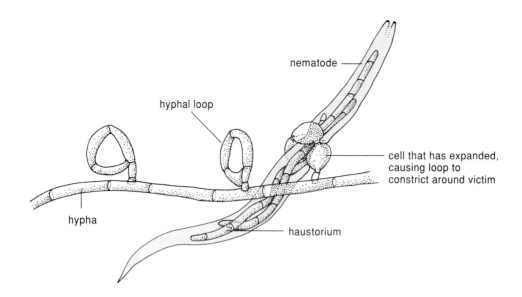

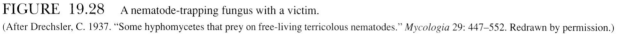

FIGURE 19.28   A nematode-trapping fungus with a victim.
(After Drechsler, C. 1937. "Some hyphomycetes that prey on free-living terricolous nematodes." *Mycologia* 29: 447–552. Redrawn by permission.)

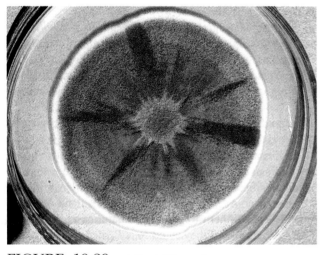

FIGURE 19.29   A *Penicillium* colony.

FIGURE 19.30   Blue cheese. The dark areas are parts of the mycelium of a species of *Penicillium* that gives the cheese its unique flavor.

they eventually were able to isolate a strain that produced more than 80 times the original quantity of penicillin. Later, when this strain was subjected to X-radiation, still other forms were produced that upped the penicillin output to 225 times that of Fleming's mold. Today, most of the penicillin produced around the world comes from descendants of that cantaloupe mold. Literally hundreds of other antibiotics effective in combating human and animal diseases have been discovered since the close of World War II, and the production of these drugs is a vast worldwide industry.

Penicillium molds are also used in other ways. Some are introduced into the milk of cows, sheep, and goats at stages in the production of "smelly" cheeses, such as blue (Fig. 19.30), Camembert, Roquefort, Gorgonzola, and Stilton. The molds produce enzymes that break down proteins and fats in the milk, giving the cheeses their characteristic flavors.

Since the early 1980s, organ transplants have been aided by the discovery and production of a "wonder drug" from an imperfect fungus found in soil. Called *cyclosporine,* the drug suppresses immune reactions that cause rejection of transplanted organs, without risking the development of leukemia and other undesirable side effects associated with other drugs.

B.

FIGURE

The l
ers of cells
tive layer (
compresse
that the ge
function. I
algal cells
medulla c
pies at lea
stances p
layer, cal
present. I
and is oft
rhizines.

Li
growth f
ships bu
identific
or emb(

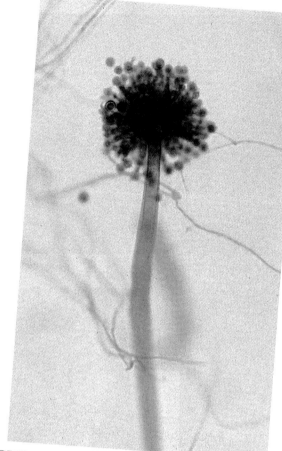

FIGURE 19.31 A conidiophore of *Aspergillus* seen with the aid of a microscope. Rows of conidiospores (also called *conidia*) are being produced at the tips of branches of the conidiophore.

A number of diseases of both humans and animals are caused by *Aspergillus* species. The diseases, called *aspergilloses,* attack the respiratory tract after the spores have been inhaled. One type thrives on and in human ears. Other diseases caused by different genera of imperfect fungi include those responsible for the widespread problems of athlete's foot and ringworm, for white piedra (a mild disease of beards and mustaches), and for tropical diseases of the hands and feet that cause the limbs to swell in grotesque fashion. One serious disease called *valley fever,* found primarily in the drier regions of the southwestern United States, usually starts with the inhalation of dust-borne spores of an imperfect fungus that produces lesions in the upper respiratory tract and lungs. The disease may spread elsewhere in the body, with sometimes fatal results.

Aspergillus flavus, which grows on moist seeds, secretes *aflatoxin,* the most potent natural carcinogen known. The toxin causes liver cancer, and no more than 50 parts per billion is allowed in human food. In humid climates, such as those of the southeastern United States and adjacent Mexico, improperly stored grain can become moist enough to support the fungus. Carcinogenic foods such as peanuts, peanut butter, and peanut-based dairy feeds may result. Dairy cattle feed is even more strictly controlled because concentrations of aflatoxin can accumulate in milk.

Two imperfect fungi show promise as biological controls of pest organisms. One has already been used with some success in controlling scale insects in Florida and other warm, humid regions. Another may be used to combat water hyacinths, which have caused serious clogging of waterways in areas of the world with mild to tropical climates.

Aspergillus is a genus of imperfect fungi whose species produce dark brown to blackish or yellow spores (Fig. 19.31). It is closely related to the *Penicillium* molds and is extensively used in industry. One or more species are used commercially for the production from sugar of citric acid, a substance for flavoring foods and for the manufacture of effervescent salts that were originally obtained from oranges. Citric acid is also used in the manufacture of inks and in medicines, and it is even used as a chicle substitute in some chewing gums.

Aspergillus fungi also produce gallic acid used in photographic developers, dyes, and indelible black ink. Other species are used in the production of artificial flavoring and perfume substances, chlorine, alcohols, and several acids. Further uses are in the manufacture of plastics, toothpaste, and soap and in the silvering of mirrors.

One species of *Aspergillus* is used in the Orient and elsewhere to ferment soybeans to make soy sauce, or *shoyu.* A Japanese food called *miso* is made by fermenting soybeans, salt, and rice with the same fungus. More than 0.5 million tons of miso are consumed annually.

LICHENS

A student who was interested in natural dyeing came to me a few years ago and asked if she could experiment with the dye potential of local plants as a special project. In the course of her experimentation, she obtained beautiful shades of yellow, brown, and green from two dozen common local plants, using simple recipes.[1] During the following summer, she extended the project to include lichens growing on the

1. Most natural dye recipes call for simmering at least 1 liter (approximately 1 quart) or 2 of loosely packed fresh or dry material covered with water in a large enamel kettle until most of the coloring appears to be in the water. This usually takes from one to several hours. The solid waste is then strained out and discarded. A *mordant* (substance that helps fibers take up dye permanently) is then often added to the liquid in amounts varying from 1 teaspoon to 0.5 cup. Commonly used mordants include alum, detergent ammonia, copper sulphate, tin (stannous chloride), and white vinegar. White wool or other fibers are washed in warm water and detergent, rinsed in warm water, and then left to soak in hot water for up to an hour. The fibers are then quickly transferred to the hot dyeing liquid and simmered for an hour or left overnight. The dyed material is then rinsed in warm water until no more dye diffuses out, washed again with detergent, and dried. Reaction of the dyed material to light can be tested by placing it in direct sunlight for several days. (See Appendix 3 for a number of sources of natural dyes.)

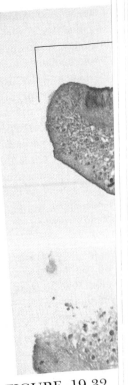

trees and rocks at a (
The rich colors she
more spectacular, wh
organisms were in th
are used in a minor

Lichens have
examples of symbio
gus and an alga ir
called a **thallus.** Th
than 1 millimeter t
feet). The alga sup
the fungus protect
produces a substar
alga, and absorbs a
ganisms. The evi
probably be more
the alga in a con
cells in some inst

There are al
algal component
terium, and a few
Three genera of
bacterium are in
species of alga m
lichen, however,
the exception of
have a club fung
press trees) that

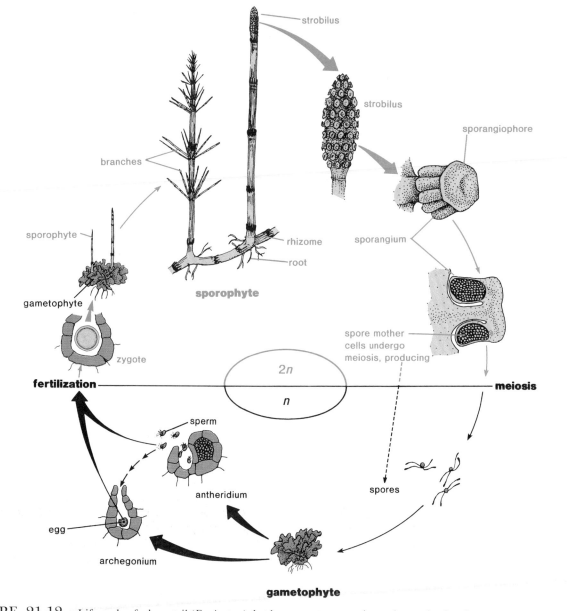

FIGURE 21.12   Life cycle of a horsetail (*Equisetum*) that has separate vegetative and reproductive shoots.

blood corpuscles." At least one or two species are known to have a mild diuretic effect (a *diuretic* is a substance that increases the flow of urine), and they have been used in the past in folk medicinal treatment of urinary and bladder disorders. Some have also been used as an antacid or an *astringent* (an astringent is a substance that arrests discharges, particularly of blood). One species was used in the treatment of gonorrhea, and others were used for tuberculosis.

At least two Native American tribes burned the stems and used the ashes to alleviate sore mouths or applied the ashes to severe burns. Members of another tribe ate the strobili of a widespread scouring rush to cure diarrhea, and still others boiled stems in water to make a shampoo for the control of lice, fleas, and mites.

At one time, the use of scouring rush stems for scouring and sharpening was widespread. They were used not only for cleaning pots and pans, but also for polishing brass, hardwood furniture and flooring, and for honing mussel shells to a fine edge. Scouring rushes are still in limited use for these purposes today. Some species of horsetails accumulate certain minerals in addition to silica. Veins of such minerals have been located beneath populations of horsetails by analyzing the plants' mineral contents. This process of analysis involves a chemical treatment of the tissues followed by the use of X-ray equipment.

In the geological past, the giant horsetails and club mosses were a significant part of the vegetation growing in vast swampy areas. In some instances, the swamps were stagnant and slowly sinking, permitting the gradual accumulation of plant remains, which, because of the lack of oxygen in the water, were not readily attacked by decay bacteria. Such circumstances, over aeons of time, were ideal for the

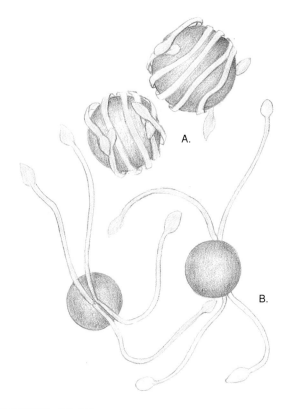

FIGURE 21.15 Reconstruction of the fossil giant horsetail, *Calamites*.

FIGURE 21.13 Horsetail spores. *A*. With elaters coiled. *B*. With elaters spread.

FIGURE 21.14 Reconstruction of a coal age (Carboniferous) forest.

(Photo by Field Museum of Natural History, Chicago)

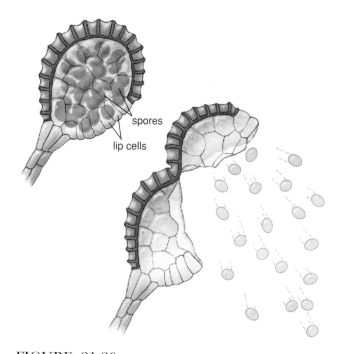

spores

lip cells

FIGURE 21.20 Release of spores from a fern sporangium. *A.* An intact sporangium. *B.* Spores being ejected as the sporangium splits; the annulus first draws back and then snaps forward.

not for the eventual substitution of alternative materials, these magnificent plants might have been totally destroyed. Some tropical hummingbirds use these hairs along with scales of other ferns to line their nests, and at one time, Polynesians used them in a form of embalming for their dead. The trunks of tree ferns have been used in the construction of small houses in the tropics. Parts of one Hawaiian species of small tree fern and the fronds of an Asian fern yield red pigments used for dyeing cloth.

The bracken fern, which is distributed worldwide, has been used and even cultivated for human food for many years, particularly in Japan and New Zealand, even though it has also long been known to be mildly poisonous to livestock. Recent research in both Europe and Japan has shown conclusively that bracken fronds fed to experimental animals produce intestinal tumors, and because of this the consumption of these fronds for food should be actively discouraged.

Indigenous peoples of many areas where ferns occur have eaten the cooked rhizomes and young fronds of various ferns. Native Americans often baked the rhizomes of sword ferns, lady ferns, and others in stone-lined pits, removed the outer layers, and ate the starchy inner material. Similarly, native Hawaiians ate the starchy core of their tree ferns as emergency food. In Asia, the oriental water fern is still sometimes grown for food in rice paddies and used as a raw or cooked vegetable. In Malaysia, a relative of the lady fern is frequently used as a vegetable.

A.

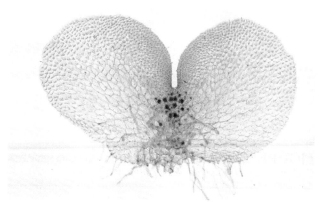

B.

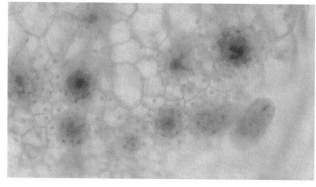

C.

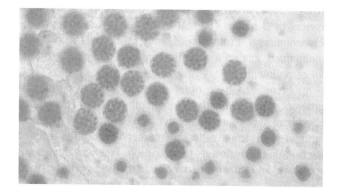

D.

FIGURE 21.21 Fern prothalli. *A.* Surface view, × 10. *B.* A prothallus as seen with the aid of a microscope, × 20. *C.* Archegonia, × 100. *D.* Antheridia × 100.

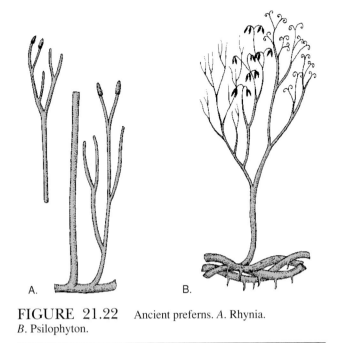

FIGURE 21.22 Ancient preferns. *A*. Rhynia. *B*. Psilophyton.

FIGURE 21.23 A fossil fern.

FIGURE 21.24 An orchid plant growing on bark.

Uses of ferns in folk medicine abound. They have been used in the treatment of diarrhea, dysentery, rickets, diabetes, fevers, eye diseases, burns, wounds, eczema and other skin problems, leprosy, coughs, stings and insect bites; as a poison antidote; and for labor pains, constipation, dandruff, and a host of other maladies. The male fern, which is more common in Europe than the United States, contains a drug that is effective in expelling intestinal worms (e.g., tapeworms). Its use for this purpose dates back to ancient times, and it is still occasionally so used, although synthetic medicines have now largely replaced it. The licorice fern, which was used by Native Americans of the Pacific Northwest in the treatment of sore throats and coughs, was also used as a flavoring agent and a sugar substitute. Members of a tribe in California chewed stalks of goldback ferns to quell toothaches and snuffed a liquid made from the fronds of the bird's-foot fern to arrest nosebleeds.

The fronds of bracken and other ferns have been used in the past for thatching houses. Anyone who has placed such fronds in compost piles knows that they break down much more slowly than the leaves of other plants. Bracken fronds are still occasionally used as an overnight bedding base by fishermen and hunters. A substance extracted from these fronds has been used in the preparation of chamois leather, and the rhizome is used in northern Europe in the brewing of ale.

The chain fern, which has large fronds up to 2 meters (6.5 feet) long, has two flexible leathery strands in the petiole and rachis of each frond. Native Americans and others have gathered the fronds for many years to strip these strands for use in basketry and weaving. They do so by gently cracking the long axis with stones to expose the strands, which are then easily removed. The glossy black petioles of the five-finger fern have also been used in intricate basketry pat-

terns by Native Americans. In Southeast Asia, the climbing fern has fronds with a rachis that may grow up to 12 meters (40 feet) long; it is still a favorite material (when available) for the weaving of baskets.

The mosquito fern, which is a floating water fern in the genus *Azolla*, forms tiny plants little bigger than duckweeds. It is found over wide areas where the climate is relatively mild. It sometimes forms such dense floating mats that it is believed to suffocate mosquito larvae, which periodically

A.

B.

FIGURE 22.6 Female pine cones. *A*. Three immature cones shortly after being produced. *B*. A mature cone.

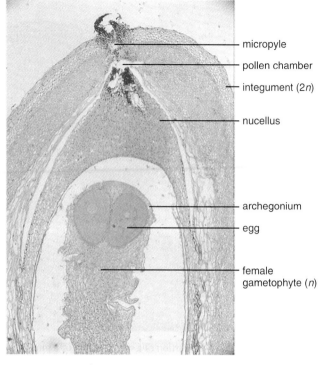

- micropyle
- pollen chamber
- integument (2*n*)
- nucellus
- archegonium
- egg
- female gametophyte (*n*)

FIGURE 22.7 A longitudinal section through part of a pine ovule.

produce more than 1 million grains, and there may be hundreds of such clusters on one tree. The grains accumulate as a fine yellow dust on cars, shrubbery, or anything else in the vicinity, and they often form an obvious scum on pools and puddles. Within a few weeks of the pollen's release, the male cones shrivel and fall from the trees.

Megaspores are produced in **megasporangia** located within **ovules** at the bases of the female cone scales. The female cones (*female strobili*—also called *seed cones*) are much larger than the male cones, becoming as much as 60 centimeters (2 feet) long in sugar pines and weighing as much as 2.3 kilograms (5 pounds) in Coulter pines. When mature, they have woody scales with inconspicuous bracts between them, arranged in a spiral around an axis. They are produced on the upper branches of the same tree on which the male cones appear (Fig. 22.6).

The *ovules* (Fig. 22.7), which occur in pairs toward the base of each scale of the immature female cones, are larger and more complex than the microsporangia of male cones. Each ovule contains a *megasporangium* embedded in multicellular nutritive tissue called the **nucellus.** The nucellus, in turn, is surrounded and enclosed by a thick, layered **integument.** The integument has a somewhat tubular channel or pore called a **micropyle** that is pointed toward the cone's central axis. One of the integument layers later becomes the **seed coat** of the seed.

A single **megaspore mother cell** within the megasporangium (nucellus) of each ovule undergoes meiosis,

---

need t
has bl
tween
been
isms

tissue
of fer

Fo

Int

Seve
cedi
that
wid
hea
how

usu
to a

grains added buoyancy, which may result in the grains being carried great distances by the wind.

Pines produce pollen grains in astronomical numbers. For example, it has been estimated that each of the 50 or more pollen cones commonly found in a single cluster may

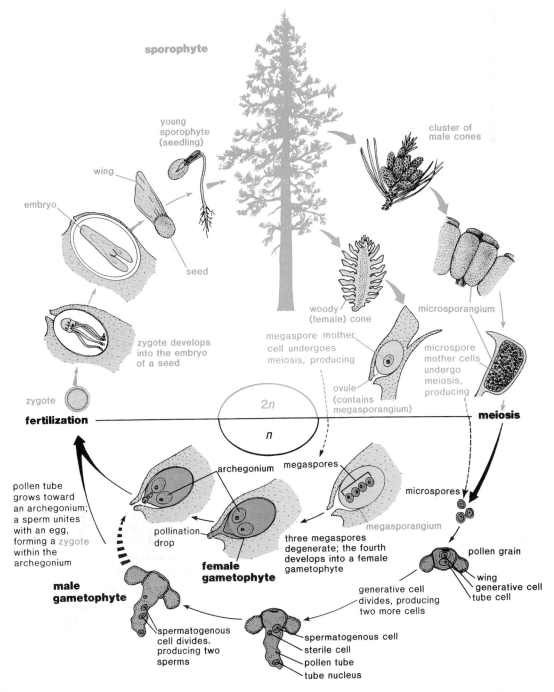

FIGURE 22.8   Life cycle of a pine.

producing a row of four relatively large **megaspores.** Three of the megaspores soon degenerate. Over a period of months, the remaining one slowly develops into a *female gametophyte,* which ultimately may consist of several thousand cells.

As gametophyte development nears completion, two to six *archegonia* differentiate at the end facing the micropyle. Each archegonium contains a single large *egg.* When a stained thin, lengthwise section of a pine ovule is examined with a microscope, only one archegonium or egg may be seen, or the micropyle may appear to be missing.

This is due to the section not having been sliced precisely through the middle of all the structures.

Female cones, which are at first usually reddish or purplish, commonly take two seasons to mature into the green and finally the brownish woody structures with which we are all familiar (Fig. 22.8). During the first spring, the immature cone scales spread apart, and pollen grains carried by the wind sift down between the scales. There they catch in sticky drops of fluid (*pollen drops*) oozing out of the micropyles. As the fluid evaporates, the pollen is drawn down through the micropyle to the top of the nucellus.

After pollination, the scales grow together and close, protecting the developing ovule. Meiosis and megaspore development don't occur until about a month after pollination. After a functional megaspore is produced, the female gametophyte and its archegonia don't mature until more than a year later.

Meanwhile, the pollen grain (immature male gametophyte) produces a **pollen tube,** which slowly grows and digests its way through the nucellus to the area where the archegonia develop. While the pollen tube is growing, two of the original four cells in the pollen grain enter it. One of these, called the **generative cell,** divides and forms two more cells, called the *sterile cell* and the *spermatogenous cell.* The spermatogenous cell divides again, producing two male gametes, or *sperms.* The pine sperms have no flagella, unlike the sperms of other organisms encountered thus far. The germinated pollen grain, with its pollen tube and two sperms, constitutes the mature *male gametophyte.* Notice that no antheridium has been formed.

About 15 months after pollination, the tip of the pollen tube arrives at an archegonium, unites with it, and discharges the contents. One sperm unites with the egg, forming a zygote. The other sperm and remaining cells of the pollen grain degenerate. The sperms of other pollen grains present may unite with the eggs of other archegonia, and each zygote begins to develop into an *embryo.* This is similar to the development of fraternal twins or triplets in animals. At a later stage, an embryo may divide in such a way as to produce the equivalent of identical twins or quadruplets in animals. Normally, however, only one embryo completes development. While this development is occurring, one of the layers of the integument hardens, becoming a *seed coat.* A thin membranous layer of the cone scale becomes a "wing" on each seed. The wing may aid in the seed's dispersal. Squirrels and other animals may also help dispersal by breaking open the cones (see Fig. 22.8).

In other species, such as the lodgepole, jack, and knobcone pines, the cones remain on the tree with the scales closed until they are seared by fire or open with old age. Sometimes, these cones are slowly buried as the cambium adds tissues that increase the girth of the branch. Seeds of such engulfed cones have been reported to germinate after they have been dug out of the stem.

Other Conifers

Some conifers don't produce woody female cones with conspicuous scales, nor do all conifers produce both male and female cones on the same tree. For example, yew (*Taxus*) and California nutmeg (*Torreya*) produce ovules singly at the tips of short axillary shoots. Each ovule is at least partially surrounded by a fleshy, cuplike covering called an **aril.** In yews, this is bright red and open at one end, giving the fleshy seed the appearance of a small red hors d'oeuvre olive with its stuffing removed (Fig. 22.9). The fleshy seeds are produced only on female trees, while the pollen-bearing strobili, or male cones, are produced only on male trees.

FIGURE 22.9 The seeds of yew (*Taxus*), which are not produced in cones, are surrounded at maturity by a red, fleshy, cuplike structure called an *aril.*

Podocarps, which are conifers of the Southern Hemisphere, are widely planted as ornamentals in regions with milder climates. Their fleshy-coated seeds, which are produced singly, are similar to those of yews, but they are not open at one end and have an additional larger appendage at the base (Fig. 22.10). The origin of these fleshy seeds is not clear and has led to speculation that yews and podocarps may have diverged from other conifers very early in the evolution of gymnosperms.

The scales of the female cones of junipers tend to be fleshy at maturity, so they look more like berries than cones. Juniper pollen, as well as that of a number of other conifers, does not have air sacs. Cypress and redwood female cone scales are flattened at the tips and narrow at the base, where they do not overlap one another as do pine cone scales.

The two California species of redwoods are both renowned for their size, height, and longevity. Coastal redwoods occasionally grow to a height of 90 meters (295 feet), and one tree in Humboldt County, California, which is 111.6 meters (366.2 feet) tall, is believed to be the tallest conifer in the world. The other species, usually referred to as *Big Tree*

FIGURE 22.10 Fleshy seeds of a podocarp (*Podocarpus*). Note the large appendage at the base of each seed.

Subdivision Pinicae, Class Ginkgoatae—*Ginkgo*

There is only one living species of *Ginkgo* (Fig. 22.11), whose name is derived from Chinese words meaning "silver apricot." The fossil record indicates *Ginkgo* and other members of its family (Ginkgoaceae) were once widely distributed, especially in the Northern Hemisphere. Despite isolated reports to the contrary, there are doubts that ginkgoes now exist anywhere they have not been cultivated, and the plant has often been called a living fossil.

Ginkgoes are often referred to as *maidenhair trees* because their notched, broad, fan-shaped leaves look like larger versions of the individual pinnae of maidenhair ferns. They are widely cultivated in the United States and are popular street trees in some areas. The leaves, which are mostly produced in a spiral on short, slow-growing spurs, have no midrib or prominent veins. Instead, hairlike veins branch dichotomously (fork evenly) and are relatively uniform in their width. They are deciduous and turn a bright golden yellow before abscission in the fall.

Ginkgo is dioecious, with a life cycle similar to that of cycads; the male and female reproductive structures are produced on separate trees, and the sperms have flagella. The mature seeds, which resemble small plums, are enclosed in fleshy seed coats. The flesh is, however, unrelated to that of true fruits. In North America, male trees (propagated from cuttings) are preferred for ornamental purposes because the seed flesh has a nauseating odor and is irritating to the skin of some individuals. However, female trees predominate in China and Korea, where seeds are considered a delicacy, and only enough males are propagated to insure pollination.

Subdivision Cycadicae—The Cycads

Cycads, which look like a cross between a tree fern and a palm, are slow-growing plants of the tropics and subtropics. They have unbranched trunks that grow more than 15 meters (50 feet) tall in a few species and have a crown of large pinnately divided leaves. Extinct members of this division, known as *cycadeoids,* were abundant during the Mesozoic era. Several of the approximately 100 known living species are presently facing extinction. Their life cycles are similar to those of conifers. Each sperm of cycads, however, has from 10,000 to 20,000 spirally arranged flagella. Cycads are dioecious; the male and female strobili, which in some species are huge (e.g., more than a meter [3 feet 3 inches] long with a weight of over 220 kilograms [100 pounds]), are produced on separate plants (Fig. 22.12). The scales of female strobili of some species are covered with feltlike or woolly hairs.

Subdivision Gneticae— The Gnetophytes

The 70 known species of *gnetophytes* are distributed among three distinctive genera. They are unique among the

or *Giant Redwood,* is confined to the western slopes of California's Sierra Nevada range. It does not grow quite as tall as the coastal redwood but exceeds it in total mass. The General Sherman tree in Sequoia National Park, for example, is 31 meters (101.5 feet) in circumference at the base and over 24 meters (79 feet) in circumference 1.5 meters (5 feet) above the ground. It weighs an estimated 5,594 metric tons (6,167 tons). There are 600,000 board feet of timber in this single tree—enough to build more than 75 five-room houses (although the wood is generally not suitable for construction purposes) or to make 20 billion toothpicks. It is over 3,500 years old.

OTHER GYMNOSPERMS

Living representatives of other gymnosperms are not as numerous or as well-known as the conifers and outwardly don't resemble them at all. In fact, some of them look more like leftover props from a science-fiction movie set. They include the **cycads,** the **ginkgoes,** and the **gnetophytes** (pronounced née-toe-fytes).

A.

B.

C.

FIGURE 22.11 *Ginkgo. A.* A mature tree in the fall. *B.* Seeds and leaves. *C.* Male strobili.

gymnosperms in having vessels in the xylem. The genus *Ephedra,* whose members are sometimes called *joint firs* (Fig. 22.13), accounts for more than half of the species. These shrubby plants, which inhabit drier regions, produce tiny leaves in twos or threes at a node. The leaves turn brown soon after they appear. The slightly ribbed branches are often whorled and are photosynthetic when they are young.

In preparation for pollination, a tubular extension, like the neck of a tiny bottle, extends into the air from the integument of an ovule. Sticky fluid oozes out of this extension, which constitutes the micropyle, and airborne pollen catches in it. Male and female strobili may be produced on the same plant or on different ones, depending on the species.

Most of the remaining species in this division are in the genus *Gnetum* (Fig. 22.14), which has not been given an English common name. Its members, which have broad leaves similar to those of flowering plants, occur in the tropics of South America, Africa, and Southeast Asia. Most are vinelike, but the best-known species is a tree that grows up to 10 meters (33 feet) tall.

The third genus, *Welwitschia,* has only one species, which is confined to the temperate Namib and Mossamedes deserts of southwestern Africa. Here the average annual rainfall is only 2.5 centimeters (1 inch), and in some years it does not rain at all. The plants carry on CAM photosynthesis and their stomata are open at night. The plants apparently survive much of the time on dew and condensate from fog that rolls in off the ocean at night.

A.

A.

B.

B.

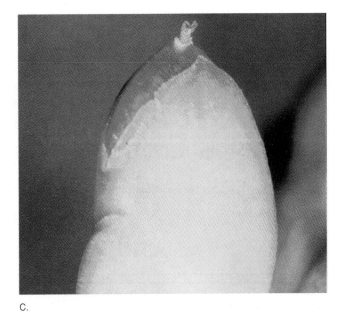

C.

FIGURE 22.12 *A.* A male cycad with a strobilus. *B.* A female cycad with a strobilus.

FIGURE 22.13 Joint fir (*Ephedra*). *A.* Part of a single plant. *B.* Male strobili, ×20. *C.* A female strobilus, ×20.

FIGURE 22.14   A climbing *Gnetum*.

Welwitschia plants are truly extraordinary in appearance. The stem, which rises only a short distance above the surface, is in the form of a large shallow cup that tapers at the base into a long taproot. At maturity, the plants, which may live to be 100 years old, have a crusty, barklike covering on the surface of the stem cup. The stems may be more than 1 meter (3 feet 3 inches) in diameter (Fig. 22.15).

Throughout their lifespan, *Welwitschia* plants usually produce only two leaves. The leaves are wide and straplike, each with a meristem at the base. The meristems constantly add to the length of the leaves, but as the leaves flap about in the wind, they become tattered and split, wearing off at the tips so that they are seldom more than 2 meters (6.5 feet) long. Male and female strobili, which are produced on separate plants, are on axes that emerge from the axils of the leaves so they appear to be growing around the rim of the stem cup.

HUMAN AND ECOLOGICAL RELEVANCE OF GYMNOSPERMS

As a group, the gymnosperms are second only to the flowering plants in their impact on our daily lives. Space does not permit a detailed account of all they contribute, but the following sections provide an overview of some of their uses, past and present.

FIGURE 22.15   A female *Welwitschia* plant in the Namibian Desert.

(Photo by Margaret Marker, in the collection of the National Botanical Institute, Kirstenbosch. Courtesy Fiona Getliffe Norris)

What do we plant when we plant the tree?
We plant the ship, which will cross the sea.
We plant the mast to carry the sails;
We plant the planks to withstand the gales—
The keel, the keelson, the beam, the knee;
We plant the ship when we plant the tree.

What do we plant when we plant the tree?
We plant the houses for you and me.
We plant the rafters, the shingles, the floors,
We plant the studding, the lath, the doors,
The beams, the siding, all parts that be;
We plant the house when we plant the tree.

What do we plant when we plant the tree?
A thousand things that we daily see;
We plant the spire that out-towers the crag,
We plant the staff for our country's flag,
We plant the shade, from the hot sun free;
We plant all these when we plant the tree.

Henry Abbey

Conifers

In the early 1970s, the late author-naturalist Euell Gibbons filmed a series of television commercials in which he mentioned uses of several wild plants for food. In one of the commercials, he rhetorically inquired if his audience had ever eaten a pine tree and added, "many parts are edible."

The edibility of parts of many conifers was known to Native Americans long before Europeans set foot on the North American continent. In fact, early explorers found large numbers of pines stripped of their bark. For centuries, the inner parts (phloem, cambium) had been used for emergency food. The Adirondack Mountains of New York are believed to have received their name from a Mohawk Indian word meaning "tree-eater," in reference to Native American use of the inner bark of eastern white pines. This material (specifically the phloem) contains sugars that make it taste sweet. Some tribes ate the material raw, some dried it and ground it to flour, and others boiled it or stored dried strips for winter food. Early settlers in New England candied strips of eastern white pine inner bark. To prevent scurvy, both they and local Native Americans drank a tea made of the needles, which are rich in vitamin C.

The seeds of nearly all pines are edible, but those of western North America include the larger and better tasting species. The protein content of those analyzed generally ranges between 15% and 30%, with much of the remainder consisting of oils.

California Indians relished gray pine seeds in particular, but even the small seeds of ponderosa pine were eaten raw or made into a meal for soups and bread. Cones of pinyons were collected by tribes of the Southwest and thrown on a fire to loosen the seeds. These were then pounded and made into cakes or soup. The soup was often fed to infants. In Siberia, people crush the seeds of Siberian white pine to obtain a nutritious oil, but its use has declined since corn and cottonseed oils have become available.

Italians and other Europeans cook *pignolias,* the seeds of the stone pine, in stews and soups. The seeds are also used in cakes and cookies, and some are exported to the United States for this purpose. Many of the so-called nuts used by commercial American bakers in cakes and confectionery are really seeds from the east Himalayan chilghoza pine. Other sources include the Mexican stone pine and a few pinyons.

Eastern white pines were often used as masts in sailing vessels. In colonial days, the royal surveyors marked certain trees for the use of the Crown, and severe penalties were imposed on colonists who ignored the ban on the use of any white pine not growing on private land. It was, however, legal for colonists to use white pines that had blown down, which gave rise to the term *windfall.* Eastern white pine wood contains less resin than that of other species and was extensively used for crates, boxes, matchsticks, furniture, flooring, and paneling. By the end of the 19th century, eastern white pines, which originally grew over vast tracts of the northeastern United States and Canada, had been decimated by wholesale logging done with no thought to conservation. Bald cypress trees in the southeastern United States met a similar fate. White pine blister rust also took its toll. Although new growth is now being promoted, most white pine lumber used today comes from large stands of western white pine in the Pacific Northwest.

The trunks of lodgepole pines are used in both the United States and Canada for telephone poles; the straight-grained wood is also used for railroad ties, mine timbers, and pulp.

Smog has severely damaged ponderosa and other native pines in California. For a number of years, the U.S. Forest Service has experimented with Afghanistan pine, a smog- and drought-resistant pine from Russia and adjacent areas, as a replacement for native trees. Growth rates in tests have been very rapid. Rapid growth is a desirable commercial feature, since it permits considerably more timber to be produced than does slow growth, but the wisdom of introducing non-native plants into natural communities is in question, since there are many examples of such activities thoroughly disrupting delicate ecological balances.

The resin produced in the resin canals of conifers is a combination of a liquid solvent called *turpentine* and a waxy substance called *rosin.* When a conifer tree is wounded or damaged by insects, resin usually covers the area, sometimes trapping the insects. Out in the air, the turpentine evaporates quickly, leaving a protective layer of rosin, which deters water loss and fungal attacks. Both turpentine and rosin are very useful products, and a large industry centered in the southern United States and in the south of France is devoted to their extraction and refinement. Turpentine is

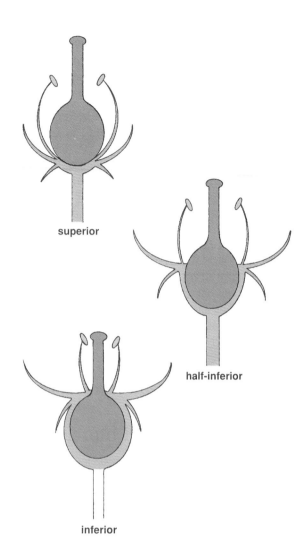

FIGURE 23.11 Ovary positions in flowers: superior (e.g., peach), half-inferior (e.g., cherry), and inferior (e.g., apple).

FIGURE 23.12 Unisexual flowers of a squash. *A.* Male. *B.* Female. Note the inferior ovary beneath the corolla.

being divided into two symmetrical halves only by a single lengthwise plane passing through the axis), as in sweet peas and orchids.

In some families, the flowers have become *imperfect* (*unisexual*). Each imperfect flower has either stamens or a pistil but not both. The Pumpkin Family (e.g., pumpkins, squashes, watermelons, cantaloupes, cucumbers) has imperfect flowers (Fig. 23.12). When both male and female imperfect flowers occur on the same plant, the species is **monoecious.** If male or female flowers occur only on separate plants, however, the species is **dioecious.**

In both the dicots and the monocots, evolutionary specialization has involved reduction and fusion of parts and a shifting of the ovary from a superior to an inferior position, as hypothetical progression is made from primitive families through intermediates to those that are advanced.

Observations on the human and ecological relevance of the flowering plants occupy literally thousands of volumes. A brief overview of several aspects of this subject is given in chapters 24 and 25.

POLLINATION ECOLOGY

When certain *consumers,* such as insects, forage for food among photosynthetic *producers,* such as plants, they often come in contact with flowers. Many insects and other animals become dusted with pollen, and as they feed or collect pollen and nectar, they unknowingly but effectively bring about pollination of the plants they visit. Throughout the evolutionary history of the flowering plants, the pollinators

A.   B.

FIGURE 23.13   Flower markings on coneflowers. *A.* In ordinary light. *B.* In ultraviolet light.

have evidently coevolved with plants. In some instances, the relationship between the two has become highly specialized.

Twenty thousand different species of bees are included among the pollinators of present-day flowering plants. By far, the best known of these are honey bees. Their chief source of food is nectar, but they also gather pollen for their larvae. The flowers that bees visit are generally brightly colored and mostly blue or yellow—rarely pure red. Pure red appears black to bees, and they generally overlook red flowers. Flowers often have lines or other distinctive markings, which may function as honey guides that lead the bees to the nectar. Bees can see ultraviolet light (a part of the spectrum not visible to humans), and some flower markings are visible only in ultraviolet light, making patterns seen by bees sometimes different from those seen by humans (Fig. 23.13).

Many bee-pollinated flowers are delicately sweet and fragrant. In contrast, flowers pollinated by beetles tend to have stronger, yeasty, spicy, or fruity odors. Beetles don't have keen visual senses, and flowers pollinated by them are usually white or dull in color. Some beetle-pollinated flowers don't secrete nectar but either furnish the insects with pollen or have food available on the petals in special storage cells, which the beetles consume.

Some flowers, including the stapelias of South Africa (Fig. 23.14), smell like rotten meat. Short-tongued flies pollinate such flowers, which tend to be dull red or brown. These plants are related to our milkweeds, although superficially they don't resemble milkweeds at all. They are often called *carrion flowers* because of their foul odor and appearance. Flies with longer tongues may also pollinate bee-pollinated flowers.

Moth- and butterfly-pollinated flowers, like bee-pollinated flowers, often have sweet fragrances. Night-flying moths visit flowers that tend to be white or yellow—colors that stand out against dark backgrounds in starlight or moonlight. Some very specialized relationships between moths and flowers occur between certain small moths and

FIGURE 23.14   A *Stapelia* (carrion flower) plant.

members of the genus *Yucca;* these relationships are discussed on the back cover of this book.

Red flowers are sometimes pollinated by butterflies, some of which can detect red colors. The nectaries of these flowers are found at the bases of corolla tubes or spurs, where only moths and butterflies with longer tongues can

Pods of a tropical cacao tree (Theobroma cacao). *The seeds are the source of chocolate.*

24 Flowering Plants and Civilization

*This chapter begins with comments on some of the problems
involved in distinguishing between fact and fancy in reported
past uses of plants. It continues with a brief discussion of
Vavilov's centers of origin of cultivated plants and a survey
of 16 well-known flowering plant families. Miscellaneous
information given for the families, which are presented in
phylogenetic sequence, includes brief comments on family
characteristics and some past, present, and possible future uses.*

Some Learning Goals

1. Give reasons for basing scientific evaluation on more than a single sampling.
2. Learn Vavilov's centers of distribution of cultivated plants, and identify several plants from each center.
3. Know characteristics of 10 flowering plant families.
4. Understand what constitutes a primitive family.
5. Know five useful plants in the Laurel, Rose, Legume, and Spurge Families.
6. Identify medicinal plants in the Poppy and Nightshade Families.
7. Construct a simple, original key to five flowering plant families.

Puccoons are herbaceous plants that grow on dry plains and slopes throughout the western United States and British Columbia. One puccoon, which has greenish-yellow flowers, has seeds that are so hard it was given the common name of *stoneseed* in Nevada and California.

Native American women of the Shoshoni tribe of Nevada reportedly drank a cold water infusion of the roots of stoneseed every day for six months to ensure permanent sterility. Biologist Clellan Ford became curious about these reports and gave extracts of stoneseed plants to mice. He found the extracts effectively eliminated the estrous cycle of the mice and decreased the weights of the ovaries, the thymus, and the pituitary glands.

Although this reported Native American use of a plant was demonstrated experimentally to have a basis in fact, distinguishing between fact and fantasy in recorded past uses of plants is often difficult, particularly if the plants have become rare or extinct. But this type of research is essential today if we're going to save potential sources of medicinal drugs and other useful plant products before they are eliminated by clearing of land and other symbols of "progress."

Primitive peoples, despite occasional misguided superstition and folklore, did cure some of the diseases they treated with plants, even though they didn't know the scientific reasons for the results. As indicated earlier, botanists who have recognized this have teamed with anthropologists, medical doctors, and interpreters to interview tribal and other primitive medical practitioners in the tropics of the Americas and Africa. They are sifting through as much of this information as possible before it is too late. Their work is already leading to useful new discoveries.

ORIGIN OF CULTIVATED PLANTS

Alphonse de Candolle, a Swiss botanist, published in 1822 a book entitled *Origin of Cultivated Plants*, based on data he gathered from many sources. He deduced that cultivated plants probably originated in areas where their wild relatives grow.

Ninety-four years later, in 1916, N. I. Vavilov, a Russian botanist, began a follow-up of de Candolle's work. During the next 20 years, he expanded on de Candolle's work and modified his conclusions. Vavilov became persuaded, as a result of his research, that most cultivated plants differ appreciably from their wild relatives. He also concluded that dispersal centers of cultivated plants are characterized by the presence of dominant genes in plant populations, with recessive genes becoming apparent toward the margins of a plant's distribution. Vavilov recognized eight centers of origin of cultivated plants, with some plants originating in more than one center. The centers (Fig. 24.1), some of which are subdivided, are as follows.

Chinese Center

The Chinese Center (1), the earliest independent center, consists of mountainous areas and adjacent lowlands of western and central China. Cultivated plants believed to have originated here include millets, soybean and several other legumes, bamboo, radish, eggplant, cucumber, some citrus, peach, apricot, walnut, persimmon, tea, some sugar canes, and hemp.

Indian Center

The Indian Center (2A), second in importance only to the Chinese Center, consists of Burma and India (exclusive of the northwest portion). Included with the cultivated plants believed by Vavilov to have originated here are rice, sorghum, mung and several other beans, gourds, yam, mango, orange and other citrus fruits, sugar cane, bowstring hemp, black pepper, betel nut, cardamon, gum arabic, henna, senna, strychnine, and rubber plant.

Indo-Malayan Center

The Indo-Malayan Center (2B) consists of Indochina, Malaysia, Java, Borneo, Sumatra, and the Philippines. Vavilov believed giant bamboo, ginger, banana, breadfruit, candlenut, coconut, clove, nutmeg, Manila hemp, and many tropical fruits originated here.

Central Asiatic Center

The Central Asiatic Center (3) consists of northwest India, Afghanistan, and adjacent Soviet provinces. Wheat, garden pea, lentil, mustard, safflower, cotton, garlic, carrot, onion, basil, pear, almond, grape, apple, and other fruits and nuts are believed to have originated here.

DICOTS

The Buttercup Family (Ranunculaceae)

Nearly all the 1,500 members of the Buttercup Family are herbaceous. The flowers, whose petals often vary in number, have numerous stamens and several to many pistils with superior ovaries (Fig. 24.2). Most have dissected leaves with no stipules and with petioles that are slightly expanded at the base. Well-known representatives include ornamental plants such as buttercup, columbine, larkspur, anemone, monkshood, and *Clematis* (Fig. 24.3). Most members of the Buttercup Family are concentrated in north temperate and arctic regions.

Columbine flowers, which have five spurred petals, resemble a circle of doves, and their name comes from *columba,* the Latin word for "dove." A blue and white species of columbine is the state flower of Colorado. Native Americans, for control of diarrhea, made a tea from boiled columbine roots, and members of at least two tribes believed columbine seeds to have aphrodisiac properties. A man would pulverize the seeds, and after rubbing them in the palms of his hands, he would try to shake hands with the woman of his choice, believing the woman would then succumb to his advances. Others crushed and moistened the seeds and applied them to the scalp to repel lice.

Most members of the family are at least slightly poisonous, but the cooked leaves of cowslips have been used for food, and the well-cooked roots of the European bulbous buttercup are considered edible. The European buttercup, in its natural state, causes blistering on the skin of sensitive individuals. East Indian fakirs are reported to deliberately blister their skin with buttercup juice in order to appear more pitiful when begging. Native Americans of the West gathered buttercup achenes, which they parched and ground into meal for bread. Others made a yellow dye from buttercup flower petals. Karok Indians made a blue stain for the shafts of their arrows from blue larkspurs and Oregon grape berries.

Goldenseal, which is still sold in health food stores, is a plant that was once abundant in the woods of temperate eastern North America. It has become virtually extinct in the wild because of relentless collecting by herb dealers. They sold the root for various medicinal uses, including remedying inflamed throats, skin diseases, and sore eyes. At least one Native American tribe mixed the pounded root in animal fat and smeared it on the skin as an insect repellent.

Monkshood yields a drug complex called *aconite,* which was once used in the treatment of rheumatism and neuralgia. Although popular as garden flowers, monkshoods are very poisonous. Death may follow within a few hours of ingestion of any part of the plant. Most species have purplish to bluish or greenish flowers, but one Asian monkshood, called *wolfsbane,* has yellow flowers. Wolf hunters in the past used to poison the animals with a juice obtained from wolfsbane roots.

FIGURE 24.2   A buttercup flower.

The Laurel Family (Lauraceae)

The Laurel Family is a primitive family whose flowers have no petals but whose six sepals are sometimes petal-like. The stamens, which occur in three or four whorls of three each, are a curiosity because the anthers open by flaps that lift up. The ovary is superior. Most of the approximately 1,000 species in this family are tropical evergreen shrubs and trees, many with aromatic leaves. The family received its name from the famous laurel cultivated for centuries in Europe. Its foliage was used by the ancient Greeks to crown victors in athletic events and later was used in the conferring of academic honors.

Several important spices come from members of this family. Powdered cinnamon is the pulverized bark of a small tree that is native to India and Sri Lanka, although it is also grown commercially elsewhere. Cassia, which is very similar, is today often sold interchangeably with cinnamon. Cinnamon oil is distilled from young leaves of the trees. Use of cinnamon and cassia dates back thousands of years. They were used in perfumes and anointing oils at the time of Moses, and other records reveal their use in Egypt at least 3,500 years ago.

Camphor has been used since ancient times. This evergreen tree, which is native to China, Japan, and Taiwan, is the main source of camphor essence still used in cold remedies and inhalants, insecticides, and perfumes. The essence is distilled from wood chips. Some American cities and towns with milder climates have been using camphor trees as street trees, because of their capacity to withstand smog.

Sassafras trees, which are native to the eastern United States and eastern Asia, also have spicy-aromatic wood. A flavoring widely used in toothpaste, chewing gum,

A.

B.

C.

FIGURE 24.3 Representatives of the Buttercup Family. *A.* Columbine. *B. Hepatica. C.* Monkshood. *D. Isopyrum.*
(*C.* Courtesy Donald E. Brink, Jr.)

D.

mouthwashes, and soft drinks used to be obtained by distillation of wood chips and bark. Sassafras tea still is considered a refreshing beverage. Sassafras is also an ingredient of some homemade root beers, and in the southern states an alcoholic beer has been made by adding molasses to boiled sassafras shoots and allowing the mixture to ferment. In Louisiana, powdered sassafras leaves (called *filé*) have been used as a thickening and flavoring agent for gumbo. It has in the past been used by country physicians for treating hypertension and for inducing a sweat in those with respiratory infections. Reports indicate that in large doses it has a narcotic-stimulant effect, and it is also reported to be carcinogenic. Most sassafras flavorings now in use are artificial.

The sweet bay, used as a flavoring agent in gravies, sauces, soups, and meat dishes, comes from the leaves of the laurel. Leaves of the related California bay (Fig. 24.4) are sometimes used as a substitute for sweet bay and for making Christmas wreaths. This tree, which is native to California,

FIGURE 24.4   A fruit and leaves of a California bay tree.

FIGURE 24.5   A prickly poppy flower.

also occurs in southwestern Oregon, where it is known as *myrtle* (true myrtles, however, belong to a different family). Its wood, which is hard, can be polished to a high luster and is used for making a variety of bowls, ornaments, and other smaller, wooden articles.

Early settlers in the West and Native Americans of the region used California bay for the relief of rheumatism, bathing in hot water to which a quantity of leaves had been added. The nutlike fruits (drupes) were roasted and used for winter food. A leaf was placed under a hat on the head to cure a headache (but even a small piece of leaf placed near the nostrils can produce an almost instant headache!). A few leaves placed on top of flour or grains in a canister will keep weevils away, and small branches have been used as chicken roosts to repel bird lice and fleas. A leaf or two rubbed on exposed skin functions as a mosquito repellent.

Avocados are also members of the Laurel Family. The fruits, which have more energy value by weight than do red meats, are rich in vitamins and iron.

The Poppy Family (Papaveraceae)

Most members of the Poppy Family are herbs distributed throughout temperate and subtropical regions north of the equator, but several poppies occur in the Southern Hemisphere, and a number are widely planted as ornamentals. Poppies, like buttercups, tend to have numerous stamens, but most have a single pistil (Fig. 24.5). Most also have milky or colored sap, and their sepals usually fall off as the flowers open. All members produce alkaloidal drugs.

Bloodroot is a pretty early flowering spring plant of eastern North American deciduous forests. A bright reddish sap, which is produced in its rhizomes, was used by some Native Americans as a facial dye, an insect repellent, and a cure for ringworm. Children today still paint their nails with it. It has a very bitter taste, except when ingested in minute amounts. The bitterness made it effective in inducing vomiting, but members of one tribe use it to treat sore throats after compensating for the bitterness by squeezing a few drops on a lump of maple sugar so that it could be held in the mouth.

Opium poppies have had a significant impact on societies of both the past and the present. Opium itself was described by Dioscorides in the first century A.D., and ancient Assyrian medical texts refer to both opium and opium poppies. Opium smoking, which does not extend back nearly as far as the use of the drug in other ways, became a major problem in China in the 1600s. Smoking opium has given way to other forms of use in recent years. The substance is obtained primarily by making small gashes in the green capsules of the poppies (Fig. 24.6). The crude opium appears as a thick, whitish fluid oozing out of the gashes and is scraped off. It contains two groups of drugs. One group contains the narcotic and addicting drugs morphine and codeine, which are best known for their widespread medicinal use as painkillers and cough suppressants. Members of the other group are neither narcotic nor addictive. They include papaverine, which is used in the treatment of circulatory diseases, and noscapine, which is used as a codeine substitute because it functions like codeine in suppressing coughs but does not have its side effects.

Heroin, a scourge of modern societies, is a derivative of morphine. It is from four to eight times more powerful than morphine as a painkiller. Less than 100 years ago, it was advertised and marketed in the United States as a cough suppressant. It is estimated that 75% of American drug

FIGURE 24.6 Immature opium poppy capsules that were gashed with a razor blade. Note the opium-containing latex oozing from the gashes.

FIGURE 24.7 Shepherd's purse.

addicts used heroin up until the early 1980s, but cocaine has now largely replaced it. The loss to society in terms of its economic, physical, and moral impact is enormous, since addicts frequently commit violent crimes to obtain the funds needed to support their habits, which often cost well over $200 per day.

The seeds of opium poppies contain virtually no opium and are widely used in the baking industry as a garnish. They also contain up to 50% edible oils, which are used in the manufacture of margarines and shortenings. Another type of oil obtained from the seeds after the edible oils have been extracted is used in soaps and paints.

The Mustard Family (Brassicaceae)

The original Latin name for the Mustard Family, still in widespread use today, was *Cruciferae*. The name describes the four petals of the flowers, which are arranged in the form of a cross. The flowers also have four sepals, usually four nectar glands, and six stamens, two of which are shorter than the other four. All members produce siliques or silicles (shown in Fig. 8.17), which are unique to the family. All 2,500 species of the family produce a pungent, watery juice, and nearly all are herbs distributed primarily throughout the temperate and cooler regions of the Northern Hemisphere.

Among the widely cultivated edible plants of the Mustard Family are cabbage, cauliflower, brussels sprouts, broccoli, radish, kohlrabi, turnip, horseradish, watercress, and rutabaga. Some edible members are also widespread weeds. The leaves of shepherd's purse (Fig. 24.7), for example, can be cooked and eaten, and the seeds can be used for bread meal. Other wild edible members are several cresses, peppergrass, sea rocket, toothwort, and wild mustard. Wild mustards are often weeds in row crops. Their leaves are sometimes sold as vegetable greens in markets.

The seeds of wild mustard, shepherd's purse, and several other members of this family produce a sticky mucilage when wet. Biologists at the University of California at Riverside discovered a potential new use for these seeds. They fed pelleted alfalfa rabbit food to mosquito larvae in water tanks, which they were using for experiments on mosquito control. They noticed that the larvae, which had to come to the surface at frequent intervals for air, often stuck to the pellets and suffocated. Curious, the workers examined the pellets under a microscope and found that they contained mustard seeds. Evidently, the field where the alfalfa had been harvested had also contained mustard plants. The scientists then tried heating the mustard seeds to kill them and found that this did not affect production of mucilage by wet seeds. It was calculated that 0.45 kilogram (1 pound) of such seeds could kill about 25,000 mosquito larvae. A few mosquito abatement districts have used the seeds effectively, but

FIGURE 24.8   A Sitka rose.

FIGURE 24.9   A raspberry.

experiments are needed to determine if there is a practical way to harvest many more seeds and control mosquitoes by such nonpolluting means on a much larger scale.

Native Americans mixed the tiny seeds of several members of this family with other seeds and grains for bread meal and gruel. To prevent or reduce sunburn, Zuni Indians applied a water mixture of ground western wallflower plants to the skin. Watercress, which is widely known as a salad plant, has had many medicinal uses ascribed to it. During the first century A.D., for example, Pliny listed more than 40 medicinal uses. Native Americans of the West Coast of the United States treated liver ailments with a diet consisting exclusively of large quantities of watercress for breakfast, abstinence from any further food until noon, and then resumption of an alcohol-free but otherwise normal diet for the remainder of the day. This was repeated until the disease, if curable, disappeared.

Dyer's woad, a European plant that has become naturalized and established in parts of North America, is the source of a blue dye that was used for body markings by the ancient Anglo-Saxons. Another member of the family, camelina, has been grown in the Netherlands for the oil that is obtained from its seeds. Camelina oil has been used in soaps and was once used as an illuminant for lamps.

The Rose Family (Rosaceae)

The Rose Family includes more than 3,000 species of trees, shrubs, and herbs distributed throughout much of the world. The flowers characteristically have the basal parts fused into a cup, with petals, sepals, and numerous stamens attached to the cup's rim (Fig. 24.8). The family is divided into subfamilies on the basis of flower structure and fruits. The flowers of one group have inferior ovaries and produce pomes for fruits. Flowers of other groups have ovaries that are superior or partly inferior and produce follicles, achenes, or drupes or clusters of drupelets.

The economic impact of members of the Rose Family is enormous, with large tonnages of stone fruits (e.g., cherries, apricots, peaches, plums), pome fruits (e.g., apples, pears), and aggregate fruits such as strawberries, blackberries, loganberries, and raspberries being grown annually in temperate regions of the world (Fig. 24.9).

Members of this family have been relevant to humans in many other ways in the past and still continue to be so. Roses themselves, for example, have for centuries been favorite garden ornamentals of countless numbers of gardeners, and the elegant fragrance of some roses delights many people. In Bulgaria and neighboring countries, a major perfume industry has grown up around the production from damask roses of a perfume oil known as *attar* (or *otto*) *of roses*. In a valley near Sofia, more than 200,000 persons are involved in the industry, whose product brought more than $2,200 per kilogram ($1,000 per pound) during the 1970s. A considerable quantity of the oil is blended with less expensive substances in the perfume industry. Perfume workers are reported rarely to develop respiratory disorders, suggesting that the plant extracts may have medicinal properties.

The fruits of wild roses, called *hips* (Fig. 24.10), are exceptionally rich in vitamin C. In fact, they may contain as much as 60 times the vitamin C of citrus fruit. Native Americans from coast to coast included rose hips in their diets (except for members of a British Columbia tribe, who believed they gave one an "itchy seat"), and it is believed that this practice contributed to scurvy being unknown among them. During World War II when food supplies became scarce in some European countries, children in particular were kept healthy on diets that included wild rose hips. The hips also contain, in addition to vitamin C, significant amounts of iron, calcium, and phosphorus. Today, many Europeans eat *Nyppon Sopa*, a sweet, thick pureé of rose hips, whenever they have a cold or influenza.

FIGURE 24.10   Mature rose hips.

After giving birth, the women of one western Native American tribe drank western black chokecherry juice to staunch the bleeding. Other tribes frequently made a tea from blackberry roots to control diarrhea. Five hundred Oneida Indians once cured themselves of dysentery with blackberry root tea, while many nearby white settlers, who refused to use "Indian cures," died from the disease. Men of certain tribes used older canes of roses for arrow shafts (presumably after removing the prickles!). Wild blackberries, raspberries, salmonberries, thimbleberries, dewberries, juneberries, and strawberries all provided food for Native Americans and early settlers, and they are still eaten today, either fresh or in pies, jams, and jellies. A spiced blackberry cordial is still a favorite for "summer complaints" in southern Louisiana. Wild strawberries are considered by many to be distinctly superior in flavor to cultivated varieties.

The Legume Family (Fabaceae)

The Legume Family is the third largest of the approximately 300 families of flowering plants, with only the Sunflower and Orchid Families having more species. Its 13,000 members, which are cosmopolitan in distribution, include many important plants. The flowers range in symmetry from radial (regular) to bilateral (irregular). The irregular flowers have a characteristic *keel* (which is a boat-shaped fusion of two petals enclosing the pistil), two *wing petals,* and a larger *banner* petal (Fig. 24.11). The stamens in such flowers are generally fused in the form of a tube around the ovary. The common feature that keeps the members together in one family is the fruit, which is a legume (shown in Fig. 8.16).

Important crop plants include peas, many kinds of beans (e.g., kidney, lima, garbanzo, broad, mung, tepary), soybeans, lentils, peanuts, alfalfa, sweet clover, licorice, and wattle. Wattle is an Australian tree that is grown commer-

cially as a source of tannins for leather tanning. Carob, which is widely used as a chocolate substitute, is also a member of this family. Several copals (hard resins used in varnishes and lacquers) are obtained from certain legume plants, as are gum arabic and gum tragacanth, which are used in mucilages, pastes, paints, and cloth printing.

Important dyes, such as indigo, logwood (used in staining tissues for microscope slides and now scarce), and woadwaxen (a yellow dye), come from different legume plants. Locoweeds, which have killed many horses, cattle, and sheep, particularly in the southwestern United States, belong to a large genus (*Astragalus*) of about 1,600 species. The poisonous principle in those species affecting livestock seems to vary in concentration according to the soil type in which the plants are growing. Other poisonous legumes are lupines, jequirity beans, black locusts, and mescal beans.

About 90% of the members of the Legume Family exhibit leaf movements, but few are as rapid as those of the sensitive plant (*Mimosa pudica*), whose leaves fold within seconds in response to a disturbance. Sensitive plants, which grow as weeds in the tropics and the deep South of the United States, are discussed in Chapter 11 and are shown in Figure 11.17. Many of the movements of other legume plant leaves are correlated primarily with day length.

Clovers were widely used in the past by gatherers of wild food plants. The leaves are difficult to digest in quantity, but the rhizomes were gathered and usually roasted or steamed in salt water and then dipped in grease before being eaten. The seeds of both clovers and vetches also were gathered and either ground for meal or cooked in a little water and eaten as a vegetable. Today, seeds of several legumes, including alfalfa and mung beans, are popular for their sprouts, which are widely used in salads and Oriental dishes. A tropical bean called *winged bean* (Fig. 24.12) has unusually high levels of protein, and all parts of the plant are edible. It is presently being grown in several widely scattered tropical and subtropical regions and also is being marketed on a limited scale in some temperate zones. It is believed to have great potential for improving the diet of undernourished peoples throughout the tropics.

The Spurge Family (Euphorbiaceae)

Although many of the members of the Spurge Family are tropical, they are widespread in temperate regions both north and south of the equator. The stamens and pistils are produced in separate flowers, which often lack a corolla and are inconspicuous. In true spurges (*Euphorbia*), the female flower is elevated on a stalk called a *gynophore* and is surrounded by several male flowers that each consist of little more than an anther. Both the female and male flowers are inserted on a cup composed of fused bracts, the cup usually having distinctive glands on the rim. This type of inflorescence is called a *cyathium* (Fig. 24.13). Sometimes, the inconspicuous flowers are surrounded by brightly colored

A.

B.

FIGURE 24.14 Cactus flowers.

giant saguaro (shown in Fig. 25.19), which can attain heights of 15 meters (50 feet) and weigh more than 4.5 metric tons (5 tons). They generally grow exceptionally slowly and, because they need so little care, make good houseplants for sunny windows.

In 1944, a marine pilot was forced to bail out of his aircraft over the desert near Yuma, Arizona. Until he was rescued five days later, he survived the intense heat and low humidity of the area by chewing the juicy pulp of barrel cacti in the vicinity. Since then, the use of cacti for emergency fluids and food has been recommended in most desert survival manuals.

Most cacti have edible fruits, and only three cacti (peyote, living rock, and hedgehog cactus) are known to be poisonous. Prickly pear fruits, which are occasionally sold in American supermarkets, taste a little like pears. Prickly pear fruits also have seeds that Native Americans of the Southwest dried and ground for flour they used in *atole,* a staple food. A good syrup is obtained from boiling the fruits of prickly pears and also those of the giant saguaros. In the past, cactus candy was made by partly drying strips of barrel cactus and boiling them in saguaro fruit syrup, but the cactus is now usually boiled in cane sugar syrup.

Native Americans of the Southwest used to scoop out barrel cacti, dry them, and use them for pots. They also mixed the sticky juice of prickly pear cacti in the mortar used in constructing their adobe huts. In Texas, a poultice of prickly pear stem was applied to spider bites. Hopi Indians chewed raw cholla cactus as a treatment for diarrhea, and the skeletons of these cacti were used for flower arrangements.

In the middle of the 19th century, Australians planted a few imported prickly pear cacti in the dry interior. These cacti found no natural enemies in their new environment and multiplied rapidly, infesting more than 24 million hectares (60 million acres) within 75 years. In 1925, in an effort to control them, Australia introduced an Argentine moth among the cacti. The moth's caterpillars, which feed on prickly pear cacti, gradually brought the plants under control, and the land was made usable again.

Another cactus parasite, the cochineal insect (related to the mealybug, a common houseplant pest), feeds on prickly pear cacti in Mexico. At one time, the insects were collected for a crimson dye they produce, which was used in lipstick and rouge before aniline dyes were introduced.

Peyote cacti are small buttonlike plants that have no spines, with roots resembling those of carrots. They contain several drugs, the best known of which is *mescaline,* a powerful hallucinogen. Dried slices of peyote have been used in native religious ceremonies in Mexico for centuries and more recently by at least 30 tribes of Native Americans. The drug gives the user a variety of hallucinations in vivid colors.

The Mint Family (Lamiaceae)

The 3,000 members of the Mint Family are relatively easy to distinguish since they have a unique combination of angular stems (which are square in cross section), opposite leaves, and bilaterally symmetrical (irregular) flowers (Fig. 24.16). Most also produce aromatic oils in the leaves and stems. The superior ovary has four parts, with each of the four divisions developing into a nutlet. Included in the family are such well-known plants as rosemary, thyme, sage (not to be confused with sagebrush, which is in the Sunflower Family), oregano, marjoram, basil, lavender, catnip, peppermint, and spearmint.

Mint oils can be distilled at home with ordinary canning equipment. Whole plants (or at least the foliage) are loosely packed to a depth of about 10 centimeters (4 inches) or more in the bottom of a large canning pot. Then a wire rack or other support is also put in the pot, and a bowl is placed in the middle on the rack. Enough water is added to

FIGURE 24.15 Cacti. *A*. Prickly pear cacti. *B*. Peyote. *C*. A barrel cactus. *D*. An organ-pipe cactus.

cover the vegetation, the pot is placed on a range, and the lid is inverted over it. The water is brought to a boil, and as it boils, ice is placed on the inverted lid. The oils vaporize, and condense when they contact the cold lid, dripping then from the low point into the bowl (Fig. 24.17). Of course, some moisture also condenses, but the oil, being lighter, floats on top. Peppermint oil is easy to collect this way and will keep for a year or two in a refrigerator.

Mint oils have been used medicinally and as an antiseptic in different parts of the world. Mohegan Indians used catnip tea for colds, and dairy farmers in parts of the midwestern United States used local mint oils to wash their milking equipment. As a result, mastitis, a common disease of dairy cattle, was seldom encountered in their herds. Horehound, a common mint weed of Europe, has become naturalized on other continents and is cultivated in France. A leaf

FIGURE 24.16   Flowers of lamb's ear mint.

FIGURE 24.17   A simple apparatus for distilling mint oil at home.

FIGURE 24.18   A peppermint plant in flower. Oil from the leaves is a source of menthol.

extract is still used in horehound candy and cough medicines. In England, it is a basic ingredient of horehound beer. Vinegar weed, also known as *blue curls,* is a common fall-flowering plant of western North America. Native Americans of the area used it in cold remedies, for the relief of toothaches, and in a bath for the treatment of smallpox. It was also used to stupefy fish.

Menthol, the most abundant ingredient of peppermint oil (Fig. 24.18), is widely used today in toothpaste, candies, chewing gum, liqueurs, and cigarettes. Most American mint is grown commercially in the Columbia River basin of Oregon and Washington. Geese are sometimes used in the mint fields to control both insects and weeds, since they do not interfere with the growth of the mint plants themselves.

Ornamental mints include salvias and the popular variegated-leaf *Coleus* plants, neither of which has typical mint oils in the foliage. *Chia* (Fig. 24.19), another relatively odorless mint, is confined to the drier areas of western North America. Native Americans parched chia seeds and used them in gruel. The seeds, which become mucilaginous when wet, were also ground into a paste that was placed in the eye to aid in the removal of dirt particles. The paste was also used as a poultice for gunshot wounds, and Spanish Californians made a refreshing drink from ground chia seeds, lemon juice, and sugar. Chia seeds reportedly contain an unidentified substance that has effects similar to those of caffeine. Before the turn of the century, one physician reported that a tablespoon of chia seeds was sufficient to sustain a man on a 24-hour endurance hike. Since that time,

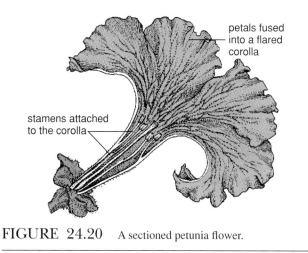

FIGURE 24.20   A sectioned petunia flower.

FIGURE 24.19   Chia.

backpackers have experimented with the seeds, and results tend to support the earlier claim. A thorough scientific investigation of the matter is needed. Chia seeds are presently sold commercially for making into a paste that is spread on clay models and then watered. The seeds sprout and resemble green hair.

The Nightshade Family (Solanaceae)

Flowers of the Nightshade Family, which is concentrated in the tropics of Central and South America, have fused petals, with the stamen filaments fused to the corolla so they appear to arise from it (Fig. 24.20). The superior ovary develops into a berry or a capsule. The more than 3,000 species of the family have alternate leaves and occur as herbs, shrubs, trees, or vines. Well-known representatives include tomato, white potato, eggplant, pepper, tobacco, and petunia.

Many nightshades produce poisonous drugs, some of which have medicinal uses. One of the best-known medicinal drug producers is the deadly nightshade of Europe. A drug complex called *belladonna* is extracted from its leaves. Belladonna, which was used in the "magic potions" of the past and also for dilating human pupils for cosmetic purposes, is now the source of several widely used drugs, including atropine, scopolamine, and hyoscyamine. Atropine is used in shock treatment, for relief of pain, to dilate eyes, and to counteract muscle spasms. Scopolamine is used as a tranquilizer, and hyoscyamine has effects similar to those of atropine. Capsicum, obtained from a pepper, is used as a gastric stimulant and is also a principal ingredient of *mace,* which is used to repel human or animal assailants. Capsaicin, derived from peppers, is used in ointments for the relief of arthritic and neuropathic pain.

Jimson weed (shown in Fig. 8.6) is also a source of medicinal drugs that have been used in the treatment of asthma and other ailments. The drugs can be fatal if ingested in sufficient quantities but have been much used in controlled amounts in Native American rituals of the past. Records indicate that users became temporarily insane but had no recollection of their activities when the effects of the drug wore off. The drug solanine is present in most if not all members of the family. Many arthritis sufferers apparently are sensitive to solanine, and a number of arthritics have reported partial relief through total avoidance of consuming members of this family (including potatoes, tomatoes, peppers, and eggplant).

Tobacco cultivation occupies more than 800,000 hectares (2 million acres) of American farmland. In its dried form, tobacco contains 1% to 3% of the drug nicotine. Nicotine is used in certain insecticides, and it is also used for killing intestinal worms in farm livestock. It is, however, an addictive drug. The evidence that human tobacco use is a primary cause of heart and respiratory diseases including lung cancer and other cancers such as those of the mouth and throat mounts almost daily. The only "benefit" it may have to humans appears to be as a killer of leeches. It is said that leeches attaching themselves to heavy smokers will drop off dead within five minutes from nicotine poisoning—a very dubious justification for continued human use!

Tomatoes are among the most popular of all "vegetables." About 18 million metric tons (20 million tons) are grown annually around the world. The plants are

FIGURE 24.21   A mechanical tomato harvester harvesting a field.

day-neutral (day lengths are discussed in Chapter 11), and even though they require warm night temperatures (16°C or 60°F) to set fruit well, they are easily cultivated in greenhouses when natural conditions are unfavorable. Most commercially grown American tomatoes are processed into juice, tomato paste, and catsup. In Italy, a small amount of edible oil is extracted from the seeds after the pulp has been removed. Most American tomatoes are grown in California, where they are harvested with special machinery developed during the 1960s when inexpensive labor became unavailable (Fig. 24.21).

The white or Irish potato is one of the most important foods grown in temperate regions of the world, with annual production estimated at well over 270 million metric tons (300 million tons). The leading producers are China, Poland, the United States, and countries of the former Soviet Union, which account for about 30% of the total. It is believed that white potatoes originated on an island off the coast of Chile and were sent back to Europe by Spanish invaders of South America in the 16th century. In the 1840s, late blight infested and destroyed the potato crop of Ireland, causing severe famine. Irish settlers subsequently emigrated to the United States, Canada, Australia, and other parts of the world.

When potato tubers are exposed to the sun, they turn green at the surface. Poisonous drugs are produced in the green areas. These have proved fatal to both animals and humans and should never be eaten.

The Carrot Family (Apiaceae)

Many members of the Carrot Family, which is widespread in the Northern Hemisphere, have savory-aromatic herbage. The flowers tend to be small and numerous and are arranged in umbels. The ovary is inferior, and the stigma is two-lobed. The petioles of the leaves, which are generally dissected, usually form sheaths around the stem at their bases. Included in the 2,000 members of the family are dill, celery, carrot, jicama, parsley, caraway, coriander, fennel, anise, and parsnip. Anise is one of the earliest aromatics mentioned in literature. It is used for flavoring cakes, curries, pastries, and candy. Pocket gophers apparently are attracted by its aroma, and some poison baits are enhanced with anise. A liqueur known as *anisette* is flavored with it.

Another liqueur, called *kümmel,* is flavored with caraway seeds, which are well-known for their use in rye and pumpernickel breads.

Some members of the Carrot Family are poisonous. Water hemlock (Fig. 24.22) and poison hemlock, which are common weeds in ditches and along streams, are deadly and have often been fatal to unwary wild-food lovers. Socrates is believed to have died as a result of ingesting poison hemlock, which should not be confused with cone-bearing hemlock trees.

Several members of the Carrot Family, such as cow parsnip, squawroot, and hog fennel, have edible roots and were used for food by Native Americans. The reader is

FIGURE 24.22 A water hemlock in flower.

advised, however, to be absolutely certain of the identity of such plants before experimenting with them.

The Pumpkin Family (Cucurbitaceae)

Although most species in the Pumpkin Family are tropical or subtropical, many occur in temperate areas of both the Northern and Southern Hemispheres. Plants are prostrate or climbing herbaceous vines with tendrils. The flowers have fused petals, and female flowers have an inferior ovary with three carpels. All are unisexual. Some species have both male and female flowers on the same plant, while others have only male or only female flowers on one plant. In male flowers, the stamens cohere to varying degrees, depending on the species. The family has about 700 members, several of which have numerous horticultural varieties.

This family includes many important edible plants, and some have been cultivated for so long that they are unknown in the wild state. Well-known members of the family include pumpkins, squashes, cucumbers, cantaloupes (Fig. 24.23 B), and watermelons. The vegetable sponge (Fig. 24.23 C), which resembles a large cucumber when it is growing, has a highly netted fibrous skeleton, which serves as a bath sponge after the soft tissues have been removed.

Gourds found in Mexican caves have been dated back to 7000 B.C. Various types of gourds (Fig. 24.23 A) serving many purposes are still grown today. Some are scooped out and used for carrying liquids or for storing food, particularly grains. South Americans drink maté, a

A.

B.

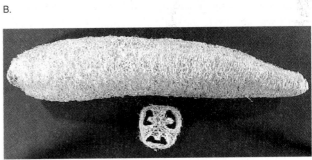

C.

FIGURE 24.23 Fruits and items associated with members of the Pumpkin Family (Cucurbitaceae). *A.* A Hawaiian ceremonial gourd with feathers attached and gourds used in South America for drinking maté. A metal straw that strains out the maté leaves is resting in one gourd. *B.* Cantaloupes. *C.* A luffa (vegetable sponge).

tea, from gourds, which are also used for several types of musical instruments. In parts of Africa, gourds are used to catch monkeys. A type with a narrow neck is scooped out and partly filled with corn or other grains. One end of a catnip, peppermint, spearmint, horehound, salvia, *Coleus,* and chia.

13. The Nightshade Family (Solanaceae) includes tomato, white potato, eggplant, pepper, tobacco, belladonna, petunia, and jimson weed (a source of hallucinogenic drugs).

14. The Carrot Family (Apiaceae) includes dill, caraway, celery, carrot, parsley, jicama, coriander, fennel, anise, parsnip, water hemlock, poison hemlock, cow parsnip, squawroot, and hog fennel.

15. The Pumpkin Family (Cucurbitaceae) includes pumpkin, squash, cucumber, cantaloupe, watermelon, vegetable sponge, gourd, melonette, and manroot.

16. The Sunflower Family (Asteraceae) includes sunflower, dandelion, lettuce, endive, chicory, Jerusalem and globe artichokes, dahlia, chrysanthemum, marigold, thistle, sagebrush, pyrethrum, balsamroot, tarweed, and yarrow.

17. The Grass Family (Poaceae) includes all cereals (e.g., wheat, barley, rye, oats, rice, corn), sugar cane, sorghum, citronella, and lemon grass.

18. The Lily Family (Liliaceae) includes lilies, asparagus, sarsaparilla, squill, meadow saffron, bowstring hemp, *Aloe,* and New Zealand flax.

19. The Orchid Family (Orchidaceae) has highly specialized flowers. It includes a species that is the source of vanilla flavoring.

Review Questions

1. To which flowering plant family does each of the following belong: Poinsettia, lupine, columbine, peach, pear, cinnamon, sarsaparilla, belladonna, peyote, horehound, rubber, gourd, jimsonweed, parsley, sorghum, asparagus, broccoli, lettuce, tomato, opium?

2. Make a list of the poisonous plants in the families discussed.

3. Which plants mentioned are or have been used for medicines?

4. Which plants mentioned have been used for tools or utensils?

5. Native Americans made extensive use of plants for a wide variety of purposes in the past. List such uses for as many plants as possible.

Discussion Questions

1. Scientific investigations often take a great deal of time and money. Is it worth the effort to check out scientifically the past uses of plants?

2. A return to exclusively herbal medicines is being advocated in some quarters. Is this a good idea? What are the pros and cons?

3. Would you expect drugs produced naturally by plants to be more effective or better than drugs produced synthetically? Why?

4. If you were asked to single out the three most important families of the 16 discussed, which would you choose? Why?

5. A number of wild edible plants were mentioned in this chapter. What would happen if a large portion of the population were to gather these wild plants as a major source of food?

Additional Reading

Advisory Committee on Technology Innovation. 1975. *Underexploited tropical plants with promising economic value.* Washington, DC: National Academy of Sciences.

Akerle, O., et al. (Eds.). 1991. *Conservation of medicinal plants.* New York: Cambridge University Press.

Brill, S., and E. Dean. 1994. *Identifying and harvesting edible and medicinal plants.* New York: Hearst Books.

Gibbons, E. 1987. *Stalking the wild asparagus.* New York: David McKay Co.

Harrington, H. D. 1974. *Edible native plants of the Rocky Mountains.* Albuquerque NM: University of New Mexico Press.

Hornsok, L. (Ed.). 1993. *Cultivation and processing of medicinal plants.* New York: Wiley & Sons.

Kirk, D. R. 1975. *Wild edible plants of the western United States,* color ed. Healdsburg, CA: Naturegraph Publishers.

Krochmal, A., and C. Krochmal. 1984. *A field guide to the medicinal plants.* New York: Times Books.

Kunkel, G. 1984. *Plants for human consumption: Annotated checklist of edible phanerogams and ferns.* Forestburgh, NY: Lubrecht & Cramer.

Lewis, W. H., and M. P. F. Elvin-Lewis. 1982. *Medical botany: Plants affecting man's health.* New York: John Wiley & Sons, Inc.

Simpson, B. B., and M. Conner-Orgarzaly. 1986. *Economic botany.* Hightstown, NJ: McGraw-Hill Book Co.

Sweet, M. 1976. *Common edible and useful plants of the West,* rev. ed. Healdsburg, CA: Naturegraph Publishers.

Turner, N. J., and A. F. Szczawinski. 1991. *Common poisonous plants and mushrooms of North America,* 2d ed. Portland, OR: Timber Press.

Van Allen Murphey, E. 1987. *Indian uses of native plants,* 3d ed. Glenwood, IL: Meyer Books.

Wijesekera, R. O. 1991. *The medicinal plant industry.* Boca Raton, FL: CRC Press.

Chapter Outline

A pond succession scene.

Ecology

25

FIGURE 25.1 Butte Sink, California. A typical wetland, as seen from the air.

Hazardous Waste

Too many members of earlier generations routinely drained on the ground or into storm drains the oil from their vehicle crankcases or took the used oil and other hazardous wastes to the dump. Highly toxic industrial wastes were also disposed of both in and at the outskirts of cities and towns. In 1996, it was estimated that some 12 million children in the United States were living less than 4 miles from hazardous waste sites. With the advent of atomic energy, radioactive wastes were sometimes insufficiently isolated from living organisms (including humans), with catastrophic results. Even when hazardous wastes aren't unceremoniously dumped, serious accidents and spills take place, with the effects sometimes lingering indefinitely, as, for example, in and around the former Soviet Union's Chernobyl atomic meltdown site.

Today, there are concerted efforts to curb the disposal of hazardous wastes and to greatly reduce the probabilities of accidents and spills. At most solid waste dumps, it is illegal to dispose of even empty latex paint cans, let alone more toxic materials, and heavy fines are levied on those found disposing of industrial wastes in an improper manner. Moneys from a U.S. government superfund are being used, with some success, to clean up selected old hazardous waste sites. The process thus far is slow and inadequate, but the increased restrictions on disposal methods and, as previously noted, the

genetic engineering of bacteria that can dismantle and render harmless many types of wastes hold promise for the future.

GLOBAL ISSUES

There are many issues, such as ozone depletion, worldwide changes in climate, and loss of biodiversity, that are global in scope and long lasting in impact. Strategies for attacking the problems emphasize the need for multinational dialogue and cooperation between specialists in fields as diverse as biology, chemistry, economics, and political science.

To comprehend the big picture, we'll first examine some basic principles and then apply the principles to global matters.

Populations, Communities, and Ecosystems

We recognize that plants, animals, and other organisms tend to be associated in various ways with one another and also with their physical environment. For example, the term *forest* is applied to **populations** (groups of individuals of the same species) of trees or other plants that form a **plant community** (unit composed of all the populations of plants occurring in a given area). The lichen and moss flora on a rock also constitute a community, as do the various seaweeds in a tidepool (Fig. 25.2). But these communities also invariably have

FIGURE 25.2   A tidepool community at low tide.

animals and other living organisms associated with them. It is preferable, therefore, to refer to the unit composed of all the populations of living organisms in a given area as a **biotic community.** Considered together, the communities and their physical environments, which interact with each other and are interconnected by physical, chemical, and biological processes, constitute **ecosystems.** Some populations, communities, and ecosystems may be microscopic in extent while others are much larger or even global.

Populations

Populations may vary in number, in density, and in the total mass of individuals. Depending on circumstances, a field biologist may investigate a population in various ways. If, for example, a conservation organization is concerned about the preservation of a rare or threatened species, the organization may simply count the **number** of individuals. If a count is not feasible, the organization may estimate population **density** (number of individuals per unit volume—e.g., five blueberry bushes per square meter). If the individuals in a population vary greatly in size or are unevenly scattered, a better estimate of the population's importance to the ecosystem may be calculated by determining the **biomass** (total mass of the individuals present). Population biology studies also include, among many components, physiological and reproductive ecology with the examination and tabulation of factors, such as pollination, seed dispersal, germination, and seedling survival and establishment in plant populations.

Communities

Communities are composed of populations of one to many species of organisms living together in the same location. Similar communities occur under similar environmental conditions, although actual species composition can vary considerably from one location to another. A community is difficult to define precisely, because species of one community may also occur in other communities. Furthermore, species of one community may have specific genetic adaptations to that community. If individuals are transplanted to a second, different community where the same species occurs, the transplanted individuals may not necessarily be able to survive alongside their counterparts, which are themselves adapted to this second community. Individuals adapted to specific communities within their overall distribution are called *ecotypes,* and areas transitional between communities are called *ecotones.*

Analysis and classification of communities is important in the preparation of maps that form the basis of activities such as land-use planning, forestry, natural resource management, and military maneuvers.

Ecosystems

Living organisms interacting with one another and with factors of the nonliving environment constitute an *ecosystem.* The nonliving factors of the environment include light, temperature, oxygen level, air circulation, fire, precipitation, energy, and soil type. The distribution of a plant species in an ecosystem is controlled mostly by temperature, precipitation, soil type, and the effects of other living organisms (biotic factors). In Mediterranean climates, such as those that occur in parts of California and Chile, nearly all precipitation occurs during the winter months and the summers are dry. This type of climate favors spring annuals that complete their life cycles by summer and evergreen shrubs that can tolerate long periods of drought. Forests may occur in areas where heavy winter snowfall soaks deeply into the soil, compensating to a certain extent for the lack of summer moisture.

The leaves and other parts of plant species that occur naturally in areas of low precipitation and high temperatures (*xerophytes*) generally are adapted to their particular environment through modifications that reduce transpiration. These modifications are discussed under "Specialized Leaves" in Chapter 7 and "Regulation of Transpiration" in Chapter 9. Plants of arid areas may also have specialized forms of photosynthesis, such as CAM photosynthesis

Decomposers

FIGURE 25.3   A food web.

(discussed in Chapter 10). Similarly, plants that grow in water (*hydrophytes*) are modified for aquatic environments.

The distribution of plant species is influenced by the mineral content of soils. For example, serpentine soils—which contain relatively high amounts of magnesium, iron, usually nickel, and chromium and low amounts of calcium and nitrogen—often support species that are not found on nearby nonserpentine soils. Biotic factors, such as competition for light, the mineral nutrients and water available, and grazing by the animal members of the biotic community, also influence the distribution of plant species.

Ideally, ecosystems sustain themselves entirely through photosynthetic activity, energy flow through food chains, and the recycling of nutrients. Organisms that are capable of carrying on photosynthesis capture light energy and convert it, along with carbon dioxide and water, to energy-storing molecules. Such organisms are called **producers.** Animals, such as cows, caribou, and caterpillars, and other organisms that feed directly on producers are called **primary consumers. Secondary consumers,** such as tigers, toads, and tsetse flies, feed on primary consumers. **Decomposers** break down organic materials to forms that can be reassimilated by the producers. The foremost decomposers in most ecosystems are bacteria and fungi.

In any ecosystem, the producers and consumers interact, forming **food chains** or interlocking *food webs,* which determine the flow of energy through the different levels (Fig. 25.3). Since most organisms have more than a single source of food and are themselves often consumed by a variety of consumers, there are considerable differences in the length and intricacy of food chains or webs.

Light energy itself, which enters at the producer level, cannot be recycled in an ecosystem. Only about 1% of the light energy falling on a temperate zone community is actually converted to organic material. As the organisms at each level respire, energy gradually dissipates as heat into the atmosphere. Additionally, parts of organisms are not consumed at subsequent levels in the food chain. For example, the energy in leaves that fall from a plant before a herbivore grazes on it is released when decomposers (bacteria and fungi) break down the leaves. It has been estimated that when cattle graze, only about 10% of the energy stored by the green plants they consume is converted to animal tissue, with most of the remaining energy dissipating as heat into the atmosphere. When we eat beef, only roughly 10% of its stored energy is utilized by our bodies to manufacture new blood cells and otherwise sustain life. The remaining energy is converted to heat. Obviously, then, in a long food chain, the final consumer gains only a tiny fraction of the energy originally captured by the producer at the bottom of the chain.

Conversely, there is proportionately much less loss of energy between levels in short food chains. Let's assume, for example, that for every 100 calories of light energy falling on lettuce plants each day, 10 calories are converted to lettuce tissue. Then suppose that the lettuce is fed to hogs. Again, only 10%, or 1 calorie, of the original energy may be converted to animal tissue. If as secondary consumers we eat

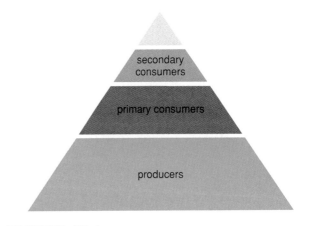

FIGURE 25.4 An energy pyramid of an ecosystem. There is much more energy at the bottom than at the top.

such pork, our bodies, in turn, may utilize 10% of the energy available in the pork, or only 0.1% of the original energy available. If, however, we eat the lettuce directly, we end up with the hog's percentage, which is 10 times more of the original solar energy than we would get if we ate the hog that ate the lettuce. (The actual amount of energy per gram of fatty hog meat is considerably higher than that found in a gram of lettuce tissue.)

From this, it follows that a vegetarian diet makes much more efficient use of solar energy than one that relies heavily on meats, and where food is scarce or humans are very abundant (as in India or Ethiopia), humans become virtual vegetarians. It also follows that, in terms of the numbers of individuals and the total mass, there is a sharp reduction at each level of the food chain. In a given portion of ocean, for example, there may be billions of microscopic algal producers supporting millions of tiny crustacean consumers, which, in turn, support thousands of small fish, which meet the food needs of scores of medium-sized fish, which are finally consumed by one or two large fish (Fig. 25.4). In other words, one large fish may very well depend on a billion tiny algae to meet its energy needs every day.

The interrelationships and interactions among the components of an ecosystem can be quite complex, but all function together in a regulatory fashion. An increase in food made available by producers can result in an increase in consumers, but the increased number of consumers reduces the available food, which then inevitably leads to a reduction of consumers to earlier levels. While cyclical in nature, the net result is sustained self-maintenance of the ecosystem. This is the basis for the so-called *Balance of Nature.*

Interactions Between Plants, Herbivores, and Other Organisms

While it is easy to see that the total mass of consumers is largely determined by the total mass of food made available by the producers, the interactions among producers themselves, between predators and prey, and between the decomposers and the other members of the ecosystem are usually more subtle. Many flowering plants produce

A.

B.

FIGURE 25.5 A species of *Acacia* that is host to ants that live in its hollow thorns. The ants attack any organisms, large or small, that come in contact with the plant. The plant provides food for the ants through nectaries at the bases of the thorns (*A*) and nitrogenous bodies at the tips of the leaflets (*B*).

roots. Many other plants produce *phytoalexins* (chemicals that kill or inhibit disease fungi or bacteria), making them resistant to various diseases (see discussion under "Use of Resistant Varieties" in Appendix 2).

Conversely, some bacteria and fungi limit higher plant growth by producing various inhibitory chemical compounds. Parasitism limits population size in other bacteria, fungi, and flowering plants. The degree of parasitism varies considerably with the organisms involved. Some of the species of the Figwort Family, which includes snapdragons and similar plants, have no chlorophyll and depend entirely on their flowering plant hosts for their energy and other nutritional needs. Other related species do have chlorophyll but apparently also require supplemental food from their hosts. Still other species often parasitize the roots of certain plants but are also capable of existing independently.

Mycorrhizal fungi are intimately associated with the roots of most woody and many other plants in such a way that both organisms derive benefit (such associations, called *mutualisms,* are discussed in Chapter 5). The fungi greatly increase the absorptive surface of the root, usually playing a major role in the absorption of phosphorus and other nutrients, while obtaining energy from root cells.

Thomas Belt, a naturalist of more than 100 years ago, first called attention to an association between tropical ants and thorny, rapidly growing species of *Acacia* (Fig. 25.5). *Acacia,* which has large hollow thorns at the base of each leaf, is host to ants that feed on sugars, fats, and proteins produced by petiolar nectaries and special bodies at the tip of each leaflet. The ants live within the hollow thorns and vigorously attack any other organisms, from insects to large animals, that come in contact with the plant. They also kill, by *girdling,* any plant that touches the *Acacia* (girdling is discussed under "bridge grafting" in Appendix 4). Experiments have shown that, when ants are removed from these *Acacia* species, the plants grow very slowly and usually soon die from insect attacks or from shading by other plants.

Large herbivorous animals, such as deer and moose, feed on a wide variety of plants, each differing in nutritional value. Each plant species also produces different combinations, types, and amounts of chemical compounds in addition to proteins, fats, and carbohydrates. Many of the chemical compounds are poisonous to the consumers, but the animals do not display symptoms of poisoning because their digestive systems are capable of breaking down the compounds and eliminating or excreting them to a limited degree. The limitations imposed by such compounds result in the consumers varying their diet, seeking familiar foods and being wary of new ones. If a plant species did not have some natural defense, such as chemical compounds or structural modifications (for example, spines), primary consumers of all kinds, from insects to elephants, would soon render that species extinct. In an ecosystem, the defenses that both producers and consumers have against each other have been developed through a process of coevolution resulting from natural selection and are maintained in delicate balance.

substances that either inhibit or promote the growth of other flowering plants. Black walnut trees, for example, produce a substance that wilts tomatoes and potatoes and injures apple trees that come in contact with black walnut

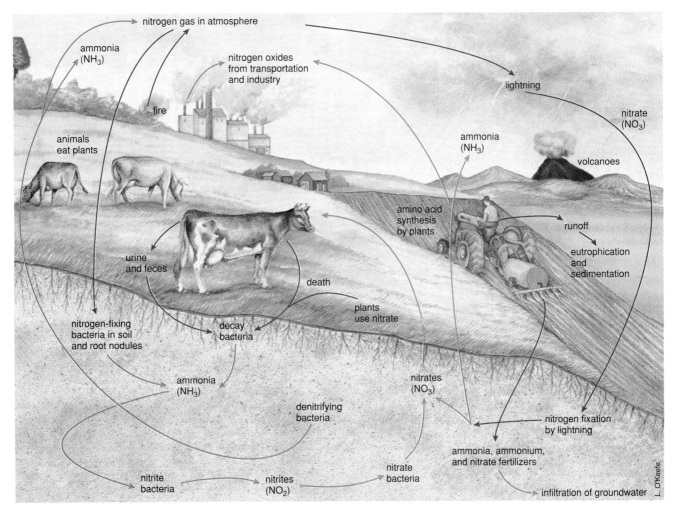

FIGURE 25.6   The nitrogen cycle.

NATURAL CYCLES

Water and elements such as nitrogen, carbon, and phosphorus are constantly recycling throughout nature. We'll use the nitrogen and carbon cycles as examples.

The Nitrogen Cycle

Much of the protoplasm of living cells consists of protein. Nitrogen, the most abundant element in our atmosphere, constitutes about 18% of the protein. There are nearly 69,000 metric tons of nitrogen in the air over each hectare of land (35,000 tons per acre), but the total amount of nitrogen in the soil seldom exceeds 3.9 metric tons per hectare (2 tons per acre) and is usually considerably less. This discrepancy results from the nitrogen of the atmosphere being chemically inert, which is another way of saying that it will not combine readily with other molecules. It is, therefore, largely unavailable to plants and animals for their use in building proteins and other substances containing nitrogen.

Most of the nitrogen supply of plants (and indirectly of animals) is derived from the soil in the form of inorganic compounds and ions taken in by the roots. These compounds and ions include those that contain nitrogen chemically combined with oxygen or hydrogen. They are produced by bacteria and fungi as these break down the more complex molecules of dead plant and animal tissues to simpler ones. Some nitrogen from the air is also *fixed*—that is, converted to ammonia or other nitrogenous compounds by various *nitrogen-fixing bacteria*. Some of these organisms gain access to various plants, particularly legumes (e.g., peas, beans, clover, alfalfa), through the root hairs, with the plant producing root nodules in which the bacteria multiply (root nodules are shown in Fig. 5.15). Others live free in the soil.

Figure 25.6 shows that there is a constant flow of nitrogen from dead plant and animal tissues into the soil and from the soil back to the plants. Decay bacteria and fungi can break down enormous quantities of dead leaves and other tissues to tiny fractions of their original volumes within a few days to a few months. If they were abruptly to cease their activities, the available nitrogen compounds would be completely exhausted and the carbon dioxide supply needed for photosynthesis seriously depleted within a few decades. Forests and prairies would die as the accumulations of shed leaves, bodies, and debris buried the living plants and shielded their leaves from the light essential to

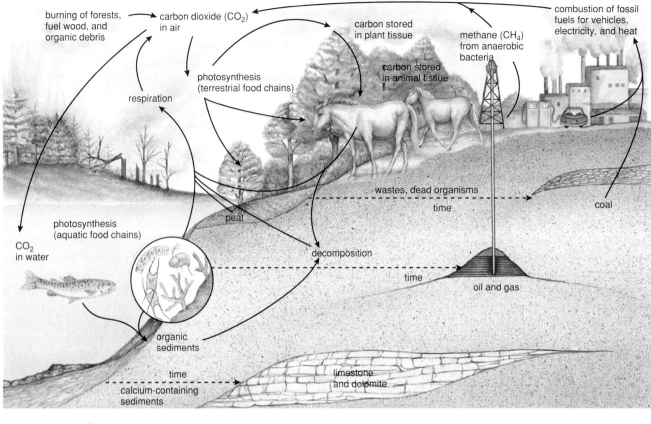

FIGURE 25.7 The carbon cycle.

photosynthetic activity. At present, even with the various bacteria involved in the nitrogen cycle functioning normally, the total amount of nitrogen in the soil is not being increased by their activities but is merely being recycled.

Significant amounts of nitrogen are continually being lost as water leaches it out or carries it away through erosion of topsoil. More is lost with each harvest, the average crop removing about 25 kilograms per hectare (25 pounds per acre) per year. This nitrogen loss from the harvesting of crops can be sharply reduced if vegetable and animal wastes are recycled and returned to the soil each year. While bacteria are decomposing tissues, they use nitrogen, and little is available until they die and release their accumulations into the soil. Accordingly, crops should not be planted in soils to which only partially decomposed materials have been added. Likewise, when sawdust, straw, or other organic mulches are spread around plants in a garden to control weeds and conserve soil moisture, less nitrogen will be available to the growing plants until the mulches have been decomposed.

Weeds and stubble are often controlled by burning. Fire, however, causes serious loss of nitrogen, which has to be replaced. It has been estimated that the annual combined loss of nitrogen from the soil in the United States from fire, harvesting, and other causes exceeds 21 million metric tons (23 million tons), and only 15.5 million metric tons (17 million tons) are replaced by natural means. To offset the net

loss, some 32 million metric tons (35 million tons) of inorganic fertilizers are applied to the soils each year. If organic matter is not added at the same time, however, this application of inorganic fertilizers, combined with the annual burning of stubble, may eventually result in the creation of a *hardpan* soil.

Hardpan develops through the gradual accumulation of salt residues, which dissolve humus and disrupt the structure of the soil, causing the clay particles to clump and also producing colloids that are impervious to moisture. In hardpan soils and others low in oxygen (e.g., flooded areas), *denitrifying bacteria* use nitrates instead of oxygen in their respiration, depleting the remaining soil nitrogen.

Precipitation returns a little nitrogen to the soil from the atmosphere, where it has accumulated as a result of the action of light on industrial pollutants, fixation by flashes of lightning, and diffusion of ammonia released through decay. The activities of nitrogen-fixing bacteria and volcanoes also contribute to the natural replenishment of nitrogen by converting it to forms that can be utilized by plants.

The Carbon Cycle

In the process of recycling nitrogen, bacteria also play a major role in recycling carbon and many other substances (Fig. 25.7). One of the two raw materials of photosynthesis, carbon dioxide, constitutes 0.03% of our atmosphere. The combined plant life of the oceans and the land masses is estimated to use about 14.5 billion metric tons (16 billion tons) of carbon obtained from carbon dioxide every year. Respiration by all living organisms constantly replaces carbon dioxide, with perhaps as much as 90% or more of it being produced by the incredible numbers of decay bacteria and fungi as they decompose tissues.

The burning of fossil fuels by the internal combustion engines of industry and transportation releases lesser but significantly increasing amounts of carbon dioxide into the air, and a small amount originates with fires and volcanic activity. At the present rate of use by photosynthetic plants, it has been calculated that all of the carbon dioxide of our atmosphere would be used up in about 22 years if it were not constantly being replenished.

Scientists have found that plant growth can be accelerated by increasing levels of carbon dioxide in the air around them, and some commercial nurseries are now pumping carbon dioxide along rows of bedding plants as a "fertilizer." Recent studies indicate that while C_3 plants respond to increased carbon dioxide levels, C_4 plants do not, suggesting that in natural communities the response of C_3 plants to higher levels of carbon dioxide may give them a competitive edge over C_4 plants.

Anaerobic bacteria produce large volumes of carbon-containing methane gas, which is discussed on page 459.

In the web of life, nutrients constantly cycle and are recycled. Carbon, nitrogen, water, phosphorus, and other molecules have for aeons been passing through cycles. Some molecules that were a part of a primeval forest that became compressed and turned to coal may have become part of another plant after the coal burned. Then the new plant may have been eaten by an animal, which, in turn, contributed molecules to a part of yet another living organism. Just think of where the molecules in your own body may have been in the past few billion years. They may well have been a part of some prehistoric seaweed, a saber-toothed tiger, a mighty dinosaur, or even all three!

SUCCESSION

After a volcano spews lava over a landscape, or after an earthquake or a landslide exposes rocks for the first time, there is initially no sign of life on the lava or rock surfaces. Within a few years or sometimes within a few months or even weeks, living organisms begin to appear, and a sequence of events known as *succession* takes place. During succession, the species of plants and other organisms that first appear gradually alter their environment as they carry on their normal activities, such as metabolism and reproduction. In time, the accumulation of wastes, dead organic material, and inorganic debris and other changes (such as of shade and water content in the habitat) favor different species, which may replace the original ones. These, in turn, modify the environment further so that yet other species become established.

Succession occurs whenever and wherever there has been a disturbance of natural areas on land or in water. It proceeds at varying rates, depending on the climate, the soils, and the animals or other organisms in the vicinity. Ecologists recognize a number of variations of two basic types of succession. *Primary succession* involves the actual formation of soil in the beginning stages, while *secondary succession* takes place in areas that previously were covered with soil and vegetation.

Primary Succession

Xerosere

One of the most universal types of primary succession, called a **xerosere,** begins with bare rocks and lava that have been exposed through glacial or volcanic activity or through landslides. Initially, the rocks are sometimes subjected to alternate thawing and freezing, at least in temperate to colder areas. Tiny cracks or flaking may occur on the surface as a result. Lichens often become established on such surfaces (Fig. 25.8). They produce acids that very slowly etch the rocks, and as they die and contribute organic matter, they are replaced by other, larger lichens. Certain rock mosses adapted to long periods of desiccation also may become established, and a small amount of soil begins to build up. This is augmented by dust and debris blown in by the wind. Eventually, enough of a mat of lichen and moss material is present to permit some ferns or even seed plants to become established, and the pace of soil buildup and rock breakdown accelerates.

If deep cracks appear in the rocks, the larger seeds may widen them further as they germinate and the roots expand in girth. It has been calculated that germinating seedlings can exert a force of up to 31.635 kilograms per square centimeter (450 pounds per square inch). Indeed, there are known instances of seedlings splitting rocks that weigh several tons (Fig. 25.9).

As soil buildup continues, larger plants take over, and eventually the vegetation reaches an equilibrium in which the associations of plants and other organisms remain the same until another disturbance takes place or climatic changes occur. Such relatively stable plant associations are referred to as **climax vegetation.**

The climax vegetation of deciduous forests in eastern North America is dominated by maples and beeches, oaks, hickories, and hemlocks or other combinations of trees. In desert regions, various cacti form a conspicuous part of the climax vegetation, while in the Pacific Northwest, large coniferous trees predominate. In parts of the Midwest, prairie grasses and other herbaceous plants form the climax vegetation, and in wet tropical regions, a complex association of trees and herbs constitutes the climax.

Occasionally, when a volcano produces ash instead of lava that buries existing landscape and associated vegetation, some of the successional stages involving lichens and mosses may be bypassed, with larger plants becoming the

FIGURE 25.8   Early stages of succession on a Hawaiian road after an earthquake produced cracks and a hole. Ferns have become established in the cracks and hole less than three years after the disruption.

FIGURE 25.9   A blue oak that grew from an acorn lodged in a small crack of the rock. Over the years, the oak has split the rock apart.

successional pioneers. This occurred following the series of ash eruptions, accumulation of debris, and mud flows of Mount St. Helens in the state of Washington during the early 1980s.

Hydrosere

Succession takes place in wet habitats as well as in drier ones, and such succession is called a **hydrosere.** In the northern parts of midwestern states, such as Michigan, Wisconsin, and Minnesota, ponds and lakes of various sizes abound. Many were left behind by retreating glaciers and often have no streams draining them. The water that evaporates from them is replaced annually by precipitation runoff. They also grow a tiny bit smaller each year as a result of succession (Fig. 25.10).

This succession often begins with algae either carried in by the wind or transported on the muddy feet of waterfowl and wading birds. The algae tend to multiply in shallow water near the margin, and with each reproductive cycle, the dead parts sink to the bottom. Floating plants, such as duckweeds, may then appear, often forming a band around the body of water just offshore (Fig. 25.11). When nutrients, oxygen, pH, and temperatures are low, peat mosses become the dominant floating plants. There are presently about 4 million square kilometers (1.5 million square miles) of peat bogs throughout the world. Next, water lilies and other rooted aquatic plants with floating leaves become established, each group of plants contributing to the organic material on the bottom, which slowly turns to muck. Cattails and other flowering plants that produce their inflorescences above the water often take root in the muck around the edges, and the accumulation of organic material accelerates.

Meanwhile, the algae, duckweeds, peat mosses, and other plants move farther out, and the surface area of

FIGURE 25.10   Succession in a pond in northern Wisconsin.
(Courtesy Robert A. Schlising)

exposed water gradually diminishes. Grasslike sedges become established along the damp margins and sometimes form floating mats as their roots interweave with one another. Dead organic material accumulates and fills in the area under the sedge mats, and herbaceous and shrubby plants then move in. As the margins become less marshy, coniferous trees whose roots can tolerate considerable moisture (e.g., tamaracks or eastern white cedars) gain a foothold, eventually growing across the entire site as the pond or lake disappears. The trees continue to aid in the formation of true soil, and in due course, the climax vegetation takes over. No visible trace of the pond or lake now remains, and the only evidence of its having been there lies beneath the surface, where fossil pollen grains, bits of wood, fossilized fish skeletons, and other material reveal the past history. Such succession may take thousands of years and has never been witnessed from start to finish. The evidence that it does occur, however, is extensive and compelling.

Under natural conditions, some stream-fed lakes and ponds eventually become filled with silt and debris, although this, too, may take thousands of years to occur. The streams that feed these lakes bring in silt, and the nutrient content of the water rises as dissolved organic and inorganic materials (particularly nitrogen and phosphorus) are

FIGURE 25.11   Duckweeds floating on a pond in early stages of succession.

FIGURE 25.15   Taiga in Alaska.

FIGURE 25.16   A scene within the eastern deciduous forest.
(Courtesy Robert A. Schlising)

marshes dot the region, which is inhabited by a variety of birds, including jays, warblers, and nuthatches. Many rodents, such as shrews and jumping mice, and larger mammals, notably moose and deer, occur. Ermine and wolverines also make their home in the taiga, and caribou overwinter in this biome.

Temperate Deciduous Forest

Most deciduous trees are broad-leaved species that shed their leaves annually during the fall and remain dormant during the shorter days and colder temperatures of winter. Most *temperate deciduous forests* (Fig. 25.16) are found, like the taiga, on large continental masses in the Northern Hemisphere. In North America, an example of this type of forest occurs from the Great Lakes region south to the Gulf of Mexico and roughly extends from the Mississippi River to the eastern seaboard, although some do not consider the eastern coastal plain to be part of the temperate deciduous forest biome. Temperatures within the area vary a great deal but normally fall below 4°C (39°F) in midwinter and rise above 20°C (68°F) in summer. The climax trees of the forest are well adapted to subfreezing temperatures as long as the cold is accompanied by precipitation or snow cover. Most of the annual precipitation, which totals between 50 and 165 centimeters (20 to 65 inches), occurs in the summer.

Some of the most beautiful of all the broad-leaved trees are found in a variety of associations in this biome. In the upper Midwest, sugar maple and American basswood predominate. Sugar maple, birch, and oak are also found to the Northeast, where they tend to be associated with the stately American beech. In the west-central part of the forest, oaks and hickories abound. Oaks also are abundant along the eastern slopes of the Appalachian Mountains, where American chestnuts were once a conspicuous part of the flora. The chestnuts have now virtually disappeared, having succumbed to chestnut blight disease, which was introduced and began killing the trees during the early 1900s. Oaks and hickories extend into the southeastern United States, where they become associated with pines and other tree species, such as the bald cypress.

Before the arrival of European immigrants, a mixture of large deciduous trees that included maples, ashes, basswood, beeches, buckeyes, hickories, oaks, tulip trees, and magnolias was found on the eastern slopes and valleys of the Appalachian Mountains. Some of the trees grew over 30 meters (100 feet) tall and had trunks up to 3 meters (10 feet) in diameter. Except for in a few protected pockets in the Great Smoky Mountains National Park, the largest trees of this rich forest have been all but eliminated through logging. A smaller-treed extension of this forest is found in an area northeast of Baton Rouge, Louisiana, western Tennessee, and Kentucky and in southern Illinois. American elms, also once a part of the forest, are rapidly disappearing as Dutch elm disease fells both wild trees and those planted along city streets. One small midwestern town, which had some 600 elms planted along its streets, found itself left with only six live trees within a year or two after Dutch elm disease struck.

A mixture of deciduous trees and evergreens occurs on the northern and southeastern borders of the forest. Hemlocks and eastern white pine are found from New England west to Minnesota and south to the Appalachians. The once vast stands of eastern white pine are now largely gone, their valuable lumber having been used for construction and other purposes. Some have been lost to still another tree disease, white pine blister rust, but scattered trees remain, particularly where they have been protected. Various pines dominate the eastern coastal plain from New Jersey to Florida and west to east Texas. Pitch pine is common in New Jersey, while some of the southern pines (e.g., long-leaf, slash,

loblolly) are now cultivated in the southeastern United States for wood pulp, turpentine, lumber, and other commercially valuable products.

During the summer, the trees of the deciduous forests form a relatively solid canopy that keeps most direct sunlight from reaching the floor. Many of the showiest spring flowers of the region (such as bloodroot, hepatica, trilliums, and violets) flower before the trees have leafed out fully and complete most of their growth within a few weeks. Other plants that can tolerate more shade, principally members of the Sunflower Family (e.g., asters and goldenrods), flower in succession in forest openings from midsummer through fall.

Animal life in the eastern deciduous forest includes red and gray foxes, raccoons, opossums, many rodents such as gray squirrels, snakes such as copperheads and black rat snakes, salamanders, and a wide variety of birds, including hawks, flickers, and mourning doves.

Grassland

Naturally occurring *grasslands* are found toward the interiors of continental masses (Fig. 25.17). They tend to intergrade with forests, woodlands, or deserts at their margins, depending on precipitation patterns and amounts. A grassland may receive as little as 25 centimeters (10 inches) of rainfall or as much as 100 centimeters (39 inches) annually. Temperatures can range from 45°C (113°F) in midsummer to −45°C (−50°F) in midwinter.

Large areas of grassland are now used for growing cereal crops (particularly corn and wheat) and for grazing cattle. Before it was destroyed, the American prairie was a remarkable sight. In Illinois and Iowa, the grasses grew over 2 meters (6 feet 6 inches) tall during an average season and another meter taller during a wet one. A dazzling display of wildflowers began before the young perennial grasses emerged in the spring and continued throughout the growing season. I have counted over 50 species of flowering plants in bloom at the same time on 1 hectare (2.47 acres) of protected grassland in the middle of spring.

Grasslands occurring in areas with a Mediterranean climate (e.g., the Great Central Valley of California), where most of the precipitation takes place during the winters, usually include *vernal pools,* some of which may be more than 50,000 years old. These are temporary accumulations of rainwater that evaporate after the rains come to an end. Their unique floras include an orderly sequence of flowering plants, some appearing initially at the pool margins, with each species forming a distinct zone or band until the water is gone (Fig. 25.18). Some species flower only in the damp soil and drying mud that remains. The seeds of many species germinate under water.

In North America, huge herds of buffalo once grazed the prairies, which grew on fertile soils, but these disappeared as the settlers cultivated more and more of the land and hunters slaughtered more and more of the large animals. By 1889, there were only 551 left, but breeding and conservation has since increased their numbers.

FIGURE 25.17   Undisturbed western grassland in the spring.

FIGURE 25.18   Zonation in a vernal pool in northern California.

(Courtesy Robert A. Schlising)

FIGURE 25.19   A desert community in the southwestern United States.

Grassland animals include cottontails, jackrabbits, gophers, mice, and pronghorns. As indicated, buffalo were once abundant. Some 15,000 of these large animals now are protected in national parks, game preserves, and private ranches. Various sparrows (e.g., vesper and savannah sparrows) and other birds still find homes in uncultivated tracts of this biome.

Desert

If you ask the average person to describe a *desert,* the response will probably include words such as *sand, heat, mirage, oasis,* and *camels* (Fig. 25.19). These features are indeed found in some of the world's large deserts that are located in the vicinity of 30 degrees latitude both north and south of the equator, but deserts occur wherever precipitation is consistently low or the soil is too porous to retain water. Most receive less than 12.5 centimeters (5 inches) of rain per year. The low humidity results in wide daily temperature ranges. On a summer day when the temperature has reached over 35°C (95°F), it will generally fall below 15°C (59°F) the same night. The light intensities reach higher peaks in the dry air than they do in areas where atmospheric water vapor filters out some of the sun's rays. Some desert plants have become adapted to these higher light intensities through the evolution of crassulacean acid metabolism (CAM) photosynthesis (discussed in Chapter 10), a process by which certain organic acids accumulate in the chlorophyll-containing parts of the plants during the night and are converted to carbon dioxide during the day. This permits much more photosynthesis to take place than would otherwise be possible, since most of the carbon dioxide of the atmosphere is excluded from the plants during the day by the stomata, which remain closed (thereby also retarding water loss).

Other adaptations of desert plants include thick cuticles, water-storage tissues in stems and leaves, leaves with a leathery texture and/or reduced in size, and even a total absence of leaves. In cacti and other succulents without functional leaves, the stems take over the photosynthetic activities of the plants. Desert perennials are adapted to the biome in various ways. Cacti and similar succulents have widespread shallow root systems that can rapidly absorb water from the infrequent rains. The water can then be stored for long periods in the interior of the stems, which are modified for such storage. Other perennials grow from bulbs that are dormant for much of the year. Annuals provide a spectacular display of color and variety, particularly during an occasional season when above-average precipitation has occurred. The seeds of the annuals often germinate after a fall or winter rain, and the plants then grow slowly or remain in a basal, circular cluster of leaves (rosette) for several months before producing flowers in the spring. Literally hundreds of different species of desert annuals may occur within a few square kilometers (1 or 2 square miles) of typical desert in the southwestern United States.

FIGURE 25.20 Coastal redwoods in northern California.

Desert animals are adapted, in many instances, to foraging at night when it is cooler. These animals include various mice and kangaroo rats, snakes (notably rattlesnakes and king snakes), chuckwallas, and lizards (e.g., gila monsters). Various thrashers, doves, and flycatchers are included in the bird life.

Mountain Forests

In the geologic past, deciduous forests extended to western North America. As the climate changed and summer rainfall was reduced, conifers largely replaced the deciduous trees, although some (e.g., maples, birches, aspens, oaks) still remain, particularly at the lower elevations. Today, coniferous forests occupy vast areas of the Pacific Northwest and extend south along the Rocky Mountains and the Sierra Nevada and California coast ranges. Isolated pockets of this biome also occur in other parts of the West, particularly toward the southern limits of the mountains.

The trees tend to be very large, particularly in and to the west of the Cascade Mountains of Oregon and Washington and on the western slopes of the Sierra Nevada. Part of the reason for the huge size of trees such as Douglas fir is the high annual rainfall, which exceeds 250 centimeters (100 inches) in some areas. The world's tallest conifers, the coastal redwoods of California (Fig. 25.20), are at low elevations between the Pacific Ocean and the California outer coast ranges. Here they apparently depend more on fog, which reduces transpiration rates, than on large amounts of rain for their size and longevity.

One of the characteristics of mountain forests is a fairly conspicuous altitudinal zonation of species. In other words, one encounters different associations of plants when proceeding from sea level up the mountainsides. At lower elevations in both the Rocky Mountains and the Sierra Nevada, the predominant conifer is ponderosa pine. At lower elevations in the northern part of the Cascades, Douglas firs, western red cedars, and western hemlocks are more common. At intermediate elevations in the Sierra Nevada, sugar pine, white fir, and Jeffrey pine take over, while at higher elevations, other species of pines, firs, and hemlocks predominate.

Most of the mountain forest biome has comparatively dry summers. This has led to frequent forest fires, even before human carelessness became a factor. Several tree species are well adapted to fires. The Douglas fir, for example, has a very thick protective bark that can be charred without transmitting sufficient heat to the interior to damage more delicate tissues, and its seedlings thrive in open areas after a fire. When the bark of the giant redwoods of the Sierra Nevada is burned, the trees are rarely killed. This has undoubtedly contributed to the great age and size of many of the trees. The cones of some of the pine trees (e.g., knobcone pine) remain closed and do not release their seeds until a fire opens them, while seeds of several other species germinate best after they have been exposed to fire. These attributes of members of the mountain forest biome, as mentioned in the section on fire ecology, have led to the practice in some of our national parks of occasionally allowing fires at higher elevations to run their natural course. The higher-than-normal incidence of fires occurring since humans came in large numbers to the forest has made it necessary, however, to control fires in most instances, even though in doing so we may be interfering with natural cycles that would otherwise occur.

Animal life in the mountain forests includes many different rodents (especially chipmunks and voles), bears, mountain lions, bobcats, mountain beavers (not related to true beavers), mule deer, and elk. Large birds such as the golden eagle and many small birds, including mountain chickadees, warblers, and juncos, are an integral part of this biome.

FIGURE 25.21 A tropical rain forest scene.
(Courtesy Mary Lane Powell)

Tropical Rain Forest

About 5% of the earth's surface, representing nearly half of the forested areas of the earth, are included in the *tropical rain forest* biome (Fig. 25.21). The biome is distributed throughout those areas of the tropics where annual rainfall amounts normally range between 200 and 400 centimeters (79 and 157 inches) and where temperatures range between 25°C and 35°C (77°F and 95°F). Although monthly rainfall amounts vary, there is no dry season and some precipitation occurs throughout all 12 months of the year, frequently in the form of afternoon cloudbursts. The humidity seldom drops below 80%.

Such climatic conditions support a diversity of flora and fauna so great that the number of species exceeds those of all the other biomes combined. The forests are dominated by broadleaf evergreen trees, whose trunks are often unbranched for as much as 40 or more meters (160 feet), with luxuriant crowns that form a beautiful dark green and several-layered canopy. The root systems are shallow and the tree trunks often buttressed (buttress roots are shown in Fig. 5.12). There are literally hundreds of species of such trees, each usually represented by widely scattered specimens.

Most of the plants of the rain forests are woody, although not all of them are evergreen. Several of the deciduous tree species shed their leaves from some branches, retain the leaves on others, and flower on yet other branches all at the same time, while branches of adjacent trees of the same species are losing their leaves or flowering at different times. Numerous hanging woody vines and even more numerous *epiphytes*—especially orchids and bromeliads—can be seen on or attached to tree branches. The epiphyte roots, which are not parasitic, have no contact with the ground. The plants are sustained entirely by rainwater that accumulates in their leaf bases and by their own photosynthesis. Traces of minerals, also necessary to the growth of the epiphytes, accumulate in the rainwater as it trickles over decaying bark and dust.

The multilayered canopy is so dense that very little light penetrates to the floor, and the few herbaceous plants that survive are generally confined to openings in the forest. Despite the lush growth, there is little accumulation of litter or humus, and the soil is relatively poor. Decomposers rapidly break down any leaves or other organic material on the forest floor, and the nutrients released by decomposition are quickly recycled or leached by the heavy rains.

A few larger animals, with adaptations for moving through the mesh of branches in the rain forest, are found on the forest floor. Such animals include peccaries, tapirs, and anteaters. Most of the great numbers of animals, however, live out their lives in the canopies. Tree frogs, with adhesive pads on their toes, are common in tropical rain forests, as are various monkeys, sloths, opossums, tree snakes, and lizards. Ants abound, and many of the extraordinary variety of other insects are adapted to their environment through excellent camouflages. Large flocks of parrots and other birds feed on the abundance of insects as well as the available fruits.

The Future of the Tropical Rain Forest Biome

In the 1960s, major plans were developed to convert the Amazon rain forest into large farms, hydroelectric plants, and mines. At the beginning of the 1990s, gold mining activities were filling the rivers with silt, and, as indicated in the discussion of the greenhouse effect, the tropical rain forests were being destroyed or damaged for commercial purposes at the rate of more than 100 acres per minute. In one study of Amazon rain forest birds conducted in 1989, it was found that a population of low-flying species in a 25-acre section of the forest fell by 75% within just six weeks after adjacent land had been cleared, and 10 of the 48 species apparently disappeared completely.

This situation has been repeated on numerous occasions, and confirmed by a project cosponsored by the Smithsonian Institution and Brazil's Institute for Research in the Amazon, which during 10 years of study demonstrated that cutting large expanses of forest into smaller pieces separated by as little as 10 meters (33 feet) of cleared land can have disastrous effects on the ecology of the entire forest, with most of the bird and other animal species originally present disappearing permanently. Very little of the biome, which is the home of more than 50% of all living species of organisms, is presently protected from commercial development. Many of the organisms are doomed to extinction, often before they have been described for the first time, and the biome will essentially have vanished within 20 years if governments and individuals do not take definitive action to halt the large-scale destruction.

John Muir, Father of America's National Park System

Today America's National Parks are overcrowded as some 270 million people enjoy their beauty each year, relaxing in the great outdoors, hiking in fields and mountains, or navigating their rivers. Whether it's Yosemite, the Grand Canyon, Acadia, or one of the many others, each park has a unique and special appeal. While these magnificent parks are the result of many people's vision and foresight, perhaps none more so than John Muir who lived from 1838 to 1914. Often called the "Father of the National Park System," he influenced presidents, members of Congress, and "just plain folks" by his love of nature and his writings about it. Always traveling and exploring various ecosystems, he visited Alaska five times, walked 1,000 miles from Indianapolis to the Gulf of Mexico, as well as walking across California's San Joaquin Valley to explore the Sierra Nevada. Later he wrote of viewing the Sierra Nevada range for the first time, "Then it seemed to me the Sierra should be called not the Nevada, or 'Snowy Range,' but the Range of Light, the most divinely beautiful of all the mountain chains I have ever seen."

Wherever he roamed in the wilderness, he noticed increasing waste and destruction from various sources, whether overgrazing by cattle and sheep or overcutting by loggers. It was his love for the American forest that

Shafts of light penetrate the coastal redwoods of Muir Woods.

Summary

1. Human populations have increased dramatically in the past few centuries, and the disruption of ecosystems by the activities directly or indirectly associated with the feeding, clothing, and housing of billions of people threatens the survival of not only humans but many other living organisms.

2. Ecology is the study of the relationships between organisms and their environment. It is a vast field with many facets. Regional ecological issues include acid rain, water contamination, destruction of wetlands, and hazardous waste.

3. Global ecological issues include populations, communities, ecosystems, interactions between organisms, and loss of biodiversity.

4. Acid rain, which damages or kills living organisms, is produced when sulphur and nitrogen compounds re-leased by the burning of fossil fuels are converted to nitric acid and sulphur dioxide by sunlight and rain.

5. Water contamination occurs when toxic wastes, pesticides, septic tanks, and fertilizers wash or leach into surface and groundwater. Bacteria may in the future be used to break down various contaminants.

6. Wetlands, which were once considered wastelands, have been drained, but are now being increasingly protected.

7. Hazardous waste sites are slowly being detoxified, but progress is slow.

8. Populations vary in numbers, in density, and in the total mass of individuals. Populations of one community may also occur in another community, and eco-types may be specifically adapted to a single community.

9. Precipitation, temperature, soils, and biotic factors play roles in determining the distributions of plant species in an ecosystem.

drove him to work tirelessly for its preservation. He wrote some 300 articles and 10 books that told of his travels and contained his naturalist philosophy. He called on everyone to "climb the mountains and get their good tidings." He wrote that the wilderness should be preserved, viewed as a treasure rather than a resource to be exploited. Through his writing, he drew attention to the destruction forests were facing and worked to find remedies.

One solution was the formation of national parks. Largely through Muir's efforts, Congress, in 1890, created Yosemite National Park. He was also personally involved in the creation of Sequoia, Grand Canyon, Mount Rainier, and Petrified Forest National Parks. He went one step further by founding the Sierra Club in 1892 and served as its first president until his death in 1914. He and his friends conceived the Sierra Club with the mission to "do something for wildness and make the mountains glad." His idea was to form an association that would work to protect Yosemite National Park.

This brought him to the attention of President Theodore Roosevelt who visited Muir in Yosemite in 1903. Together with Muir, under the trees of the park, Roosevelt's innovative conservation programs began to take form.

John Muir's words and life inspire environmental activism today; they also teach the importance of enjoying nature and protecting it. Muir wrote a warning that is just as timely today as it was 100 years ago.

The axe and saw are insanely busy, chips are flying thick as snowflakes and every summer thousands of acres of priceless forests, with their underbrush, soil, springs, climate, scenery and religion, are vanishing away in clouds of smoke, while, except in the national parks, not one forest guard is employed. Any fool can destroy trees. They cannot run away; and if they could, they would still be destroyed—chased and hunted down as long as fun or a dollar could be got out of their bark hides, branching horns or magnificent bole backbones. Few that fell trees plant them; or would planting avail much towards getting back anything like the noble primeval forests.[1]

A living memorial to John Muir is located 12 miles north of San Francisco, Muir's home in his later years. Muir Woods is a 560-acre national monument consisting of a grove of majestic coastal redwoods along a canyon floor. The towering redwoods and lush canyon ferns growing along Redwood Creek make this a forest of tranquility. Only a 30-minute drive from San Francisco, Muir Woods is visited by over 1.5 million people each year. He declared that this grove of coastal redwoods was, "the best tree-lovers monument that could possibly be found in all the forests of the world."

John Muir—lover of forests, roamer of the wilderness, inspired writer, and fighter for conservation.

1. *Atlantic Monthly*, no. 80, 1897.

10. Xerophytes and hydrophytes have modifications of leaves and other organs.

11. Species distribution is influenced by soil mineral content and biotic factors, such as competition for light, nutrients, and water.

12. Ecosystems ideally sustain themselves entirely through photosynthetic activity, energy flow through food chains, and the recycling of nutrients. Producers carry on photosynthesis; consumers feed on producers. Primary consumers feed directly on producers, and secondary consumers feed on primary consumers.

13. Decomposers break down organic materials to forms that can be reassimilated by the producers. In any ecosystem, the producers and consumers comprise food chains, which determine the flow of energy through the different levels. Food chains vary in length and intricacy and, because of their interconnections, form food webs.

14. Energy itself is not recycled in an ecosystem; it gradually escapes in the form of heat as it passes from one level to another. In a long food chain, the final consumer gains only a tiny fraction of the energy originally captured by the producer at the bottom of the chain.

15. When producers increase the amount of food available, consumers increase correspondingly; this reduces the available food, resulting in a self-maintaining ecosystem. The composition of an ecosystem may be influenced by its living components through the secretion of growth-inhibiting substances, the secretion of growth-promoting substances, and parasitism.

16. Some nitrogen from the air is fixed by the nitrogen-fixing bacteria found in legumes and other plants. In the nitrogen cycle, there is a constant flow of nitrogen from dead plant and animal tissues into the soil and from the soil back to the plants.

17. Water leaches nitrogen from the soil and carries it away when erosion occurs. Other nitrogen is lost from harvesting crops, but the loss can be reduced if wastes

are decomposed and annually returned to the soil. Fire also causes nitrogen loss.

18. Replacement of nitrogen loss by the application of chemical fertilizers can eventually create hardpan by altering the soil structure. Bacteria also recycle carbon and other substances, such as water and phosphorus.

19. Succession occurs whenever there has been a disturbance of natural areas on land or in water.

20. Primary succession involves the formation of soil in the beginning stages, while secondary succession takes place in areas previously covered with vegetation. A xerosere is a primary succession in which bare rock or lava is converted to soil through the activities of lichens, plants, and physical forces in an orderly progression of events over a period of time.

21. Climax vegetation becomes established at the conclusion of succession and remains until or unless a disturbance disrupts it.

22. A hydrosere is initiated in a wet habitat and culminates in a climax vegetation. As a lake or other body of water is filled in with silt and debris, eutrophication facilitating the growth of algae and other organisms occurs.

23. Secondary succession, which proceeds more rapidly than does primary succession, may take place if soil is present. It may occur after fires.

24. The greenhouse effect refers to a global rise in temperature due to carbon dioxide, methane, and other gases preventing the sun's radiant heat from escaping back into space.

25. The carbon dioxide and methane levels in the earth's atmosphere and the earth's temperature have been rising. It is predicted that polar ice will, as a result, melt and flooding of low-lying coastal areas will occur.

26. In the stratosphere, sunlight converts methane, chlorofluorocarbons, and halons into active compounds that destroy the ozone shield that protects us from intense ultraviolet radiation.

27. Loss of biodiversity has serious consequences, including the potential loss of means with which to combat crop diseases.

28. Biotic communities considered on a global or at least a continental scale are called *biomes*. Major biomes of North America include tundra, taiga, temperate deciduous forests, grasslands, deserts, mountain forests, and tropical rain forests.

29. Tundra is found primarily above the Arctic Circle. It includes few trees and many lichens and grasses and is characterized by the presence of permafrost (permanently frozen soil) below the surface. Alpine tundra occurs above timberline in mountains below the Arctic Circle.

30. Taiga is dominated by coniferous trees, with birches, aspens, and willows in the wetter areas. Many perennials and few annuals occur in taiga.

31. Temperate deciduous forests are dominated by trees such as sugar maples, American basswoods, beeches, oaks, and hickories, with evergreens such as hemlocks and eastern white pines toward the northern and southeastern borders.

32. Grasslands are found toward the interiors of continental masses. Those located in Mediterranean climatic zones usually include vernal pools with unique annual floras. Many grasslands have been converted to agricultural use.

33. Deserts are characterized by low annual precipitation and wide daily temperature ranges, with plants adapted both in form and metabolism to the environment.

34. Mountain forests occupy much of the Pacific Northwest and extend south along the Rocky Mountains and California mountain ranges. The primarily coniferous tree species tend to be in zones determined by altitude. The forests have mostly dry summers, and some of the species associated with it have thick bark that protects the trees from fires, which occur frequently in this type of climate. Other species (e.g., knobcone pine) depend on fires for normal distribution and germination of seeds.

35. The tropical rain forests constitute nearly half of all forest land and contain more species of plants and animals than all the other biomes combined. Numerous woody plants and vines form multilayered canopies, which permit very little light to reach the floor. The soils are poor, with nutrients released during decomposition being rapidly recycled. The biome is being destroyed so rapidly that it will disappear within 20 years if the destruction is not halted.

Review Questions

1. How is acid rain formed?

2. What are the principal threats to global drinking water sources?

3. Distinguish among a plant community, a biotic community, an ecosystem, and a biome.

4. What is the difference between a primary consumer and a secondary consumer?

5. How are short food chains more efficient in the use of solar energy than are long ones?

6. Why is the diet of larger herbivorous animals varied?

7. What types of checks and balances exist in an ecosystem?

8. How are nitrogen and carbon recycled in nature?

9. What is primary succession as opposed to secondary succession?

10. What are general characteristics of climax vegetation as compared with nonclimax vegetation?

11. What is the greenhouse effect?

12. What is the ozone shield, and what significant role does it play? What threats are there to its existence?

13. Characterize desert and tundra biomes.

14. What is the function of fire in grassland and mountain forest biomes?

15. What kinds of plants predominate in tropical rain forests?

16. Why does the loss of the tropical forest biome have very serious implications for the future?

Discussion Questions

1. Humans have disrupted ecosystems almost everywhere they have established themselves, at least in industrialized countries. Do you believe that humans could also improve an ecosystem? Explain.

2. If a vegetarian diet makes more efficient use of solar energy, should we all strive to become vegetarians for this reason? Explain.

3. Besides the hypothetical example given (in which it was observed that eating lettuce directly was a more efficient use of solar energy than feeding it to hogs and then eating the pork), can you think of other ways in which food chains might be shortened?

4. Peanut butter has many of the nutrients needed in human nutrition. On the basis of what you have learned about the diet of animals in an ecosystem, do you think it would be a good idea to live on peanut butter and water as a means of saving money? Explain.

5. Could succession take place in an abandoned swimming pool? Explain.

6. Fire has been a natural phenomenon in several biomes for thousands of years, and most plant species are adapted to it in various ways in the ecosystem where it occurs regularly. Should we then not extinguish forest and grassland fires when they occur? Explain.

7. Much of the tropical rain forest biome is located in third world countries, which have precarious economies. Bearing that in mind, what could be done to halt the destruction of this priceless and vital part of our ecosystem?

Additional Reading

Barbour, M., et al. 1987. *Terrestrial plant ecology,* 2d ed. Menlo Park, CA: Benjamin/Cummings Publishing Co., Inc.

Barbour, M. G., and W. D. Billings. 1990. *North American terrestrial vegetation.* New York: Cambridge University Press.

Dafni, A. 1993. *Pollution ecology: A practical approach.* New York: Oxford University Press.

Daubenmire, R. 1978. *Plant geography.* San Diego, CA: Academic Press, Inc.

Fowden, L., et al. 1993. *Plant adaptation to environmental stress.* New York: Chapman and Hall.

Glenn-Lewin, D. C., et al. (Eds.). 1992. *Plant succession: Theory and prediction.* New York: Chapman and Hall.

Koopowitz, H., and H. Kaye. 1983. *Plant extinction: A global crisis.* Washington, DC: Stone Wall Press.

Kormondy, E. J. 1984. *Concepts of ecology,* 3d ed. Englewood Cliffs, NJ: Prentice-Hall.

Kozlowski, T. T., and C. E. Ahlgren (Eds.). 1974. *Fire and ecosystems.* San Diego, CA: Academic Press, Inc.

Rice, E. L. 1983. *Allelopathy,* 2d ed. San Diego, CA: Academic Press, Inc.

Silvertown, J. 1987. *Introduction to plant population ecology,* 2d ed. New York: Halsted Press.

Whitmore, T. C. 1990. *An introduction to tropical rain forests.* Fair Lawn, NY: Oxford University Press.

Appendix 1

Scientific Names of Organisms Mentioned in the Text

This is an alphabetical list of the organisms whose scientific names may not be mentioned in the text. Common and scientific names of organisms mentioned in Appendices 2 through 4 are provided within the respective appendices.

COMMON NAMES AND SCIENTIFIC NAMES OF ORGANISMS

COMMON NAME	SCIENTIFIC NAME	COMMON NAME	SCIENTIFIC NAME
Aardvark	*Orycteropus* spp.	Algae, yellow-green	members of Class Xanthophyceae, Division Chrysophyta, Kingdom Protoctista
Abrasives, horsetail source of	*Equisetum* spp.		
Absinthe liqueur, source of ingredients	*Pimpinella anisum, Artemisia absinthium,* and others	Almond	*Prunus amygdalus*
Aconite, source of	*Aconitum* spp.	Aloe juice, source of	*Aloe barbadensis, A. ferox, A. vera,* and others
Afghanistan pine	*Pinus eldarica*	'Ama'uma'u	*Sadleria cyatheoides*
Agar, source of	*Gelidium* spp. and others	Amaryllis	*Amaryllis* spp.
Agave	*Agave* spp.	Amoeba	*Amoeba proteus* and others
Air plant	*Kalanchoë* spp.	Amoeba, fungal internal parasites of	*Cochlonema verrucosum* and others
Alder	*Alnus* spp.		
Alfalfa	*Medicago sativa*	Amoeba, fungal trappers of	*Dactylella* spp. and others
Alfalfa caterpillar	*Colias philodice*	Anabaena	*Anabaena* spp.
Algae	members of Kingdom Protoctista—all divisions	Anemone	*Anemone* spp.
Algae, brown	members of Division Phaeophyta, Kingdom Protoctista	Angelica	*Angelica archangelica*
		Anise	*Pimpinella anisum*
		Annatto	*Bixa orellana*
Algae, coralline	*Bossiella* spp., *Corallina* spp., *Lithothamnion* spp., and others	Ant	*Formica* spp. and others
		Anteater	*Myrmecophaga jubata*
Algae, flatworm	*Platymonas* spp.	Ants, bullhorn *Acacia*	*Pseudomyrmex ferruginea*
Algae, golden-brown	members of Class Chrysophyceae, Division Chrysophyta, Kingdom Protoctista	Aphid	*Anuraphis* spp., *Aphis* spp., and others
		Apple	*Pyrus malus*
		Apricot	*Prunus armeniaca*
Algae, green	members of Division Chlorophyta, Kingdom Protoctista	Arbor-vitae	*Thuja occidentalis*
		Archaebacteria	members of Division Archaebacteria, Subkingdom Archaebacteriobionta, Kingdom Monera
Algae, platelike colonial green	*Pediastrum* spp., *Chaetopeltis* spp., and others		
		Arrowroot, Florida, source of	*Zamia floridana*
Algae, red	members of Division Rhodophyta, Kingdom Protoctista	Artichoke, globe	*Cynaria scolymus*
		Artichoke, Jerusalem	*Helianthus tuberosus*
		Arum Lily Family	Araceae
Algae, snowbank	*Chlamydomonas nivale* and others	Ash, Oregon	*Fraxinus latifolia*
		Ash, blue	*Fraxinus quadrangulata*
Algae, sponge	*Chlorella* spp., *Zoochlorella* spp.	Ash, white	*Fraxinus americana*

COMMON NAME	SCIENTIFIC NAME	COMMON NAME	SCIENTIFIC NAME
Fungi—used in manufacturing birth control pills	*Rhizopus nigricans, R. arrhizus*	**Fungus, white piedra**	*Trichosporon beigeli*
Fungi, bracket	*Fomes* spp., *Daedalea* spp., and others	**Fungus—used in manufacturing yellow food-coloring agent**	*Blakeslea trispora*
Fungi, brown fruit rot	*Monolinia fruticola* and others	**Funori, source of**	*Gloiopeltis* spp.
Fungi, cap-thrower	*Pilobolus* spp.	**Fur, green algae that inhabit animal**	*Trentepohlia* spp.
Fungi, cheese	*Penicillium camembertii* (for Camembert cheese), *P. roquefortii* (for blue, Gorgonzola, Roquefort, and Stilton cheeses)	**Gentian, source of**	*Gentiana* spp.
		Geranium	*Geranium* spp., *Pelargonium* spp.
		Geranium Family	Geraniaceae
		Gila monster	*Heloderma suspectum*
Fungi, citric acid-producing	*Aspergillus niger* and others	**Ginger**	*Zingiber officinale* and others
Fungi, flavor-producing	*Aspergillus* spp.	**Ginseng, source of**	*Panax* spp.
Fungi, hallucinogenic	*Amanita muscaria, Conocybe* spp., *Panaeolus* spp., *Psilocybe* spp., and others	**Giraffe**	*Giraffa camelopardalis*
		Gladiola	*Gladiolus* spp.
		Gloeocapsa	*Gloeocapsa* spp.
		Goat	*Capra* spp.
Fungi, horse dung	*Pilobolus* spp.	**Golden-brown algae**	members of Class Chrysophyceae, Division Chrysophyta, Kingdom Protoctista
Fungi, industrial alcohol-producing	*Aspergillus* spp.		
Fungi, insect-parasitizing	members of Order Laboulbeniales, Class Ascomycetes, Division Eumycota, Kingdom Fungi, and others	**Golden chain tree**	*Laburnum anagyroides*
		Goldenrod	*Solidago* spp.
		Goldenseal	*Hydrastis canadensis*
		Goldenweed	*Haplopappus gracilis*[4]
Fungi, jelly	*Auricularia* spp., *Exidia* spp., *Tremella* spp., and others	**Goose**	*Branta* spp. and others
		Gooseberry	*Ribes* spp.
Fungi, meat-tenderizing	*Thamnidium* spp.	**Goosefoot Family**	Chenopodiaceae
Fungi, ringworm	*Epidermophyton* spp., *Mircosporium* spp., *Trichophyton* spp.	**Gopher**	*Geomys* spp., *Thomomys* spp.
		Gopher plant	*Euphorbia lathyrus*
		Gopher, pocket	*Geomys bursarius* and others
Fungi, shelf—*see* Fungi, bracket		**Gourd**	*Lagenaria siceraria* and others
Fugi, *shoyu*	*Aspergillus oryzae, A. soyae*	**Grape**	*Vitis* spp.
Fungi—used in silvering of mirrors	*Aspergillus* spp.	**Grapefruit**	*Citrus paradisi*
		Grass	*Bromus* spp. and others[5]
Fungi—used in manufacturing soap	*Penicillium* spp.	**Grass, Bermuda**	*Cynodon dactylon*
		Grass, crested wheat	*Agropyron cristatum*
Fungi, soil	*Fusarium* spp. and others	**Grass Family**	Poaceae (Gramineae)
Fungi, soy sauce	*Aspergillus oryzae, A. soyae*	**Grass, Indian**	*Sorghastrum nutans*
Fungi, *sufu*	*Actinomucor elegans, Mucor* spp.	**Grass, pampas (Fig. 7.4)**	*Cortaderia selloana*
		Grass tree	*Xanthorrhea* spp.
Fungi, *teonanacatl* (sacred)	*Conocybe* spp., *Panaeolus* spp., *Psilocybe* spp., and others	**Greenbrier**	*Smilax* spp.
		Ground pine	*Lycopodium* spp.
		Ground pine—used for baby powder	*Lycopodium clavatum*
Fungus, bracket (Fig. 19.13 C)	*Polyporus sulphureus*	**Ground pine—used to arrest bleeding**	*Lycopodium clavatum*
Fungus, bracket (Fig. 19.19)	*Phacolus* sp.		
Fungus, chlorine-assimilating	*Aspergillus terreus*	**Ground pine, fossil relative of**	*Baragwanathia* spp., *Drephanophycus* spp., *Protolepidodendron* spp., and others
Fungus, cup (Fig. 19.7)	*Caloscypha fulgens*		
Fungus, "foolish seedling"	*Gibberella fujikuroi*		
Fungus, *miso*	*Aspergillus oryzae*	**Ground pine—used as intoxicant**	*Lycopodium selago*
Fungus—used in producing plastics	*Aspergillus terreus*		
Fungus, *tempeh*	*Rhizopus oligosporus*		
Fungus—used in manufacturing toothpaste	*Aspergillus niger*		

[4]Species with four chromosomes per cell.

[5]About 4,500 species of the Grass Family, Poaceae (Gramineae), in all.

COMMON NAME	SCIENTIFIC NAME	COMMON NAME	SCIENTIFIC NAME
Ground pine—used for ornaments	*Lycopodium clavatum,* *L. complanatum,* and others	Horsetail, giant	*Equisetum telmateia*
Ground pine—used to reduce fevers	*Lycopodium clavatum*	Horsetail, Hopi Indian flour source	*Equisetum laevigatum*
Guava	*Psidium guajava*	Horsetail—used as shampoo	*Equisetum hyemale*
Gum arabic, source of	*Acacia senegal* and others	Horsetail, treelike fossil	*Calamites* spp.
Gum tragacanth, source of	*Astragalus echidenaeformis, A. gossypinus, A. gummifer,* and others	Horsetail—used as water source	*Equisetum telmateia*
Guppy	*Lebistes reticulatus*	Hot springs, blue-green bacteria of	*Bacillosiphon induratus, Synechococcus* spp., and others
Hawk	*Buteo* spp., *Falco* spp., and others	"Human hair" slime mold	*Stemonitis* spp.
Hazelnut	*Corylus* spp.	Hummingbird	*Archilocus* spp. and others
Heath	*Erica* spp. and others	Hummingbird, Oasis (Fig. 23.15)	*Rhodopis vesper*
Heath Family	Ericaceae		
Hemlock, eastern	*Tsuga canadensis*	Hummingbirds, ferns used by (for nest material)	*Cyathea arborea, Lophosoria quadripinnata, Nephelea mexicana*
Hemlock, mountain	*Tsuga mertensiana*		
Hemlock, poison	*Conium maculatum*	Hummingbirds, tropical	*Chlorostilbon maugaeus* and others
Hemlock, water	*Cicuta* spp.		
Hemlock, western	*Tsuga heterophylla*	Hyacinth	*Hyacinthus* spp.
Hemp	*Cannabis sativa*	Hyacinth, grape	*Muscari* spp.
Hemp, Manila	*Musa textilis*	Hyacinth, water	*Eichhornia crassipes*
Hemp, Mauritius	*Furcraea gigantea*	Hyssop	*Hyssopus officinalis*
Henbit	*Lamium amplexicaule*	Ice plant	*Mesembryanthemum* spp.
Henna	*Lawsonia inermis*	India, toxic blue-green bacteria of	*Lyngbya majuscula*
Hepatica	*Hepatica* spp.		
Hepatica (Fig. 1.8; Fig. 24.3 C)	*Hepatica americana*	Indian pipe	*Monotropa* spp.
		Indian warrior	*Pedicularis densiflora*
Hickory	*Carya* spp.	Indigo	*Indigofera* spp.
Hog	*Sus scrofa* and others	Insects—*see individual entries*	
Hog fennel	*Lomatium* spp.		
Holly, American	*Ilex opaca*	Insects, fern used for treating stings and bites of	*Adiantum capillus-veneris*
Honeybee	*Apis mellifera*		
Hoopoe	*Upupa africana*		
Hop hornbeam	*Ostrya virginiana*	Ipecac, source of	*Cephaelis ipecacuanha*
Hops	*Humulus lupulus*	Iris	*Iris* spp.
Horehound	*Marrubium vulgare*	Iris, butterfly	*Moraea* sp.
Hornwort	*Anthoceros* spp.	Iris Family	Iridaceae
Horse	*Equus caballus*	Ironwood, South American	*Krugiodendron ferreum*
Horseradish	*Rorippa armoracia*	Ivy, Boston	*Parthenocissus tricuspidata*
Horsetail	*Equisetum* spp.	Ivy, English	*Hedera helix*
Horsetail (Fig. 21.10 A)	*Equisetum hyemale*	Ivy, poison	*Toxicodendron radicans*
Horsetail (Fig. 21.10 B)	*Equisetum telmateia*	Jacaranda	*Jacaranda* spp.
Horsetail(s)—used as abrasive	*Equisetum* (all spp.)	Jaeger	*Stercorarius* spp.
		Jicama	*Pachyrhizus* spp.
Horsetail(s)—used as astringent	*Equisetum arvense, E. debile,* and others	Jimson weed	*Datura* spp.
		Jimson weed (Fig. 8.6)	*Datura stramonium*
Horsetail(s)—used for treating burns	*Equisetum hyemale* and others	Jojoba	*Simmondsia californica*
		Joshua tree	*Yucca brevifolia*
Horsetail—used for treating diarrhea	*Equisetum hyemale*	Jumping mouse	*Zapus hudsonius, Napaeozapus insignis*
Horsetail(s)—used as diuretic	*Equisetum arvense, E. debile,* and others	Junco	*Junco* spp.
		Junco, slate-colored	*Junco hyemalis*
Horsetail—used for treating dysentery	*Equisetum hyemale*	Juneberry	*Amelanchier* spp.
		Juniper	*Juniperus* spp.
Horsetail, field	*Equisetum arvense*	Juniper, dwarf	*Juniperus communis* and others
Horsetail(s), fossil	*Equisetites* spp., *Hyenia* spp., *Sphenophyllum* spp., and others		
		Jute	*Corchorus* spp.

COMMON NAME	SCIENTIFIC NAME	COMMON NAME	SCIENTIFIC NAME
Turtle	*Chelydra* spp., *Chrysemys* spp., and others	**Wattle**	*Acacia decurrens* and others
Twinflower	*Linnaea borealis*	**Weaver birds**	*Anaplectes* spp., *Hyphantoris* spp., and others
Ulothrix	*Ulothrix* spp.	**Webworm, fall**	*Hyphantria cunea*
Ultraviolet light, flowers seen in (Fig. 23.13)	*Rudbeckia* sp.	**Welwitschia**	*Welwitschia mirabilis*
		Whale, sperm	*Physeter catodon*
Unicorn plant	*Proboscoidea* spp.	**Wheat**	*Triticum* spp. and their hybrids
Venus flytrap	*Dionaea muscipula*		
Vetch	*Vicia* spp.	**Wheel tree**	*Trochodendron aralioides*
Vetchling, yellow	*Lathyrus aphaca*	**Whisk fern**	*Psilotum* spp.
Vinegar weed	*Trichostema* spp.	**Whisk fern, fossil relatives of**	*Asteroxylon* spp., *Psilophyton* spp., *Rhynia* spp., and others
Violet	*Viola* spp.		
Violet, African	*Saintpaulia* spp.		
Violet, gold	*Viola douglasii*	**Whisk fern, living relatives of**	*Tmesipteris* spp.
Virginia creeper	*Parthenocissus quinquefolia*		
Virus[15]		**White pine blister rust**	*Cronartium ribicola*
Vole	*Microtus* spp. and others	**Willow**	*Salix* spp.
Wahoo	*Euonymus alata* and others	**Willow Family**	Salicaceae
Wake-robin	*Trillium* spp.	**Window leaves, plants with**	*Fenestraria* spp. and others
Wallflower, western	*Erysimum capitatum*	**Wintergreen oil, sources of**	*Gaultheria procumbens* and others
Walnut	*Juglans* spp.		
Walnut, black	*Juglans nigra, J. hindsii*	**Wisteria**	*Wisteria sinensis* and others
Warbler	*Dendroica* spp. and others	**Witch hazel**	*Hamamelis virginiana*
Watercress	*Nasturtium officinale*	**Woad, dyer's**	*Isatis tinctoria*
Water fern, oriental	*Ceratopteris thalictroides*	**Woadwaxen**	*Genista tinctoria*
Water molds	member of Division Oomycota, Subkingdom Mastigobionta, Kingdom Protoctista	**Wolfsbane**	*Aconitum vulparia*
		Wolverine	*Gulo luscus*
		Woodpecker	*Dendrocopus* spp. and others
		Wormwood	*Artemisia absinthium*
Water net	*Hydrodictyon* spp.	**Wounds, ferns used for treating**	*Lygodium circinatum, Ophioglossum vulgatum*
Water silk	*Spirogyra* spp.		
Water weed	*Elodea* spp.	**Yam**	*Dioscorea* spp.
Water weed, yellow	*Ludwigia repens*	**Yareta**	*Azorella yareta*
Watermelon	*Citrullus vulgaris*	**Yarrow, American**	*Achillea lanulosum*
		Yarrow, European	*Achillea millefolium*
		Yeast	*Saccharomyces* spp.
		Yew	*Taxus* spp.
		Yucca	*Yucca* spp.
		Zebra	*Equus zebra* and others
		Zinnia	*Zinnia elegans* and others

[15]Depending on the classification used, viruses may not have a scientific name. Many are named after the disease they cause; e.g., tobacco mosaic virus causes tobacco mosaic disease. One classification attempts to give them at least a Latin prefix, so the virus for warts is *Papavovirus;* for smallpox, *Poxvirus;* for polio, *Picornavirus;* for measles and mumps, *Paramyxovirus.*

Appendix 2

Biological Controls

If you were to ask the average farmer or backyard gardener how to control a particular insect or plant pest, you might be given the name of some poisonous spray or bait that has proved "effective" in the past. Evidence that spraying with such substances yields only temporary results, however, has been mounting for many years, and the spraying is frequently followed by even larger invasions of pests. Also, the residues of poisonous sprays often accumulate in the soil and disrupt the microscopic living flora and fauna essential to the soil's health. The problem is compounded and the ecology further upset when large amounts of inorganic fertilizers are added. As increasing numbers of people become aware of the devastating effects of pesticides and herbicides on the environment, they have turned to **biological controls** as an alternative to the use of poisonous sprays. To the surprise of some, such controls are often more effective than traditional controls.

Poisonous sprays often promote pest invasions because the sprays usually kill beneficial insects along with the undesirable ones. In addition, the pests, through mutations, often become resistant to the sprays. In undisturbed natural areas, weeds are never a problem, and even though pests may be present they seldom destroy the ecological community. Why is this so? You'll recall from Chapter 25 that all members of a community are in ecological balance with one another. The plants produce a variety of substances that may either repel or attract insects, inhibit or promote the growth of other plants, and generally contribute to the health of the community as a whole.

Virtually all insects have their own pests and diseases, as do most other living organisms. Each pest ensures, at least indirectly, that the various species of a community are perpetuated. This principle of nature can be applied, to a certain extent, to farming and gardening. The following are some general and specific biological controls that either are now in widespread use or are in various tests showing promise for the future.

GENERAL CONTROLS

Establishment of Beneficial Insects

Ladybugs (Family Coccinellidae)
The small and often colorful beetles called *ladybugs,* and particularly their larval stages, consume large numbers of aphids, thrips, insect eggs, weevils, and other pests. They are obtainable from various commercial sources (e.g., Bozeman Bio-Tech, P.O. Box 3146, Bozeman, MT 59772; Arbico, P.O. Box 4247, Tucson, AZ 85738), but, if given a chance, they probably will establish themselves without being imported. When obtained from outside of the local area, they should be placed in groups at the bases of plants on which pests are present, preferably in the early evening after watering.

Lacewings (Families Chrysopidae and Hemerobiidae)
Lacewings are slow-flying, delicate-winged insects that consume large numbers of aphids, mealybugs, and other pests. They lay their eggs on the undersides of leaves, each egg being borne at the tip of a slender stalk. The larvae consume the immature stages of leafhoppers, bollworms, caterpillar eggs, mites, scale insects, thrips, aphids, and other destructive pests. Commercial sources include Arbico, P.O. Box 4247, Tucson, AZ 85738, and All Pest Control, 6030 Grenville Lane, Lansing, MI 48910.

Appendix 3

Useful Plants and Poisonous Plants

Wild Edible Plants	Medicinal Plants	Dye Plants
Words of Caution	Hallucinogenic Plants	Additional Reading
Poisonous Plants	Spice Plants	

WILD EDIBLE PLANTS

Words of Caution

Literally thousands of native and naturalized plants, or at least parts of them, have been used for food and other purposes by Native Americans and the immigrants who came later from other quarters of the globe. Table A3.1 has been compiled from a variety of sources; I have had opportunities to sample only a fraction of these plants myself and thus cannot confirm the edibility of all of the plants listed. *The reader is cautioned to be certain of the identity of a plant before consuming any part of it.* Cow parsnip (*Heracleum*

lanatum) and water hemlocks (*Cicuta* spp.), for example, resemble each other in general appearance, but although cooked roots of cow parsnip have been used for food for perhaps many centuries, those of water hemlocks are very poisonous and have caused many human fatalities.

As was indicated in Chapter 21, many species of organisms are now on rare and endangered species lists, and a number of them are doomed to extinction within the next few years. Although the wild edible plants discussed here are not included on such lists, it might not take much indiscriminate gathering to endanger their existence as well. Because of this, one should exercise the following rule of thumb: *Never reduce a population of plants by more than 10% when collecting them for any purpose!*

Table A3.1

WILD EDIBLE PLANTS

PLANT	SCIENTIFIC NAME	USES
Amaranth	*Amaranthus* spp.	Young leaves used like spinach; seeds ground with others for flour.
Arrow grass	*Triglochin maritima*	Seeds parched or roasted. (***Caution***: *The plant is otherwise poisonous.*)
Arrowhead	*Sagittaria latifolia*	Tubers used like potatoes.
Balsamroot	*Balsamorhiza* spp.	Whole plant edible, especially when young, either raw or cooked.
Basswood	*Tilia* spp.	Fruits and flowers ground together to make a paste that can serve as a chocolate substitute; winter buds edible raw; dried flowers used for tea.
Bedstraw	*Galium aparine*	Roasted and ground seeds make a coffee substitute.
Beechnuts	*Fagus grandifolia*	Seeds used like nuts; oil extracted from seeds for table use.
Biscuit root	*Lomatium* spp.	Roots eaten raw or dried and ground into flour; seeds edible raw or roasted.
Bitterroot	*Lewisia rediviva*	Outer coat of the bulbs should be removed to eliminate the bitter principle; bulbs are then boiled or roasted.
Blackberry (wild)	*Rubus* spp.	Fruits edible raw and in pies, jams, and jellies.
Black walnut	*Juglans nigra*	Nut meats highly edible.
Bladder campion	*Silene cucubalus*	Young shoots (less than 5 cm tall) cooked as a vegetable.
Blueberry	*Vaccinium* spp.	Fruits edible raw, frozen, and in pies, jams, and jellies.
Bracken fern	*Pteridium aquilinum*	Young uncoiling leaves (fiddleheads) cooked like asparagus; rhizomes also edible but usually tough. (***Caution***: *Recent evidence indicates that frequent consumption of bracken fern can cause cancer of the intestinal tract.*)

Table A3.1

WILD EDIBLE PLANTS

PLANT	SCIENTIFIC NAME	USES
Broomrape	*Orobanche* spp.	Entire plant eaten raw or roasted.
Bulrush (Tule)	*Scirpus* spp.	Roots and young shoot tips edible raw or cooked; pollen and seeds also edible.
Butternut	*Juglans cinerea*	Nut meats edible.
Caraway	*Carum carvi*	Young leaves in salads; seeds for flavoring baked goods and cheeses.
Cattail	*Typha* spp.	Copious pollen produced by flowers in early summer is rich in vitamins and can be gathered and mixed with flour for baking; rhizomes can be cooked and eaten like potatoes.
Chicory	*Cichorium intybus*	Leaves edible raw or cooked; dried, ground roots make a coffee substitute.
Chokecherry	*Prunus virginiana*	Fruits make excellent jelly or can be cooked with sugar for pies and cobblers.
"Coffee" (wild)	*Triosteum* spp.	Berries dried and roasted make a coffee substitute.
Common chickweed	*Stellaria media*	Plants cooked as a vegetable.
Corn lily	*Clintonia borealis*	Youngest leaves can be used as a cooked vegetable.
Cow parsnip	*Heracleum lanatum*	Roots and young stems cooked.
Cowpea	*Vigna sinensis*	"Peas" and young pods cooked as a vegetable (plant naturalized in southern United States).
Crab apple	*Pyrus* spp.	Jelly made from fruits.
Crowberry	*Empetrum* spp.	Fruits should first be frozen then cooked with sugar.
Dandelion	*Taraxacum officinale*	Leaves rich in vitamin A; dried roots make a coffee substitute; wine made from young flowers.
Dock	*Rumex* spp.	Leaves cooked like spinach; tartness of leaves varies from species to species and sometimes from plant to plant—tart forms should be cooked in two or three changes of water.
Douglas fir	*Pseudotsuga menziesii*	Cambium and young phloem edible; tea made from fresh leaves.
Elderberry	*Sambucus* spp.	Fresh flowers used to flavor batters; fruits used in pies, jellies, wine. (*Caution*: Other parts of the plant are poisonous.)
Evening primrose	*Oenothera hookeri, O. biennis,* and others	Young roots cooked.
Fairy bells	*Disporum trachycarpum*	Berries edible raw.
Fennel	*Foeniculum vulgare*	Leaf petioles edible raw or cooked.
Fireweed	*Epilobium angustifolium*	Young shoots and leaves boiled as a vegetable.
Ginger (wild)	*Asarum* spp.	Rhizomes can be used as substitute for true ginger.
Gooseberry	*Ribes* spp.	Berries edible cooked, dried, or raw; make excellent jelly.
Grape (wild)	*Vitis* spp.	Berries usually tart but can be eaten raw; make good jams and jellies.
Grass	Many genera and species	Seeds of most can be made into flour; rhizomes of many perennial species can be dried and ground for flour.
Greenbrier	*Smilax* spp.	Roots dried and ground; refreshing drink made with ground roots, sugar, and water.
Groundnut	*Apios americana*	Tubers cooked like potatoes.
Hawthorn	*Crataegus* spp.	Fruits edible raw and in jams and jellies.
Hazelnut	*Corylus* spp.	Nuts edible raw or roasted.
Hickory	*Carya* spp.	Nuts edible.
Highbush cranberry	*Viburnum trilobum*	Fruits make excellent jellies and jams.
Huckleberry	*Vaccinium* spp.	Berries eaten raw or in jams and jellies.
Indian paintbrush	*Castilleja* spp.	Flowers of many species edible. (*Caution*: In certain soils, plants absorb toxic quantities of selenium.)
Indian pipe	*Monotropa* spp.	Whole plant edible raw or cooked.
Juniper	*Juniperus* spp.	"Berries" dried, ground, and made into cakes.
Labrador tea	*Ledum* spp.	Tea made from young leaves.
Lamb's quarters	*Chenopodium album*	Leaves and young stems cooked as a vegetable.
Licorice	*Glycyrrhiza lepidota*	Roots edible raw or cooked.
Mallow	*Malva* spp.	Leaves and young stems used as vegetable (use only small amounts at one time).
Manzanita	*Arctostaphylos* spp.	Berries eaten raw, used in jellies or pies, or made into "cider."

PLANTS ASSOCIATED WITH MEDICINAL USES

PLANT
Angel's trump
Azalea
Baneberry
Belladonna
Black cherry
Black locust
Black snakero
Buckeye
Caladium
Carolina jessa
Castor bean
Chinaberry
Daphne
Death camas
Dieffenbachia
Duranta
English ivy
False hellebor
Foxglove
Golden chain
Jequirity bean
Jimson weed
Lantana
Lily of the vall
Lobelia
Mistletoe
Monkshood
Moonseed
Mountain laur
Mushrooms
Nightshade
Oleander
Poke
Rhododendron
Rhubarb
Rubber vine
Sandbox tree
Tansy
Tung tree
Water hemlock
White snakero
Yellow oleande
Yew

PLANT	SCIENTIFIC NAME	USES
Ginkgo	*Ginkgo biloba*	Evidence that concentrated leaf extract improves oxygen-carrying capacity of capillaries, especially those of the brain, and improves memory.
Ginseng	*Panax* spp.	Considered a general panacea, especially in the Orient.
Goldenseal	*Hydrastis canadensis*	Rhizome source of alkaloidal drugs used in treatment of inflamed mucous membranes; also used as tonic.
Goldthread	*Coptis groenlandica*	Native Americans boiled plant and gargled the liquid for sore or ulcerated mouths.
Gotu kola	*Cola nitida*	Seeds contain up to 3.5% caffeine and 1% theobromine which may lessen fatigue; related to kola nuts of cola fame.
Green hellebore	*Helleborus viridis*	Plant extract used to treat hypertension (drug now synthesized); Thompson Indians used in small amounts to treat syphilis.
Green tea	*Camellia sinensis*	Unfermented leaves source of polyphenols, which appear to reduce incidence of cancers in regular users.
Hemlock	*Tsuga* spp.	Native Americans made tea of inner bark to treat colds and fevers.
Horehound	*Marrubium vulgare*	Extract from dried tops of plants used in lozenges for relief of sore throats and colds.
Horsetail	*Equisetum* spp.	Plants boiled in water; liquid used as a delousing shampoo or as a gargle for mouth ulcers.
Indigo (wild)	*Baptisia tinctoria*	Native Americans boiled plant and used liquid as an antiseptic for skin sores.
Ipecac	*Cephaelis ipecacuana*	Drug from roots and rhizome used to treat ameobic dysentery; also used as an emetic.
Jimson weed	*Datura* spp.	Drugs atropine, hyoscyamine, stramonium, and scopolamine obtained from seeds, flowers, and leaves. Stramonium used for knockout drops and in treatment of asthma (*see* Deadly nightshade).
Joe-pye weed	*Eupatorium purpureum*	Dried root said to prevent formation of gallstones.
Joshua tree	*Yucca brevifolia*	Cortisone and estrogenic hormones made from sapogenins produced in the roots.
Juniper	*Juniperus* spp.	Tea of "berries" drunk by Zuni Indian women to relax muscles following childbirth.
Kansas snakeroot	*Echinacea angustifolia*	Dried roots used as antiseptic in treatment of sores and boils, periodontal disease, and sinus drainage problems.
Kava	*Piper methysticum*	Leaf tea used as sedative, muscle relaxant, and pain reliever.
Licorice	*Glycyrrhiza glabra*	Rhizomes source of licorice used in cough drops and for the soothing of inflamed mucous membranes.
Lily of the valley	*Convallaria majalis*	All parts of plant contain a heart stimulant similar to digitalis.
Lobelia	*Lobelia inflata*	Drug lobeline sulphate, obtained from dried leaves, used in antismoking preparations and in treatment of respiratory disorders.
Mandrake	*Mandragora officinarum*	Extracts of plant used in folk medicine as painkiller (drugs hyoscyamine, podophyllin, and mandragorin have been isolated; podophyllin used experimentally in treatment of paralysis).
Manroot	*Marah* spp.	Native Americans used oil from seeds to treat scalp problems and crushed roots for relief from saddle sores.
Marginal fern	*Dryopteris marginalis*	Rhizomes contain oleoresin, used in expulsion of tapeworms from intestinal tract.
Marijuana	*Cannabis sativa*	Tetrahydrocannabinol obtained from resinous hairs in inflorescences; ancient medicinal drug of China.
Mayapple	*Podophyllum peltatum*	Podophyllin, obtained from roots, used experimentally in treatment of paralysis; dried root power used on warts.
Maypop	*Passiflora incarnata*	Dried leaves used as sedative; Native Americans used juice as treatment for sore eyes.
Mesquite	*Prosopis glandulosa*	Native Americans mixed dry leaf powder with water and used liquid to treat sore eyes.

Table A3.4

PLANTS ASSOCIATED WITH MEDICINAL USES

PLANT	SCIENTIFIC NAME	USES
Mexican yam	*Dioscorea floribunda*	Tuberous roots produce up to 10% diosgenin, a precursor of progesterone and cortisone, and are source of DHEA (dihydroepiandrosterone), a complex hormone naturally produced by humans. DHEA levels decline with aging; there is some evidence that controlled DHEA supplementation in older persons retards some aspects of aging.
Milk thistle	*Silybum marianum*	Silymarin extracted from plants has antioxidant properties that appear to be especially beneficial to liver.
Milkweed	*Asclepias syriaca*	Quebec Indians promoted temporary sterility by drinking infusion of pounded roots.
Mistletoe	*Phoradendron flavescens*	Mistletoe.
Monkshood	*Aconitum napellus*	Source of aconite once used in treatment of rheumatism and neuralgia.
Mulberry	*Morus rubra*	Rappahannock Indians applied milky latex of leaf petioles to scalp for ringworm.
Mullein	*Verbascum thapsus*	Native Americans smoked leaves for respiratory ailments and asthma; flowers once widely used in cough medicines.
Onion (wild)	*Allium* spp.	Bulbs eaten to prevent scurvy; Cheyenne Indians applied bulbs in poultice to boils.
Opium poppy	*Papaver sominferum*	Morphine and codeine obtained from latex of immature fruits.
Oregon grape	*Berberis aquifolium*	Bark tea drunk by Native Americans to settle upset stomach; used in strong doses for treatment of venereal diseases.
Pacific yew	*Taxus brevifolia*	*Taxol*, promising anticancer agent, extracted from bark.
Pansy (wild)	*Viola* spp.	Plants ground up and applied to skin sores.
Pennyroyal	*Mentha pulegium*	Native Americans used leaf tea in small amounts for relief of headaches and flatulence and to repel chiggers. (Toxic in larger amounts.)
Persimmon	*Diospyros virginiana*	Liquid from boiled fruit used as astringent. Fruits have high beta carotene content; leaves said to be high in vitamin C.
Peyote	*Lophophora williamsii*	Alcoholic extract of plant used as antibiotic.
Pine	*Pinus* spp.	Pycnogenols extracted from bark have powerful antioxidant properties.
Pinkroot	*Spigelia marilandica*	Powdery root very effective in expulsion of roundworms from intestinal tract.
Pipssisewa	*Chimaphila umbellata*	Native Americans steeped plant in water and used liquid to draw out blisters.
Pitcher plant	*Sarracenia purpurea*	Native Americans used root widely as smallpox cure (records indicate it was effective).
Plantain	*Plantago ovata* and other spp.	Seed husks (known as *psyllium*) widely used in bulking laxatives.
Pleurisy root	*Asclepias tuberosa*	Liquid from roots boiled in water used in treatment of respiratory problems.
Prickly ash	*Zanthoxylum americanum*	Bark and berries widely used by Native Americans for toothache (pieces inserted in cavities); liquid infusion drunk for venereal diseases.
Purple coneflower	*Echinacea purpurea*	Plant extracts used to boost the immune system.
Quassia	*Picraea excelsa; Quassia amara*	Wood extract used as pinworm remedy and as insecticide.
Rauwolfia	*Rauwolfia serpentina*	Reserpine obtained from roots; drug used in treatment of mental illness and in counteracting effects of LSD.
Saffron (meadow)	*Colchicum autumnale*	Drug colchicine from corms used in past for treatment of gout and back disc problems but now used mostly for experimental doubling of chromosome numbers in plants.
Sarsaparilla	*Aralia nudicaulis*	Cough medicines made from roots.
Sassafras	*Sassafras albidum*	Tea of root bark used to induce sweating; used externally as liniment.
Self-heal	*Prunella vulgaris*	Native Americans applied plants in poultices to boils.
Seneca snakeroot	*Polygala senega*	Bark boiled in water and then liquid applied to snakebites. Taken internally as abortifacient; used in cough remedy.

Table A3.4

PLANTS ASSOCIATED WITH MEDICINAL USES

PLANT	SCIENTIFIC NAME	USES
Senna	*Cassia senna* and other spp.	Leaf extract used as purgative.
Skeleton weed	*Lygodesmia juncea*	Widely used by Native American women to increase milk flow.
Skullcap	*Scutellaria laterifolia*	Dried plant used as anticonvulsive in treatment of epilepsy and as sedative.
Slippery elm	*Ulmus fulva*	Dried inner bark, which contains an aspirinlike substance, used to soothe inflamed membranes.
Spicebush	*Lindera benzoin*	Berries, buds, and bark brewed for tea used to reduce fevers.
Spruce	*Picea* spp.	Cree Indians ate small immature female cones for treatment of sore throat.
Squills	*Urginea maritima*	Bulbs of red variety are source of heart stimulant; bulbs of white variety are widely used as rodent killer.
Stoneseed	*Lithospermum ruderale*	Shoshoni women reported to have drunk cold water infusion of roots daily for six months to ensure permanent sterility (experiments with mice suggest substance to the reports).
Strophanthus	*Strophanthus* spp.	Seeds are major source of cortisone and also source of heart stimulant.
Strychnine plant	*Strychnos nox-vomica*	Strychnine extracted from seeds widely used as insect and animal poison; principal ingredient in blowgun darts used by South American aborigines. Minute amounts stimulate the central nervous systems and relieve paralysis.
Sumac	*Rhus* spp. (especially *R. glabra*)	Native Americans applied leaf decoction as remedy for frostbite; fruits and liquid made from leaves applied to poison ivy rash and gonorrhea sores. Root chewed for treatment of mouth ulcers.
Sweet flag	*Acorus calamus*	Boiled root applied to burns; root chewed for relief of colds and toothache.
Sweet gum	*Liquidambar styraciflua*	Bud balsam used to treat chigger bites; balsam also used in insect fumigating powders.
Sword fern	*Polystichum munitum*	Boiled rhizome used by Native Americans to treat dandruff; sporangia applied to burns.
Tamarind	*Tamarindus indica*	Fruit pulp used as laxative.
Turmeric	*Curcuma longa*	Rhizome extracts used to lower cholesterol levels and to prevent blood clots.
Valerian	*Valeriana septentrionalis*	Pulverized plant applied to wounds.
Velvet bean	*Mucuna* spp.	Seeds contain L-dopa used in treatment of Parkinson's disease.
Virginia snakeroot	*Aristolochia serpentaria*	Native Americans used tea of plant for reducing high fevers.
Wahoo	*Euonymus atropurpureus*	Bark steeped in water; liquid has digitalislike effect on heart.
Watercress	*Rorippa nasturtium-aquaticum*	Some evidence that daily consumption reduces development of lung cancer in smokers.
Western wallflower	*Erysimum capitatum*	Zuni Indians ground plant with water and applied it to skin to prevent sunburn.
Willow	*Salix* spp.	Chickasaw Indians snuffed infusion of roots as remedy for nosebleed; Pomo Indians boiled bark in water and applied liquid for relief of skin itches. Fresh inner bark contains salicin, an aspirinlike compound used to reduce fevers.
Wintergreen	*Gaultheria procumbens*	Oil from leaves used as folk remedy for body aches and pains.
Witch hazel	*Hamamelis virginiana*	Oil distilled from twigs used as external medicine.
Wormseed	*Chenopodium ambrosioides*	Oil from seeds used to expel intestinal worms.
Wormwood	*Artemisia* spp.	Yokia Indians made tea from leaves to treat bronchitis; other Native Americans used tea as cold remedy.
Yarrow	*Achillea millefolium*	Native Americans used plant infusion for treating wounds, earaches, and burns.
Yellow lady's slipper	*Cypripedium calceolus*	Dried root used for relief of insomina and as sedative.
Yellow nut grass	*Cyperus esculentus*	Paiute Indians pounded tubers with tobacco leaves and applied mass in wet dressing for treatment of athlete's foot.
Yerba santa	*Eriodictyon californicum*	Native Americans smoked leaves or drank leaf tea for treatment of colds or asthma.

HALLUCINOGENIC PLANTS

Although a few hallucinogenic substances produced by animals have been isolated and some have been synthesized, the majority of known hallucinogens are produced by plants. Table A3.5 is not a complete list, but it includes the better-known sources. Refer to *Additional Reading* for further information.

SPICE PLANTS

The word *spice* describes any aromatic plant or part of a plant used to flavor or season food; spices are also used to add scent or flavor to manufactured products (Table A3.6). Although spices have no nutritional value, they add a pleasurable zest to meals, and before food preservation was possible, they helped make unappealing but still edible food palatable.

The value placed on spice plants was responsible for changing the course of Western civilization as a principal motive behind the voyages of discovery.

Table A3.5

HALLUCINOGENIC SUBSTANCES PRODUCED BY PLANTS

PLANT	SCIENTIFIC NAME	PART USED	PRINCIPAL ACTIVE SUBSTANCE
Ajuca	*Mimosa hostilis*	Roots	Nigerine
Belladonna	*Atropa belladonna*	Leaves	Hyoscyamine, scopolamine
Caapi	*Banisteriopsis caapi*	Wood	Harmine
Canary broom	*Cytisus canariensis*	Seeds	Cytisine
Catnip	*Nepeta cataria*	Leaves	Unknown
Cohoba	*Piptadena peregrina*	Seeds (snuff)	Tryptamines
Coral bean	*Erythrina* spp.	Seeds	Unknown
Cubbra borrachera	*Methysticodendron amnesianum*	Leaves	Scopolamine
Ergot fungus	*Claviceps purpurea*	Rhizomorph	Ergine (LSD)
Fly agaric	*Amanita muscaria*	Mushroom cap	Ibotenic acid, muscimol
Henbane	*Hyoscyamus* spp.	Leaves	Hyoscyamine, scopolamine
Iboga	*Tabernanthe iboga*	Root bark	Ibogaine
Jimson weed	*Datura* spp.	All parts	Scopolamine
Kava kava	*Piper methysticum*	Root (large amounts of beverage produce hallucinations)	Myristicinlike compound
Mace	*Myristica fragrans*	Aril of seed	Myristicin
Mescal bean	*Sophora secundiflora*	Seeds	Cytisine
Morning glory	*Ipomoea violacea*	Seeds	Ergine
Nutmeg	*Myristica fragrans*	Seeds	Myristicin
Ololiuqui	*Rivea corymbosa*	Seeds	Tubicoryn
Peyote	*Lophophora williamsii*	Stems	Mescaline
Psilocybe mushrooms	*Psilocybe* spp, *Concocybe* spp., *Panaeolus* spp., and others	All parts	Psilocybin, psilocin
Rape dos Indios	*Maquira sclerophylla*	Dried plant (snuff)	Unknown
San Pedro	*Trichocereus pachanoi*	Stems	Mescaline
Sassafras	*Sassafras albidum*	Root bark (large amounts of tea)	Safrole
Sweet flag	*Acorus calamus*	Dried root	Asarone, β-asarone
Syrian rue	*Peganum harmala*	Seeds	Harmine
Vygie	*Mesembryanthemum expansum*	All parts	Mesebrine
Wood rose	*Argyreia nervosa*	Seeds	Ergoline alkaloids
Yakee (Parica)	*Virola* spp.	Resin from inner surface of freshly removed bark (snuff)	Tryptamine
Yohimbehe	*Corynanthe* spp.	Bark	Yohimbine

Table A3.6

PLANTS USED TO SEASON OR FLAVOR

SPICE	SCIENTIFIC NAME OF PLANT	PARTS USED; REMARKS	PRINCIPAL SOURCE
Allspice	*Pimenta dioica*	Powdered dried fruit.	Jamaica
Almond	*Prunus amygdalus*	Oil from seed used for flavoring baked goods.	Mediterranean; U.S.
Angelica	*Angelica archangelica*	Stems candied; oil from seeds and roots used in liqueurs.	Europe; Asia
Anise	*Pimpinella anisum*	Oil distilled from fruits used for flavoring.	Widely cultivated
Arrowroot	*Maranta arundinacea*	Powdered root used in milk puddings, baked goods.	South America
Asafoetida	*Ferula asafoetida*	Powdered gum from stems and roots used in minute quantities with fish.	Middle East
Balm (Melissa)	*Melissa officinalis*	Oil from leaves used in beverages; leaves used as food flavoring.	U.S.; Mediterranean
Basil	*Ocimum basilicum*	Leaves used in meat dishes, soups, sauces.	Mediterranean
Bay	*Laurus nobilis*	Leaves used in soups, sauces.	Europe
Bell pepper	*Capsicum frutescens*	Dried diced fruit used in chip dips, salad dressings.	Widely cultivated
Bergamot	*Monarda didyma*	Leaves used with pork. (*Note*: A perfume oil obtained from a variety of orange—*Citrus aurantium* var. *bergamia* — is also called *bergamot*.)	North America (*Monarda*); Italy (*Citrus*)
Black pepper	*Piper nigrum*	Dried fruits used as condiment.	India; Indonesia
Borage	*Borago officinalis*	Leaves used as beverage flavoring.	England
Burnet	*Sanguisorba minor*	Used in soups and casseroles.	Eurasia
Calamus	*Acorus calamus*	Powdered rhizome used for flavoring.	Europe; Asia; North America
Capers	*Capparis spinosa*	Flower buds used for flavoring relishes, pickles, sauces.	Mediterranean
Caraway	*Carum carvi*	Seeds used in breads, cheeses; seed oil used in the liqueur kümmel.	North America; Europe
Cardamom	*Elletaria cardamomum*	Dried fruit and seeds used for flavoring baked goods. (*Note*: Several false cardamoms—*Amomum* spp.—are sold commercially.)	India; Sri Lanka; Central America
Cassia	*Cinnamomum cassia*	Powdered bark used as cinnamon substitute.	Southeast Asia
Cayenne pepper	*Capsicum* spp.	Powdered dried fruits used in chili powder, Tabasco sauce.	American tropics
Celery	*Apium graveolens*	Seeds used in celery salt, soups.	Europe; U.S.
Chervil	*Anthriscus cerefolium*	Used as parsley substitute.	Europe; Near East
Chives	*Allium schoenoprasum*	Leaves and bulbs used with sour cream, butter.	Widely cultivated
Chocolate	*Theobroma cacao*	Ground seeds used for flavoring.	Africa; South America
Cilantro	*Coriandrum sativum*	Leaves used in avocado dip and with poultry.	Europe
Cinnamon	*Cinnamomum zeylanicum*	Ground bark used for flavoring baked goods; oil from leaves used as flavoring, clearing agent.	Seychelles; Sri Lanka
Citrus	*Citrus* spp.	Fruits, especially rinds, source of flavoring oil.	Mediterranean; South Africa; U.S.
Cloves	*Syzgium aromaticum*	Dried flower buds used as flavoring.	Zanzibar

Table A3.6

PLANTS USED TO SEASON OR FLAVOR

SPICE	SCIENTIFIC NAME OF PLANT	PARTS USED; REMARKS	PRINCIPAL SOURCE
Coffee	*Coffea arabica*	Roasted seeds source of mocha-coffee flavoring.	Tropics
Coriander	*Coriandrum sativum*	Ground seed used in German frankfurters, curry powders.	Mediterranean
Cubebs	*Piper cubeba*	Dried fruits used as seasoning.	East Indies
Cumin	*Cuminum cyminum*	Ground seeds used with meats, pickles, cheeses, curry.	Mediterranean
Curry		Spicy condiment containing several ingredients, such as turmeric, cumin, fenugreek, and zedoary.	India
Dill	*Anethum graveolens*	Seeds used in pickling brines; leaves used for seasoning meat loaves, sauces.	Europe; Asia
Dittany	*Origanum dictamnus*	Leaves used as seasoning for poultry, meats.	Crete
Eucalyptus	*Eucalyptus* spp.	Oil from leaves used in toothpastes, flavoring agents.	Australia
Fennel	*Foeniculum vulgare*	Seeds used in baked goods.	Europe
Fenugreek	*Trigonella foenumgraecum*	Oil distilled from seeds used in pickle, chutney, curry powders, imitation maple flavoring.	Widely cultivated
Filé (*see* **Sassafras**)			
Garlic	*Allium sativum*	Fresh or dry bulbs used for meat seasonings.	Widely cultivated
Ginger	*Zingiber officinale* and others	Dried rhizomes used for flavoring many foods and drinks.	India; Taiwan
Grains of paradise	*Afromomum melegueta*	Seeds used to flavor beverages and medicines.	West Africa
Hops	*Humulus lupulus*	Dried inflorescences of female plants used in brewing beer.	Europe; North America
Horseradish	*Rorippa armoracia*	Grated fresh root used as condiment.	Europe; North America
Juniper	*Juniperus* spp.	"Berries" used to season beef roasts, poultry, sauces.	North America
Licorice	*Glycyrrhiza glabra*	Dried rhizome and root used to flavor pontefract cakes, candies.	Middle East
Lovage	*Ligusticum scoticum*	Stems candied; seeds used in pickling sauces; celery substitute.	Europe
Mace	*Myristica fragrans*	Aril of seed used for flavoring beverages, foods.	Grenada; Indonesia; Sri Lanka
Marigold	*Tagetes* spp.	Petals substituted for saffron in rice dishes, stews.	Widely cultivated
Marjoram	*Marjorana hortensis*	Leaves used in stews, dressings, sauces.	Mediterranean
Mustard	*Brassica* spp.	Ground seeds used in meat condiment.	Europe; China
Nasturtium	*Tropaeolum majus*	Flowers, seeds, leaves used in salads.	Widely cultivated
Nutmeg	*Myristica fragrans*	Seeds used for flavoring foods, beverages.	Grenada; Indonesia; Sri Lanka
Oregano	*Origanum vulgare* and others	Leaves used as seasoning with poultry, meats.	Europe
Paprika (*see* **Cayenne pepper**)			
Parsley	*Petroselinum sativum*	Leaves used as meat garnish and flavoring in sauces.	Widely cultivated
Pepper, black	*Piper nigrum*	Dried drupes ground for condiment.	Madagascar; India

Table A3.6

PLANTS USED TO SEASON OR FLAVOR

SPICE	SCIENTIFIC NAME OF PLANT	PARTS USED; REMARKS	PRINCIPAL SOURCE
Peppermint	*Mentha piperita*	Oil from leaves used for food, drink, dentifrice flavoring (much commercial menthol is derived from *Mentha arvensis* grown in Japan).	U.S.; countries of the former Soviet Union
Pimiento	*Capsicum* spp.	Bright red fruits of a cultivated variety of pepper used in stuffing olives and in cold meats, cheeses.	Central and South America
Poppy	*Papaver somniferum*	Seeds used in baking.	Widely cultivated
Rosemary	*Rosmarinus officinalis*	Oil from leaves used in perfumes, soaps.	Mediterranean
Rue	*Ruta graveolens*	Flavoring for fruit cups, salads.	Europe
Saffron	*Crocus sativus*	Dried stigmas used to flavor oriental-style dishes.	Spain; India
Sage	*Salvia officinalis*	Leaves used in poultry and meat dressings.	Yugoslavia
Sarsaparilla	*Smilax* spp.	Roots source of flavoring for beverages, medicines.	American Tropics
Sassafras	*Sassafras albidum*	Bark and wood yield flavoring for beverages, toothpaste, gumbo.	U.S.
Savory (summer)	*Satureia hortensis*	Leaves used in green bean and bean salads, in lentil soup, with fish.	Mediterranean
Savory (winter)	*Satureia montana*	Leaves used as seasoning in stuffings, meat loaf, stews.	Europe
Scallion	*Allium fistulosum*	Leaves used in wine cookery, soups.	Widely cultivated
Sesame	*Sesamum indicum*	Seeds used in baking.	Asia
Shallot	*Allium ascalonicum*	Bulbs and leaves used in Colbert butter, wine cookery.	Widely cultivated
Southernwood	*Artemisia abrotanum*	Leaves used to flavor cakes.	Europe
Star anise	*Illicium verum*	Fruits used in candy and cough drops.	China
Stonecrop	*Sedum acre*	Dried leaves (ground) used as pepper substitute.	Europe
Tarragon	*Artemisia dracunculus*	Leaves and flowering tops used in pickling sauces.	Europe
Thyme	*Thymus vulgaris*	Leaves used in meat and poultry dishes, soups, sauces.	Widely cultivated
Tonka bean	*Dipteryx* spp.	Seeds source of flavoring for tobacco; vanilla substitute (now largely synthesized).	American tropics
Turmeric	*Curcuma longa*	Rhizomes powdered and used in curry powders, meat flavoring.	India; China
Vanilla	*Vanilla planifolia*	Flavoring extracted from fruits; used in foods, drinks.	Malagasay Republic
Wintergreen	*Gaultheria procumbens*	Oil from leaves and bark used as flavoring for confections, toothpaste.	U.S.
Zedoary	*Curcuma zedoaria*	Dried rhizome used in liqueurs, curry powders.	India

DYE PLANTS

In the recent and ancient past, dyes from many different plants were used to color cotton, linen, and other fabrics. Since the middle of the 19th century, however, natural dyes have been almost completely replaced in industry by synthetic dyes, and today the use of natural dyes is largely confined to the work of individual hobbyists.

Any reader interested in experimenting with natural dyes is encouraged not only to choose those plant materials included in Table A3.7 but to try any local plants available. The experimenter will soon find that quite unexpected colors may be derived from plants, as the colors of fresh flowers, bark, or leaves often bear little relationship to the colors of the dyes. For methods of dyeing, see the footnote given under the heading of Lichens in Chapter 19 and references in *Additional Reading*.

Table A3.7

PLANT SOURCES OF NATURAL DYES

PLANT OR DYE	SCIENTIFIC NAME OF PLANT SOURCE	REMARKS
Acacia	*Acacia* spp.	Brown dyes from bark and fruits.
Alder	*Alnus* spp.	Brownish dyes from bark.
Alkanet	*Alkanna tinctoria*	Red dye from roots.
Annatto	*Bixa orellana*	Yellow or red dye from pulp surrounding seeds.
Bamboo	*Bambusa* spp.	Light green dye from leaves.
Barberry	*Berberis vulgaris*	Grayish dye from leaves.
Barwood	*Baphia nitida*	Purplish dyes from wood.
Bearberry	*Arctostaphylos uva-ursi*	Yellowish dye from leaves.
Bedstraw	*Galium* spp.	Light reddish brown dyes from roots.
Birch	*Betula* spp.	Light brown to black dyes from bark.
Black cherry	*Prunus serotina*	Red dye from bark; gray to green dyes from leaves.
Black walnut	*Juglans nigra*	Rich brown dye from bark; brown dye from walnut hulls.
Bloodroot	*Sanguinaria canadensis*	Red dye from rhizomes.
Blueberry	*Vaccinium* spp.	Blue to gray dye from mature fruits (tends to fade).
Bougainvillea	*Bougainvillea* spp.	Light brownish dyes from floral bracts.
Brazilwood	*Caesalpinia* spp.	Reddish dyes from wood.
Buckthorn	*Rhamnus* spp.	Green dyes from fruits.
Buckwheat	*Fagopyrum esculentum*	Blue dye from stems.
Buckwheat (wild)	*Eriogonum* spp.	Dark gold, pale yellow, and beige dyes from stems and flowers.
Buffaloberry	*Shepherdia argentea*	Red dye from fruit.
Butternut	*Juglans cinerea*	Yellow to grayish brown dyes from fruit hulls.
Cocklebur	*Xanthium strumarium*	Dark green dye from stems and leaves.
Coffee	*Coffee arabica*	Light brown dye from ground, roasted seeds.
Cudbear (Archil)	*Rocella* spp. (lichen)	Red dye obtained by fermentation of thallus.
Cutch	*Acacia* spp.; *Uncaria gambir*	Brown to drab green dyes from stem gums.
Dock	*Rumex* spp.	Light brown dyes from stems and leaves.
Dogwood	*Cornus florida*	Red dye from bark; purplish dye from root.
Doveweed	*Eremocarpus setigerus*	Light to olive green dye from entire plant.
Dyer's rocket	*Reseda luteola*	Orangish dye from all parts.
Elderberry	*Sambucus* spp.	Blackish dye from bark; purple, blue, or dark brown dyes from fruits.
Eucalyptus	*Eucalyptus* spp.	Beige dyes from bark.
Fennel	*Foeniculum vulgare*	Yellow dyes from shoots.
Fig	*Ficus carica*	Green dyes from leaves and fruits.
Fustic	*Chlorophora tinctoria*	Yellow, bright orange, and greenish dyes from heartwood.
Gamboge	*Garcinia* spp.	Yellow dye from resins that ooze from cuts made on stems.
Giant reed	*Arundo donax*	Pale yellow dye from leaves.
Grape	*Vitis* spp.	Bright yellow to olive green dyes from leaves.
Hawthorn	*Crataegus* spp.	Pink dye from ripe fruits.
Hemlock	*Tsuga* spp.	Reddish brown dye from bark.
Henna	*Lawsonia inermis*	Orange dye from shoots and leaves.
Hickory	*Carya tomentosa*	Yellow dye from bark.
Hollyhock	*Althaea rosea*	Purplish black dye from flower petals.
Horsetail	*Equisetum* spp.	Tan dyes from all green parts.
Indigo	*Indigofera tinctoria*	Bright blue dyes from leaves.
Kendall green (*see* **Woadwaxen**)		
Larkspur	*Delphinium* spp.	Blue dyes from petals.
Lichens	Many genera and species	Many lichens yield brilliant shades of yellows, golds, and browns with various mordants.
Litmus	*Rocella tinctoria*	Famous pink-to-blue pH indicator dye from thallus.
Logwood	*Haematoxylon campechianum*	Dark blue purple dye from heartwood.
Lokao	*Rhamnus* spp.	Green dye from wood.

Table A3.7

PLANT SOURCES OF NATURAL DYES

PLANT OR DYE	SCIENTIFIC NAME OF PLANT SOURCE	REMARKS
Lupine	*Lupinus* spp.	Greenish dyes from flowers.
Madder	*Rubia tinctorium*	Bright red dye from roots.
Madrone	*Arbutus menziesii*	Brown dye from bark.
Manzanita	*Arctostaphylos* spp.	Beige to dull yellow dyes from dried fruits.
Maple	*Acer* spp.	Pink dye from bark.
Marsh marigold	*Caltha palustris*	Yellow dye from petals.
Milkweed	*Asclepias speciosa*	Pale yellow dyes from leaves.
Morning glory	*Ipomoea violacea*	Gray green dye from blue flowers.
Mullein	*Verbascum thapsus*	Gold dyes from leaves.
Oak	*Quercus* spp.	Yellow dye from bark.
Onion	*Allium cepa*	Reddish brown dyes from dry outer bulb scales of red onions; yellow dyes from similar parts of yellow onions.
Oregon grape	*Berberis aquifolium*	Yellow dyes from roots.
Osage orange	*Maclura pomifera*	Yellow, gray, and green dyes from fruits; yellow orange dye from wood.
Peach	*Prunus persica*	Green dyes from leaves.
Poke	*Phytolacca americana*	Red dyes from mature fruits.
Pomegranate	*Punica granatum*	Dark gold dye from fruit rinds.
Prickly lettuce	*Lactuca serriola*	Green dye from leaves.
Privet	*Ligustrum vulgare*	Yellow green dye from leaves; deep gray dye from berries.
Quercitron	*Quercus velutina*	Bright yellow dye from bark.
Rhododendron	*Rhododendron* spp.	Tan dyes from leaves.
Safflower	*Carthamus tinctorius*	Reddish dye from flower heads.
Saffron	*Crocus sativus*	Powerful yellow dye from stigmas.
Sage	*Salvia officinalis*	Yellow dye from shoots.
Sandalwood	*Pterocarpus santalinus*	Red dye from wood.
Sappanwood	*Caesalpinia sappan*	Red dye from heartwood.
Sassafras	*Sassafras albidum*	Orange brown dye from bark.
Scotch broom	*Cytisus scoparius*	Yellow dye from all parts of plant.
Smoke tree	*Cotinus coggyria*	Orange yellow dye from wood (dye sometimes called *young fustic*).
Smooth sumac	*Rhus glabra*	Grayish brown dye from bark.
St. John's wort	*Hypericum* spp.	Light brownish dyes from leaves.
Tansy	*Tanacetum* spp.	Yellow, green dyes from leaves.
Toyon	*Heteromeles arbutifolia*	Reddish brown dyes from leaves.
Turmeric	*Curcuma longa*	Orangish dye from rhizome.
Woad	*Isatis tinctoria*	Blue dye from leaves.
Woadwaxen	*Genista tinctoria*	Yellow dye from all parts.
Yerba santa	*Eriodictyon californicum*	Rich dark brown dyes from leaves.

ADDITIONAL READING

Adrosko, R. J. 1971. *Natural dyes and home dyeing.* New York: Dover Publications.

Bliss, A. 1986. *North American dye plants,* rev. ed. New York: Boulder, Co: Juniper House.

Furst, P. E. 1992. *Mushrooms: Psychedelic fungi,* rev. ed. Edgemont, PA: Chelsea House Publications.

Graedon, J., and T. Graedon. 1995. *The people's guide to deadly drug interactions.* New York: St. Martin's Press.

Lewis, W. H., and M. P. F. Elvin-Lewis. 1977. *Medical botany.* New York: John Wiley & Sons, Inc.

Loewenfeld, C. 1989. *Herb gardening: Why and how to grow herbs.* Winchester, MA: Faber and Faber.

Merory, J. 1968. *Food flavorings: Composition, manufacture and use,* 2d ed. New York: Chemical Publishing Co.

Schultes, R. E., and A. Hoffman. 1980. *The botany and chemistry of hallucinogens,* 2d ed. Springfield, IL: Charles C. Thomas, Publishers.

Spoerke, D. G., Jr. 1992. *Herbal medications,* 2d ed. Santa Barbara, CA: Woodbridge Press Publishing Company.

Spoerke, D. G., Jr., and S. Smolinske (Eds.). 1990. *Toxicity of houseplants.* Boca Raton, FL: CRC Press.

Vogel, V. J. 1990. *American Indian medicine.* Norman, OK: University of Oklahoma Press.

See also the *Additional Reading* entries in Chapter 24.

Appendix 4

Houseplants and Home Gardening

GROWING HOUSEPLANTS

If sales volume is an indication, houseplants have never been more popular in the United States than they are now. Many are easy to grow and will brighten windowsills, planters, and other indoor spots for years if a few simple steps are followed to ensure their health and vigor.

Water

Houseplants are commonly overwatered, resulting in the unnecessary development of rots and diseases (see Table A4.1). As a rule, the surface of the potting medium should be dry to the touch before watering, but the medium should not be allowed to dry out completely unless the plant is dormant. Care should be taken, particularly during the winter, that the water is at room temperature. Rainwater, if available, is preferred to tap water, particularly if the water has a high mineral content or is chlorinated. Broad-leaved plants should periodically have house dust removed with a damp sponge (never use detergents to clean surfaces—they remove pro-

tective waxes). Many plants benefit from a daily misting with water, particularly in heated rooms.

Containers

In time, plants may develop too extensive a root system for the pots in which they are growing (becoming what is commonly called *root bound*). Nutrients in the potting medium may become exhausted, salts and other residues from fertilizers and water may build up to the point of inhibiting growth, or the plants themselves may produce substances that accumulate until they interfere with the plant's growth. To resolve these problems, periodically repot the plants and, if necessary, divide them at the time of repotting.

Temperatures

Most houseplants don't thrive where the temperatures are either too high or too low (see Table A4.1). In general, they tend to do best with minimum temperatures of about 13°C (55°F) and maximum temperatures of about 29°C (84°F). Many houseplants that prefer

T̲able A4.1

COMMON AILMENTS OF HOUSEPLANTS

PROBLEM SYMPTOMS	POSSIBLE CAUSES
Wilting or collapse of whole plant	Lack of water; too much heat; too much water or poor drainage resulting in root rot.
Yellowish or pale leaves	Insufficient light; too much light; microscopic pests (especially spider mites); too much or too little fertilizer.
Brown, dry leaves	Humidity too low; too much heat; poor air circulation; lack of water.
Tips and margins of leaves brown	Mineral content of water too high; drafts; too much sun or heat; too much or too little water.
Ringed spots on leaves	Water too cold.
Leaves falling off	Improper watering or water too cold; excessive use of fertilizer or wrong fertilizer; too much sun or, if lower leaves drop only, too little light.
Stringy growth	Insufficient light; too much fertilizer.
Base of plant soft or rotting	Overwatering.
No flowers or flower buds drop	Too much or too little light; night temperatures too high.
Water does not drain	Drain hole plugged; potting mixture has too much clay.
Mildew present	Fungi present—arrest with sulphur dust.

COMMON PESTS	CONTROLS*
Aphids	Wash off under faucet or spray with soapy water (not detergent); pyrethrum or rotenone sprays also effective.
Mealybugs	Remove with cotton swabs dipped in alcohol; spray with Volck oil.
Scale insects	Remove by hand; spray with Volck oil.
Spider mites	Use sprays containing small amounts of xylene (as soon as possible—spider mites multiply very rapidly).
Thrips	Spray with pyrethrum/rotenone or Volck oil sprays.
White flies	Spray with soapy water or pyrethrum/rotenone sprays every 4 days for 2 weeks (only the adults are susceptible to the sprays).

*For additional controls, see the biological controls in Appendix 2.

---

warmer temperatures while actively growing also benefit from a rest period at lower temperatures after flowering.

Light

Next to overwatering, insufficient light is the most common contributor to the decline or death of houseplants (see Table A4.1). This does not mean that houseplants do better in direct sunlight—such light frequently damages them—but filtered sunlight (as, for example, through a muslin curtain) is usually better for the plant than the light available in the middle of the room. Plants can also thrive in artificial light of appropriate quality. Ordinary incandescent bulbs have too little light of blue wavelengths, and ordinary fluorescent tubes emit too little red light. A combination of the two, however, works very well. Generally, the wattage of the incandescent bulbs should be only one-fourth that of the fluorescent tubes in such a combination. Several types of fluorescent tubes specially balanced to imitate sunlight are also available.

Humidity

Dry air is hard on most houseplants. The level of humidity around the plants can be raised by standing the pots in dishes containing gravel or crushed rock to which water has been added. The humidity level can also be raised through the use of humidifiers, which come in a variety of sizes and capacities. Daily misting, as mentioned previously, can also help.

COMMON HOUSEPLANTS

Here is an explanation of the symbols given with each plant:

Water

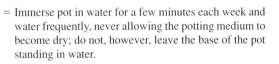 = Give little water (applies primarily to cacti and succulents; these plants store water in such a way that the soil can be completely dry for a week or two without their being adversely affected).

 = Water regularly but not excessively; wait until the potting medium surface is dry to the touch before watering.

 = Immerse pot in water for a few minutes each week and water frequently, never allowing the potting medium to become dry; do not, however, leave the base of the pot standing in water.

= Give little to regular water.

 = Give regular to frequent water.

Minerals in hard water are taxing on houseplants, and commercial water softeners do not improve water for the plants. Use rainwater or filtered water if at all possible; otherwise, repot more often.

Temperature

🕯 = cool; maximum 13°C–16°C (55°F–61°F); minimum 5°C–7°C (41°F–45°F)

🕯 = cool to medium; maximum 18°C–21°C (65°F–70°F); minimum 10°C–13°C (50°F–55°F)

🕯 = medium; maximum 30°C (86°F)

🕯 = medium to warm

🕯 = warm

Many houseplants are native to the tropics, where they thrive under year-round warm temperatures, while cacti and succulents prefer cool winters. The closer a houseplant's environment is to its natural environment, the better the plant will grow (see Table A4.2).

Light

▮ = Needs shading or indirect daylight.

▯ = Needs bright light but needs to be screened from direct sunlight.

▯ = Needs direct sunlight.

▮ = Needs shading from bright light.

▯ = Needs bright to direct light.

As mentioned previously, improper lighting is second only to over-watering as a cause of problems for houseplants; generally, they are given too little light and, occasionally, too much. A south-facing windowsill may be ideal for certain plants in midwinter but excessively bright in midsummer; conversely, a north-facing windowsill may have enough light for certain plants in midsummer but not in midwinter. Adjustable screens permit manipulation of daylight to suit the plants involved.

Humidity

☁ = Will tolerate dry air.

☁ = Needs dry to regular air.

☁ = Will tolerate the air in most houses provided it is mist-sprayed occasionally.

☁ = Needs regular to humid air.

☁ = Needs high humidity; use a humidifier if possible.

Virtually all plants with 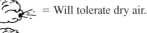 or 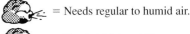 symbols benefit from having a pan of gravel with water beneath the pot.

Potting Medium

🔨 = Requires a porous, slightly acid medium that drains immediately.

🔨 = Requires a loam that is slightly alkaline (e.g., a mixture of sand and standard commercial potting medium).

🔨 = Requires a peaty potting mixture and acid fertilizer.

Table A4.2

ENVIRONMENTS SUITABLE FOR COMMON HOUSEPLANTS

PLANT	SCIENTIFIC NAME	ENVIRONMENTAL REQUIREMENTS	REMARKS
Aechmea	*Aechmea fasciata*		*See* Bromeliad; produces side shoots that should be propagated as main plant dies after flowering.
African lily	*Agapanthus* spp.		Do not repot until pot is full; keep cool in winter.
African violet	*Saintpaulia* spp.		Let rest under cooler conditions after flowering; dislikes cold water.
Agave	*Agave* spp.		Keep cool and dry in winter.
Algerian ivy	*Hedera canariensis*		Resembles variegated English ivy.
Aloe	*Aloe* spp.		Keep cool and dry in winter.

Table A4.2

ENVIRONMENTS SUITABLE FOR COMMON HOUSEPLANTS

PLANT	SCIENTIFIC NAME	ENVIRONMENTAL REQUIREMENTS	REMARKS
Aluminum plant	*Pilea cadierei*		Plants do not usually survive long in houses.
Amaryllis	*Amaryllis* spp.		Let leaves die back in fall; put bulb in cool, dark place until early spring; then repot, water, and fertilize weekly.
Anthurium	*Anthurium* spp.		If it grows without flowering, try putting it in cooler location for a few weeks.
Aphelandra	*Aphelandra squarrosa*		Mist-spray frequently; fertilize regularly.
Aralia (*see* **Fatsia**)			
Asparagus fern	*Asparagus plumosus*		Not a true fern; repot annually; fertilize weekly.
Aspidistra	*Aspidistra* spp.		Sometimes called "cast iron plant" because it can stand neglect.
Aucuba	*Aucuba japonica*		Must be kept cool in winter.
Avocado	*Persea* spp.		Easily propagated from seed; provides good greenery but will not produce fruit indoors.
Azalea	*Rhododenron* spp.		Needs acid fertilizer; avoid warm locations.
Bamboo (dwarf)	*Bambusa angulata*		Needs good air circulation, bright light.
Begonia	*Begonia* spp.		Easily propagated from leaves; repot regularly.
Bilbergia	*Bilbergia* spp.		*See* Bromeliad; tough plant that can tolerate some neglect.
Birdcatcher plant	*Pisonia umbellifera*		Sticky exudate on fruits attracts birds in the plant's native habitat of New Zealand; strictly a foliage plant in houses.
Bird of paradise plant	*Strelitzia reginae*		Can be grown outdoors in milder climates.
Bird's nest fern	*Asplenium nidus*		Produces a spongelike mass of roots at base; requires much water and regular fertilizing.
Black-eyed Susan	*Thunbergia alata*		Annual climbing vine; grow from seed.

Table A4.2

ENVIRONMENTS SUITABLE FOR COMMON HOUSEPLANTS

PLANT	SCIENTIFIC NAME	ENVIRONMENTAL REQUIREMENTS	REMARKS
Bloodleaf	*Iresine herbstii*		Easily propagated.
Boston fern	*Nephrolepis exaltata*		Needs regular watering and fertilizing.
Bromeliads	many species		These plants absorb virtually all their water and nutrients through their leaves; they should not be placed in regular potting soil, nor should they be watered with high calcium content water. They produce offshoots that should be propagated as main plant dies after flowering.
Cacti	many species		Contrary to popular belief, these slow-growing plants should not be grown in pure sand. Add some humus to potting mixture and withhold water in winter; keep cool in winter.
Caladium	*Caladium* spp.		Must have high humidity; keep root at 18°C (65°F) in pot during winter.
Calceolaria	*Calceolaria herbeohybrida*		Discard after flowering.
Calla lily	*Zantedeschia aethiopica*		After flowering, allow plant to dry up; repot in fall and start over.
Cape jasmine	*Gardenia jasminoides*		Needs night temperatures below 22°C (72°F) to initiate flowering; needs cool temperatures in winter.
Carrion flower	*Stapelia* spp.		Cactuslike plants with foul-smelling flowers.
Century plant (*see* Agave)			
Chinese evergreen	*Aglaonema costatum*		Needs warm temperatures and much water all year.
Chrysanthemum	*Chrysanthemum* spp.		Plants may be artificially dwarfed through use of chemicals; flowering initiated by short days.
Cineraria	*Senecio cruentus*		Needs cool temperatures; discard after flowering.

Table A4.2

ENVIRONMENTS SUITABLE FOR COMMON HOUSEPLANTS

PLANT	SCIENTIFIC NAME	ENVIRONMENTAL REQUIREMENTS	REMARKS
Cliff brake	*Pellaea rotundifolia*		Hanging basket fern; needs minimum temperature above 10°C (50°F).
Coffee	*Coffea arabica*		Handsome foliage plant; self-pollinating variety will produce fruit.
Coleus	*Coleus blumei*		To control size, restart plants from cuttings annually.
Copperleaf	*Acalypha wilkesiana*		Seldom survives average house environment for long.
Corn plant	*Dracaena massangeana*		Uses much water when large; easy to grow.
Croton	*Codiaeum* spp.		Needs constant high humidity and bright light.
Crown of thorns	*Euphorbia milii* and *E. splendens*		Deviation from watering routine may result in loss of leaves, but plant generally recovers.
Cyclamen	*Cyclamen* spp.		Fertilize weekly; keep cool; withhold water after flowering for few weeks, then start over.
Donkey tail	*Sedum morganianum*		Keep cool and dry in winter.
Dracaena	*Dracaena* spp.		Many kinds—all easy to grow and tolerant of some neglect.
Dumbcane	*Dieffenbachia* spp.		Needs regular fertilizing; keep away from small children (poisonous).
Dwarf banana	*Musa cavendishii*		Will produce small edible bananas if given enough light, water, and humidity.
Dwarf cocos palm	*Microcoleum weddelianum*		Keep temperature above 18°C (65°F) at all times.
Echeveria	*Echeveria* spp.		Keep cool in winter.
English ivy	*Hedera helix*		Needs cool temperatures to grow at its best.
False aralia	*Dizygotheca elegantissima*		Benefits from frequent mist-spraying.
Fatshedera	*Fatshedera lizei*		Climbing plant.
Fatsia	*Fatsia japonica*		Also called *Aralia*.

ENVIRONMENTS SUITABLE FOR COMMON HOUSEPLANTS

PLANT	SCIENTIFIC NAME	ENVIRONMENTAL REQUIREMENTS	REMARKS
Ferns	many species	🖼️	Water regularly; propagate from spores or runners.
Figs:			
Climbing fig	*Ficus pumila*	🖼️	Damp-sponge leaves regularly.
Fiddleleaf fig	*Ficus lyrata*		Damp-sponge leaves regularly.
Weeping fig	*Ficus benjamina*		Damp-sponge leaves regularly.
Fingernail plant	*Neoregelia* spp.	🖼️	*See* Bromeliad; name from red tips of leaves.
Fittonia	*Fittonia* spp.	🖼️	Strictly terrarium plants—humidity too low elsewhere.
Flame violet	*Episcia cupreata*	🖼️	Add charcoal and peat to potting medium.
Flowering maple	*Abutilon striatum*	🖼️	Needs bright light to flower.
Fuchsia	*Fuchsia* spp.	🖼️	Soil must be alkaline.
Gardenia (*see* **Cape jasmine**)		🖼️	
Geranium	*Pelargonium* spp.		Make cuttings annually and discard parent plants each fall; keep cool through winter. Available with scents of orange, rose, or coconut.
Gloxinia	*Sinningia speciosa*	🖼️	Fertilize heavily and water frequently; after flowering, withhold water and keep bulb cold for a few weeks.
Goldfish plant	*Columnea* spp.	🖼️	Pot in mixture of leaf mold, fern bark, peat moss, and charcoal; use only rainwater or filtered water.
Grape ivy	*Rhoicissus rhomboidea*	🖼️	Tolerates low light better than most plants.
Haworthia	*Haworthia* spp.	🖼️	Aloelike plants that need minimum temperatures above 10°C (50°F) in winter.
Hen and chickens	*Sempervivum tectorum*	🖼️	Keep cool in winter.
Hibiscus	*Hibiscus rosa-sinensis*	🖼️	Fertilize weekly.

Table A4.2

ENVIRONMENTS SUITABLE FOR COMMON HOUSEPLANTS

PLANT	SCIENTIFIC NAME	ENVIRONMENTAL REQUIREMENTS	REMARKS
Hippeastrum (*see* **Amaryllis**)			
Holly fern	*Cyrtomium falcatum*		Relatively tough fern; keep cool in winter.
Houseleek	*Sempervivum* spp.		Keep cool in winter.
Hydrangea	*Hydrangea* spp.		Prune after flowering; keep cool in winter. Pink-flowering plant can be converted to blue by changing the soil to acid and vice versa.
Impatiens	*Impatiens* spp.		Exceptionally easy to propagate from cuttings.
Ivy-arum	*Rhaphidophora aurea*		Can tolerate some neglect.
Jade plant	*Crassula argenta* and others		Keep cool in winter.
Kaffir lily	*Clivia miniata*		Save the plant's energy by removing flowers as they wither.
Kalanchoë	*Kalanchoë* spp.		Withhold water and fertilizer for a few weeks after flowering, then repot and start over.
Lantana	*Lantana camara*		Fertilize twice a month; can be espaliered.
Madagascar jasmine (*see* **Stephanotis**)			
Maidenhair fern	*Adiantum* spp.		Mist-spray regularly.
Meyer fern	*Asparagus densiflora* var. *meyeri*		Fertilize regularly; repot annually.
Moneywort	*Lysimachia nummularia*		Hanging pot plant; needs bright light.
Moonstones	*Pachyphytum* spp.		Keep cool in winter.
Moses in the cradle	*Rhoeo* spp.		Can tolerate some neglect.
Mother-in-law's tongue (*see* **Sansevieria**)			
Mother of thousands	*Saxifraga* spp.		Also called *Saxifrage;* hanging plant. Plantlets formed on runners can be removed and grown separately.

ENVIRONMENTS SUITABLE FOR COMMON HOUSEPLANTS

PLANT	SCIENTIFIC NAME	ENVIRONMENTAL REQUIREMENTS	REMARKS
Neanthe palm	*Chamaedorea elegans*		Sometimes called *Parlor palm;* stays less than 1 meter tall.
Norfolk Island pine	*Araucaria* spp.		Needs cold temperatures 2°C–3°C (36°F–38°F) in winter.
Octopus tree	*Schefflera arboricola*		Does best under cool conditions.
Oleander	*Nerium oleander*		Keep pot cool in winter for better flowering; keep away from small children (poisonous).
Orchid	thousands of species	no single set of environmental conditions applies	Contrary to popular belief, the common *Cattleya* and related orchids do not need high temperatures and humidity; most can get along with a minimum temperature of 13°C (56°F) at night and a minimum humidity of 40%. Most need bright light. They should never be placed in soil; pot them in sterilized pots with chips of fir bark or shreds of tree fern bark. *See Additional Reading* for culture references.
Oxtongue	*Gasteria* spp.		Can tolerate some neglect; needs a cool and relatively dry winter to flower.
Palms	many species		Use deep pots; fertilize regularly.
Parlor palm	*Howea fosteriana*		One of the easiest palms to grow.
Peperomia	*Peperomia* spp.		Many kinds; keep warm, humid; fertilize regularly.
Persian violet	*Exacum affine*		Needs good air circulation.
Philodendron	*Philodendron* spp.		Relatively tough plants; repot each spring.
Piggyback plant	*Tolmiea menziesii*		Plantlets formed on leaves can be separated and propagated.

ENVIRONMENTS SUITABLE FOR COMMON HOUSEPLANTS

PLANT	SCIENTIFIC NAME	ENVIRONMENTAL REQUIREMENTS	REMARKS
Pineapple	*Ananas comosus*		*See* Bromeliad; easily grown from the top of a pineapple. If plant has not flowered after one year, enclose it in a plastic bag with a ripe apple for a few days (ethylene from apple should initiate flowering); no temperature below 15°C (59°F).
Pink polka dot plant	*Hypoestes sanguinolenta*		Susceptible to diseases and pests.
Pittosporum	*Pittosporum* spp.		Put several cuttings in one pot for bushy appearance.
Poinsettia	*Euphorbia pulcherrima*		After flowering, let plant dry under cool conditions until it loses its leaves; then restart.
Prayer plant	*Maranta leuconeura*		Name derived from fact that leaves fold together in evening.
Primrose	*Primula* spp.		Needs much water; does well outside in cool weather.
Purple tiger	*Calathea amabilis*		Use pots that are broader than they are deep.
Rosary plant	*Ceropegia woodii*		Hanging pot plant whose potting medium must drain well or plant will not survive.
Rubber plant	*Ficus elastica*		Do not overwater!
Sago palm	*Cycas revoluta*		Very slow growing (not a palm but a gymnosperm); never allow to dry out, but be certain water drains.
Sansevieria	*Sansevieria* spp.		Perhaps the toughest of all houseplants—nearly indestructible.
Satin pothos	*Scindapsus pictus*		Basket or pot plant.
Screw pine	*Pandanus* spp.		If given space, can become large; mist-spray often.
Selaginella (*see* **Spike moss**)			

Table A4.2

ENVIRONMENTS SUITABLE FOR COMMON HOUSEPLANTS

PLANT	SCIENTIFIC NAME	ENVIRONMENTAL REQUIREMENTS	REMARKS
Sensitive plant	*Mimosa pudica*		Leaves fold when touched; does not usually last more than a few months in most houses.
Shrimp plant	*Beloperone guttata*		Winter temperatures should be above 15°C (59°F).
Spathe flower	*Spathiphyllum wallisii*		Prefers warm winters and even warmer summers.
Spider plant	*Chlorophytum comosum*		Plantlets formed at tips of stems can be propagated separately.
Spiderwort	*Tradescantia* spp.		Easy to grow; do not overwater.
Spike moss	*Selaginella* spp.		Can become a weed in greenhouses.
Splitleaf philodendron	*Monstera deliciosa*		Plant adapts to various indoor locations quite well.
Sprenger fern	*Asparagus densiflora* var. *sprengeri*		Fertilize weekly; repot annually.
Staghorn fern	*Platycerium* spp.		Tough plant; immerse in water weekly.
Stephanotis	*Stephanotis floribunda*		Use very little fertilizer; keep cool in winter but water sparingly.
Stonecrop	*Sedum* spp.; *Crassula* spp.		Keep cool in winter.
Stove fern	*Pteris cretica*		Water and fertilize regularly.
String-of-pearls	*Senecio rowellianus*		Keep cool in winter.
Sundew	*Drosera* spp.		Sterilize pots; grow only on sphagnum moss.
Syngonium	*Syngonium podophyllum*		Repot annually in spring.
Tillandsia	*Tillandsia* spp.		*See* Bromeliad; best-known species is called *Spanish moss*.
Ti plant	*Cordyline terminalis*		Needs high humidity; seems to do better with other plants in pot.
Treebine	*Cissus antarctica*		Plant dislikes acid potting medium.

Table A4.2

ENVIRONMENTS SUITABLE FOR COMMON HOUSEPLANTS

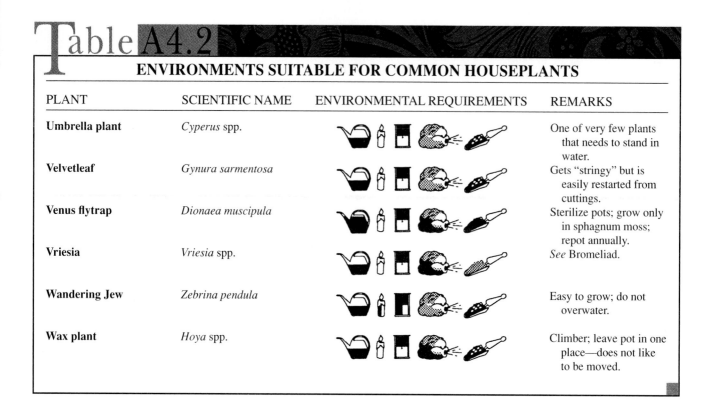

PLANT	SCIENTIFIC NAME	ENVIRONMENTAL REQUIREMENTS	REMARKS
Umbrella plant	*Cyperus* spp.		One of very few plants that needs to stand in water.
Velvetleaf	*Gynura sarmentosa*		Gets "stringy" but is easily restarted from cuttings.
Venus flytrap	*Dionaea muscipula*		Sterilize pots; grow only in sphagnum moss; repot annually.
Vriesia	*Vriesia* spp.		*See* Bromeliad.
Wandering Jew	*Zebrina pendula*		Easy to grow; do not overwater.
Wax plant	*Hoya* spp.		Climber; leave pot in one place—does not like to be moved.

GROWING VEGETABLES

Seed Germination

Many gardeners germinate larger seeds (e.g., squash, pumpkin) in damp newspaper. Soak a few sheets of newspaper in water for a minute and then hang them over a support for about 15 minutes or until the water stops dripping. Then line up the seeds in a row on the newspaper, wrap them, and place the damp mass in a plastic bag. Tie off the bag or seal it and place it in a warm (not hot!), shaded location (the floor beneath a running refrigerator is an example). Depending on the species, germination should occur within 2 to several days.

Tiny seeds (e.g., carrots, lettuce) may be mixed with clean sand before sowing to bring about a more even distribution of the seed in the rows.

Transplanting

Roots should be disturbed as little as possible when seedlings or larger plants are transplanted. Even a few seconds' exposure to air will kill root hairs and smaller roots. They should be shaded (e.g., with newspaper) from the sun and transplanted late in the day or on a cool, cloudy day if at all possible. To minimize the effects of transplanting, immediately water the seedlings or plants in their new location with a dilute solution of vitamin B/hormone preparation (e.g., Superthrive).

Bulb or Other Plants with Food-Storage Organs

All plants with food-storage organs (e.g., beets, carrots, onions) develop much better in soil that is free of lumps and rocks. If possible, the areas where these plants are to be grown should be dug to a minimum depth of 30 centimeters (12 inches) and the soil sifted through a 0.7 centimeter (approximately 0.25 inch) mesh before planting. Obviously, such a procedure is not always practical, but it can yield dramatic results.

Cutworms

Cutworms forage at or just beneath the surface of the soil. Their damage to young seedlings can be minimized by placing a collar around each plant. Tuna cans with both ends removed make effective collars when pressed into the ground to a depth of about 1 to 2 centimeters (0.4 to 0.8 inch).

Protection Against Cold

Some seedlings can be given an earlier start outdoors if plastic-topped coffee cans with the bottoms removed are placed over them. The plastic lids can be taken off during the day and replaced at night during cool weather. Conical paper frost caps can serve the same purpose.

Watering

Proper watering promotes healthy growth. It is much better to water an area thoroughly (e.g., for 20 to 30 minutes) every few days than to wet the surface for a minute or two daily. Shallow daily watering promotes root development near the surface, where midsummer heat can damage the root system. Conversely, too much watering can leach minerals out of the topsoil.

Fertilizers

Manures, bone and blood meals, and other organic fertilizers, which release the nutrients slowly and do not "burn" young plants, are preferred. Plants will utilize minerals from any available source, but in the long run the plants will be healthier and subject to fewer problems if they are not given sudden boosts with liquid chemical fertilizers.

Pests and Diseases

Biological controls are listed in Appendix 2. If sprays must be used, biodegradable substances, such as rotenone and pyrethrum, should be used.

COMMON VEGETABLES AND THEIR NUTRITIONAL VALUES

The nutritional values (NV) given are per 100 grams (3.5 ounces), edible portion, as determined by the U.S. Department of Agriculture.

Asparagus

NV (spears cooked in water): 20 calories; protein 2.2 gm; fats 0.2 gm; vit. A 900 I.U.; vit. B1 0.16 mg; vit. B2 0.18 mg; niacin 1.5 mg; vit. C 26 mg; fiber 0.7 gm; calcium 21 mg; phosphorus 50 mg; iron 0.6 mg; sodium 1 mg; potassium 208 mg.

Asparagus can be started from seed, but time until the first harvest can be reduced by a year or two if planting begins with one-year-old root clusters of healthy, disease-resistant varieties (e.g., "Mary Washington"). Asparagus requires little care if appropriate preparations are made before planting in the permanent location. Seeds should be sown sparsely and the seedlings thinned to about 7.5 centimeters (3 inches) apart. Before transplantation the following spring, dig a trench 30 to 60 centimeters (12 to 24 inches) deep and about 50 centimeters (20 inches) wide in an area that receives full sun, usually along one edge of the garden. If the soil is heavy, place crushed rock or gravel on the bottom of the trench to provide good drainage. Add a layer of steer manure about 10 centimeters (4 inches) thick, followed by about 7.5 centimeters (3 inches) of rich soil that has been prepared by thorough mixing with generous quantities of steer manure and bone meal. Place root clusters about 45 centimeters (18 inches) apart in the trench and cover with about 15 centimeters (6 inches) of prepared soil (be sure not to allow root clusters to dry out). As the plants grow, gradually fill in the trench. If one-year-old roots are planted, wait for 2 years before harvesting tips; if 2-year-old roots are planted, some asparagus may be harvested the following year. In all cases, no harvesting should be done after June, so that the plants may build up reserves for the following season. Cut shoots below the surface but well above the crown before the buds begin to expand. Cut all stems to the ground after they have turned yellow later in the season.

Beans

String or Snap Beans

NV (young pods cooked in water): 25 calories; protein 1.6 gm; fats 0.2 gm; vit. A 540 I.U.; vit. B1 0.07 mg; vit. B2 0.09 mg; niacin 0.5 mg; vit. C 12 mg; fiber 1 gm; calcium 50 mg; phosphorus 37 mg; iron 0.6 mg; sodium 4 mg; potassium 151 mg.

String or snap beans are warm-weather plants, although they can be grown almost anywhere in the United States. Wait until all danger of frost has passed and the soil is warm. Prepare the soil, preferably the previous winter, by digging to a depth of 30 centimeters (12 inches) and mixing in aged manure and bone meal. Pulverize the soil just before sowing; if soil has a low pH, add lime. Plant seeds thinly in rows about 40 to 50 centimeters (16 to 20 inches) apart; thin plants to 10 centimeters (4 inches) apart when the first true leaves have developed. Beans respond unfavorably to a very wet soil—do not overwater! In areas with hot summers, beans also prefer some light shade, particularly in midafternoon. As the bean plants grow, nitrogen-fixing bacteria invade the roots and supplement the nitrogen supply. Early vigorous growth can be enhanced by inoculating the seeds with such bacteria, which are available commercially in a powdered form. To maintain a continuous supply of green beans, plant a new row every two to three weeks during the growing season until two months before the first predicted frost. Cultivate regularly to control weeds, taking care not to damage root systems. Do not harvest or work with beans while the plants are wet, as this may invite disease problems.

Pole Beans

NV similar to those of string beans.

Soil preparation and cultivation are the same as for string beans. Plant beans in hills around poles that are not less than 5 centimeters (2 inches) in diameter and are at least 2 meters (6.5 feet) tall. As beans twine around their supports, it helps to tie them to the support with plastic tape as they grow. If harvested before the pods are mature, pole beans will produce over a longer period of time than will bush varieties.

Lima Beans

NV (immature seeds cooked in water): 111 calories; protein 7.6 gm; fats 0.5 gm; vit. A 280 I.U.; vit. B1 0.18 mg; vit. B2 0.10 mg; niacin 1.3 mg; vit. C 17 mg; fiber 1.8 gm; calcium 47 mg; phosphorus 121 mg; iron 2.5 mg; sodium 1 mg; potassium 422 mg.

Lima beans take longer to mature than other beans and are more sensitive to wet or cool weather. They definitely need warm weather to do well. Prepare soil and cultivate as for string beans.

Soybeans

NV (dry, mature seeds): 403 calories; protein 34.1 gm; fats 17.7 gm; vit. A 80 I.U.; vit. B1 1.10 mg; vit. B2 0.31 mg; niacin 2.2 mg; vit. C[1]; fiber 4.9 gm; calcium 226 mg; phosphorus 554 mg; iron 8.4 mg; sodium 5 mg; potassium 1,677 mg.

Broad (Fava) Beans

NV (dry, mature seeds): 338 calories; protein 25.1 gm; fats 1.7 gm; vit. A 70 I.U.; vit. B1 0.5 mg; vit. B2 0.3 mg; niacin 2.5 mg; vit. C[1]; fiber 6.7 gm; calcium 47 mg; phosphorus 121 mg; iron 2.5 mg; sodium 1 mg; potassium 422 mg.

Unlike other beans, broad beans need cool weather for their development. Sow as early as possible (in mild climates they may be sown in the fall, as they can withstand light frosts). Since the plants occupy a little more space than do bush beans, plant in rows about 0.9 to 1.0 meter (3 feet or more) apart and thin to about 20 centimeters (8 inches) apart in the rows. After the first pods mature, pinch out the tips to promote bushier development. To most palates, broad beans do not taste as good as other types of beans.

Beets

NV (cooked in water): 32 calories; protein 1.1 gm; fats 0.1 gm; vit. A 20 I.U.; vit. B1 0.03 mg; vit. B2 0.04 mg; vit. C 6 mg; fiber

---

[1]Values not available. *See also* Broad (Fava) Beans.

0.8 gm; calcium 14 mg; phosphorus 23 mg; iron 0.5 mg; sodium 43 mg; potassium 208 mg.

Beets will grow in a variety of climates but do best in cooler weather. They can tolerate light frosts and can be grown on a variety of soil types, although they prefer a sandy loam supplemented with well-aged organic matter. As with any bulb or root crop, they develop best in soil that is free of rocks and lumps. Beet "seeds" are really fruits containing several tiny seeds. Plant them in rows 40 to 60 centimeters (16 to 24 inches) or more apart and thin to about 10 centimeters (4 inches) apart in the rows after germination. After harvesting, the beets will keep in cold storage for up to several months. The leaves, if used when first picked, make a good substitute for spinach.

Broccoli

NV (spears cooked in water): 26 calories; protein 3.1 gm; fats 0.3 gm; vit. A 2,500 I.U.; vit. B1 0.09 mg; vit. B2 0.2 mg; niacin 0.8 mg; fiber 1.5 gm; calcium 88 mg; phosphorus 62 mg; iron 0.8 mg; sodium 10 mg; potassium 267 mg.

Broccoli is a cool-weather plant that will thrive in any good prepared soil, providing that it has not been heavily fertilized just prior to planting (fresh fertilizer promotes rank growth). The plants can stand light frosts and are planted in both the spring and fall in areas with mild climate. Although broccoli may continue to produce during the summer, most growers prefer not to keep the plants going during warm seasons because of the large numbers of pest insects they may attract.

Sow seeds indoors and transplant outdoors after danger of killing frosts has passed. Place plants about 0.9 to 1.0 meter (3 feet or more) apart and keep well watered. Keep area weeded and pests under control. Harvest heads (bundles of spears) while they are still compact. Smaller heads will develop very shortly after the first harvest; if these are removed regularly, the plants will continue to produce for some time, although the heads become smaller as the plants age.

Cabbage

NV (raw): 24 calories; protein 1.3 gm; fats 0.2 gm; vit. A 130 I.U.; vit. B1 0.05 mg; vit. B2 0.05 mg; niacin 0.3 mg; vit. C 47 mg; fiber 1 gm; calcium 49 mg; phosphorus 29 mg; iron 0.4 mg; sodium 20 mg; potassium 233 mg.

Growth requirements of cabbage are similar to those of broccoli.

Carrots

NV (raw): 42 calories; protein 1.1 gm; fats 0.2 gm; vit. A 11,000 I.U.; vit. B1 0.6 mg; vit. B2 0.5 mg; niacin 0.6 mg; vit. C 8 mg; fiber 1 gm; calcium 37 mg; phosphorus 36 mg; iron 0.7 mg; sodium 47 mg; potassium 341 mg.

Carrots are hardy plants that can tolerate a wide range of climate and soils, but the soil must be well prepared: free of rocks and lumps and preferably not too acid. The seeds are slow to germinate. Plant in rows a little more than 30 centimeters (12 inches) apart and thin seedlings to about 5 centimeters (2 inches) apart in the rows. Weed the rows regularly until harvest. Carrots keep well in below-ground storage containers when freezing weather arrives.

Cauliflower

NV (cooked in water): 22 calories; protein 2.3 gm; fats 0.2 gm; vit. A 60 I.U.; vit. B1 0.09 mg; vit. B2 0.08 mg; niacin 0.6 mg; vit. C 55 mg; fiber 1 gm; calcium 21 mg; phosphorus 42 mg; iron 0.7 mg; sodium 9 mg; potassium 206 mg.

Growth requirements of cauliflower are similar to those of broccoli except that heavier fertilizing is required. As cauliflower heads develop, protect them from the sun by tying the larger leaves over the tender heads. Harvest while the heads are still solid.

Corn

NV (cooked sweet corn kernels): 83 calories; protein 3.2 gm; fats 1 gm; vit. A 400 I.U. (yellow varieties; white varieties have negligible vit. A content); vit. B1 0.11 mg; vit. B2 0.10 mg; niacin 1.3 mg; vit. C 7 mg; fiber 0.7 gm; calcium 3 mg; phosphorus 89 mg; iron 0.6 mg; sodium trace; potassium 165 mg.

There are several types of corn (e.g., popcorn, flint corn, dent corn), but sweet corn is the only type grown to any extent by home gardeners. It can be grown in any location where there is at least a 10-week growing season and warm summer weather.

Corn does best in a fertile soil, which should be prepared by mixing with compost and liberal amounts of chicken manure or fish meal. Since corn is wind-pollinated, it can be helpful to grow the plants in several short rows at right angles to the prevailing winds rather than in a single long row. For best results, use only fresh seeds and plant in rows 60 centimeters (24 inches) apart for dwarf varieties and 90 centimeters (36 inches) apart for standard varieties. Thin to 20 to 30 centimeters (8 to 12 inches) apart in the rows after the plants have produced three to four leaves. Cultivate frequently to control weeds. The corn is ready to harvest when the silks begin to wither.

Cucumber

NV (raw, with skin): 15 calories; protein 0.9 gm; fats 0.1 gm; vit. A 250 I.U.; vit. B1 0.03 mg; vit. B2 0.04 mg; niacin 0.2 mg; vit. C 11 mg; fiber 0.6 mg; calcium 25 mg; phosphorus 27 mg; iron 1.1 mg; sodium 6 mg; potassium 160 mg.

Until drought- and disease-resistant varieties were developed in recent years, cucumbers were considered rather temperamental plants to grow. The newer varieties are no more difficult to raise than are most other common vegetables.

The soil should be a light loam—neither too heavy nor too sandy. It should be mixed with well-aged manure and compost and heaped into small mounds about 2 meters (6.5 feet) apart. Five to six seeds should be planted in each mound about 2.5 centimeters (1 inch) below the surface in the middle of the spring. When the plants are about 1 decimeter (4 inches) tall, thin to three plants per mound. Cultivate regularly and, to promote continued production, pick all cucumbers as soon as they attain eating size.

Eggplant

NV (cooked in water): 19 calories; protein 1 gm; fats 0.2 gm; vit. A 10 I.U.; vit. B1 0.05 mg; vit. B2 0.04 mg; niacin 0.5 mg; vit. C 3 mg; fiber 0.9 gm; calcium 11 mg; phosphorus 21 mg; iron 0.6 mg; sodium 1 mg; potassium 150 mg.

Eggplant is strictly a hot-weather plant that is sensitive to cold weather or dry periods and needs heavy fertilizing. Since seedling development is initially slow, plant the seeds indoors about 2 months before the plants will be set out, which should be about 5 to 6 weeks after the last average date of frost.

Eggplants do best in enriched sandy soils that are supplemented with additional fertilizer once a month. Never permit them to dry out. Place the seedlings about 70 to 80 centimeters (28 to 32 inches) apart in rows 0.9 to 1.0 meter (3 feet or more) apart. Some staking of the plants may be desirable. The fruits are ready to harvest when they have a high gloss. They are still edible after greenish streaks appear and the gloss diminishes, but they are not as tender at this stage.

Lettuce

NV (crisp, cabbage-head varieties): 13 calories; protein 0.9 gm; fats 0.1 gm; vit. A 330 I.U.; vit. B1 0.06 mg; vit. B2 0.06 mg; niacin 0.3 mg; vit. C 6 mg; fiber 0.5 gm; calcium 20 mg; phosphorus 22 mg; iron 0.5 mg; sodium 9 mg; potassium 175 mg.

NV (leaf varieties): 18 calories; protein 1.3 gm; fats 0.3 gm; vit. A 1,900 I.U.; vit. B1 0.05 mg; vit. B2 0.08 mg; niacin 0.4 mg; vit. C 18 mg; calcium 68 mg; phosphorus 25 mg; iron 1.4 mg; sodium 9 mg; potassium 264 mg.

This favorite salad plant comes in a wide variety of types and forms, all of which do better in cooler weather, although a few of the leaf types (e.g., oak leaf) can tolerate some hot periods. As long as the individual plants are given room to develop and the soil is not too acid, most varieties can be grown on a wide range of soil types.

Since lettuce can stand some frost, sow the seeds outdoors as early in the spring as the ground can be cultivated. Do not cover the seeds with more than a millimeter or two of soil—they need light to germinate. Mix the soil with a well-aged manure and a general-purpose fertilizer a week or two before sowing. Plant seedlings about 30 centimeters (12 inches) apart in rows 30 to 40 centimeters (12 to 16 inches) apart. For best results, do not allow the soil to dry out and plant only varieties suited to local conditions. The most common crisp, cabbage-head varieties found in produce markets will not form heads in hot weather, and many others will bolt (begin to flower) during hot weather and longer days. Cultivate weekly between rows to promote rapid growth and to control weeds.

Onion

NV (raw): 38 calories; protein 1.5 gm; fats 0.1 gm; vit. A 40 I.U. (yellow varieties only); vit. B1 0.03 mg; vit. B2 0.04 mg; niacin 0.2 mg; vit. C 10 mg; fiber 0.6 gm; calcium 27 mg; phosphorus 36 mg; iron 0.5 mg; sodium 10 mg; potassium 157 mg.

These easy-to-grow vegetables do best in fertile soils that are free of rocks and lumps, are well drained, and are not too acid or sandy.

Onions may take several months to mature from seed. The viability of the seed decreases rapidly after the first year. Bulb formation is determined by day length rather than by the total number of hours in the ground. Because of these characteristics of onions, most gardeners prefer to purchase *sets* (young plants that already have a small bulb) from commercial growers, although green or bunching onions are still easily grown from seed.

Plant the sets upright 6.0 to 7.5 centimeters (2.5 to 3.0 inches) apart in rows and firm in place. They will need little care except for weeding, watering, and occasional shallow cultivation until harvest about 14 weeks later. The onions are mature when the tops fall over. After pulling them from the ground, allow them to dry in the shade for 2 days. Then remove the tops 2 to 3 centimeters (about 1 inch) above the bulbs and spread out the bulbs to continue curing for 2 to 3 more weeks. After this, store them in sacks or other containers that permit air circulation until needed.

Peas

NV (green, cooked in water): 71 calories, protein 5.4 gm; fats 0.4 gm; vit. A 540 I.U.; vit. B1 0.28 mg; vit. B2 0.11 mg; niacin 2.3 mg; vit. C 20 mg; fiber 2 gm; calcium 23 mg; phosphorus 99 mg; iron 1.8 mg; sodium 1 mg; potassium 196 mg.

Peas are strictly cool-weather plants that generally produce poorly when the soil becomes too warm. Seeds should be planted in the fall or very early spring. As is the case with beans, peas receive a better start if the seeds are inoculated with nitrogen-fixing bacteria (see discussion of beans) at planting time. Prepare the soil by mixing thoroughly with liberal amounts of aged manure and bone meal. Plant the seeds about 2.5 centimeters (1 inch) deep in heavy soil or 5 centimeters (2 inches) deep in light, sandy soil, about 2.5 centimeters (1 inch) apart in single rows for dwarf bush varieties or 15 centimeters (6 inches) apart in double files for standard varieties, with intervals of 80 to 90 centimeters (32 to 36 inches) between the rows. After germination, thin the plants to 10 centimeters apart. Place support wires, strings, or chicken wire between the rows at the time of planting; peas will not do well without such support.

Green peas should be picked while still young and cooked or frozen immediately, as the sugars that make them sweet are converted into starch within 2 to 3 hours after harvest.

Peppers

NV (raw sweet or bell peppers): 22 calories; protein 1.2 gm; fats 0.2 gm; vit. A 420 I.U.; vit. B1 0.08 mg; vit. B2 0.08 mg; niacin 0.5 mg; vit. C 128 mg; fiber 1.4 gm; calcium 9 mg; phosphorus 22 mg; iron 0.7 mg; sodium 13 mg; potassium 213 mg.

Peppers, like eggplants, are strictly hot-weather plants for most of their growing season, but unlike eggplants they actually do better toward the end of their season if temperatures have moderated somewhat. Sweet or bell peppers are closely related and have similar cultural requirements.

Plant seeds indoors 8 to 10 weeks before the outdoor planting date, which is generally after the soil has become thoroughly warm. They will grow in almost any sunny location in a wide variety of soils, but to obtain the large fruits seen in produce markets fertilize the plants heavily and water regularly. Plant seedlings 50 to 60 centimeters (20 to 24 inches) apart in rows that are 60 to 90 centimeters (24 to 36 inches) apart. Sweet peppers can be harvested at almost any stage and are still perfectly edible after they have turned red.

Potatoes

NV (baked in skin): 93 calories; protein 2.6 gm; fats 0.1 gm; vit. A trace; vit. B1 0.10 mg; vit. B2 0.04 mg; niacin 1.7 mg; vit. C 20 mg; fiber 0.6 gm; calcium 9 mg; phosphorus 65 mg; iron 0.7 mg; sodium 4 mg; potassium 503 mg.

Potatoes grow best in a rich, somewhat acid, well-drained soil that has had compost or well-aged manure added to it. They are subject to several diseases, and it is advisable to use disease-free seed potatoes purchased from a reputable dealer. Two to 3 weeks before the average date of the last spring frost, plant the seed potatoes whole or cut into several pieces, making sure that each piece has at least one eye. Place the potato pieces about 30 centimeters (12 inches) or more apart at a depth of about 12 to 15 centimeters (5 to 6 inches) in rows 0.9 to 1.0 meter (about 3 feet) apart. Later plantings are feasible. Spread a thick mulch (e.g., straw) over the area after planting to keep soil temperatures down and to retain soil moisture.

Potatoes are ready for harvest when the tops start turning yellow, but they may be left in the ground for several weeks after that if the soil is not too wet. After harvest, wash the potatoes immediately and place them in a dry, cool, dark place until needed. If left exposed to light, the outer parts of the potato turn green; poisonous substances are produced in these tissues, and such potatoes should be discarded.

Spinach

NV (raw): 26 calories; protein 3.2 gm; fats 0.3 gm; vit. A 8,100 I.U.; vit. B1 0.10 mg; vit. B2 0.20 mg; niacin 0.6 mg; vit. C 51 mg; fiber 0.6 gm; calcium 93 mg; phosphorus 51 mg; iron 3.1 mg; sodium 71 mg; potassium 470 mg.

Spinach is a cool-season crop that goes to seed as soon as the days become long and warm. It should be planted in the fall or early spring. If protected by straw or other mulches, it will overwinter in the ground in most areas and be ready for use early in the spring. Spinach has a high nitrogen requirement and reacts negatively to acid soils. It is otherwise easy to grow. Mix the soil thoroughly with aged manure and bone meal before planting. Plant seedlings 3 to 5 centimeters (1 to 2 inches) apart in rows 40 to 50 centimeters (16 to 20 inches) apart. Keep the plants supplied with adequate moisture and their growing area free of weeds. Harvest the whole plant when a healthy crown of leaves develops.

Squash

NV (cooked zucchini): 12 calories; protein 1 gm; fats 0.1 gm; vit. A 300 I.U.; vit. B1 0.05 mg; vit. B2 0.08 mg; niacin 0.8 mg; vit. C 9 mg; fiber 0.6 gm; calcium 25 mg; phosphorus 25 mg; iron 0.4 mg; sodium 1 mg; potassium 141 mg.

All varieties of squash are warm-weather plants, and all are targets of a variety of pests. Thorough preparation of the soil before planting pays dividends in production and in the health of the plants. Mix compost and aged manure with the soil and heap the soil in small hills about 1.2 meters (4 feet) apart from one another. Plant four to five seeds in each hill and thin the seedlings to three after they are about 10 centimeters (4 inches) tall. Summer squashes (e.g., zucchini) mature in about 2 months, while winter squashes (e.g., acorn) can take twice as long to mature. Summer squashes should be harvested while very young—they can balloon, seemingly overnight, into huge fruits. Winter squashes should be harvested before the first frost; only clean, undamaged fruits will store well. Keep such squashes laid out in a cool, dry place and not piled on top of one another. Check them occasionally for the development of surface fungi.

Tomatoes

NV (raw, ripe): 22 calories; protein 1.1 gm; fats 0.2 gm; vit. A 900 I.U.; vit. B1 0.06 mg; vit. B2 0.04 mg; niacin 0.7 mg; vit. C 23 mg;

fiber 0.5 gm; calcium 13 mg; phosphorus 27 mg; iron 0.5 mg; sodium 3 mg; potassium 244 mg.

These almost universally used fruits are easy to grow providing one understands a few basic aspects of their cultural requirements:

1. Many tomato plants normally will not initiate fruit development from their flowers when night temperatures drop below 14°C (57°F) or day temperatures climb above 40°C (104°F). For the earliest yields, seeds may be germinated indoors several weeks before the plants are to be placed outside, but little is accomplished by transplanting before the night temperatures begin to remain above 14°C (57°F); in addition, some growers insist that plants given an early start indoors do not always do as well later as those germinated outdoors.

2. Tomatoes require considerably more phosphorus in proportion to nitrogen from any fertilizers added to the soil where they are to be grown. Many inexperienced gardeners make the mistake of giving the plants lawn or general-purpose fertilizers that are proportionately high in nitrogen. As a result, the plants may grow vigorously but produce very few tomatoes. Give tomatoes bone meal, tomato food (Magamp is an excellent commercial slow-release preparation), or steer manure mixed with superphosphate.

3. Tomato plants seem more susceptible than most to soil fungi and to root-knot nematodes. The damage caused by these organisms may not become evident until the plants begin to bear. Then the lower leaves begin to wither, and yellowing progresses up the plant or there seems to be a general loss of vigor and productivity. Using disease- and nematode-resistant varieties (usually indicated by the letters V, F, and N on seed packets) is by far the simplest method of controlling these problems. Another effective control involves dipping the seedling roots in an emulsion of 0.25% corn oil in water when transplanting; experiments have shown that the corn oil greatly reduces root-knot nematode infestation.

4. Many garden varieties of tomatoes need to be staked to keep fruits off the ground where snails and other organisms can gain easy access to them. When using wooden stakes, be sure they have a diameter of 5 centimeters (2 inches) or more, and tie the plants securely to the stakes with plastic tape. Thinner stakes are likely to break or collapse when the plants grow to a height of 2 meters (6.5 feet) or more. Some growers prefer to use heavy wire tomato towers instead of stakes.

5. Hornworms and tomato worms almost invariably appear on tomato plants during the growing season. They can virtually strip a plant and ruin the fruits if not controlled. Fortunately control is simple and highly effective with the use of *Bacillus thuringiensis*, which was discussed in Appendix 2.

6. The eating season for garden tomatoes can be extended for about 2 months past the first frost if all the green tomatoes are picked before frost occurs. Place the tomatoes on sheets of newspaper on a flat surface in a cool, dry place, where they will ripen slowly a few at a time. Generally, the taste of tomatoes ripened in this way is superior to that of hothouse tomatoes sold in produce markets. Be sure when picking the green tomatoes to handle them very gently, as they bruise very easily, and molds quickly develop in the bruised areas.

PRUNING

A good gardener or orchardist makes a habit of pruning trees, shrubs, and other plants regularly for a variety of reasons. He or she may wish to improve the quality and size of the fruits and flowers, restrict the size of the plants, keep the plants healthy, shape the shrubbery, or generally get more from the plants.

Except for spring-flowering ornamental shrubs, which should be pruned right after flowering, most maintenance pruning is done in the winter when active growth is not taking place. It usually involves removal of portions of stems, but it can also involve roots. When a terminal bud is removed, the axillary or lateral bud just below the cut will usually develop into a branch, and a bushier growth will result. Some gardeners pinch off terminal buds routinely to encourage such growth. The following sections provide a few generalities and specifics pertaining to several types of plants.

Fruit Trees

When young fruit trees are first planted, all except four or five stems and any damaged roots should be pruned off. The remaining stems should be cut back so that there is one central leader about 1 meter (3 feet) tall, with shorter side branches facing in different directions (Fig. A4.1). When cutting the stems, be careful to cut in such a way that the axillary or lateral bud just below the cut is facing outward. Each succeeding year, prune back new growth to within a few centimeters of the previous year's cut. Remove any dead or diseased branches, and prune stems that have grown so that they are rubbing against each other. Cut out the central leader the second year in peach trees so that the interior of the tree is left relatively open. Regularly remove any sucker shoots that develop from the base or along the trunk of the tree.

Grapevines

There are several methods of pruning grapevines, depending on the type of vine and the circumstances under which the vines are being grown. In general, grapevines should be pruned heavily in late winter for best fruit production. After allowing a central trunk to develop, cut back each shoot, regardless of its length, so that no more than three axillary buds remain. Exceptions to this rule involve situations in which the vines are trained on arbors or wires, when the shoots initially may be allowed to grow longer. Even then, however, after the desired training has taken place (Fig. A4.2), pruning should be heavy for best results.

Roses

Rose bushes should be pruned heavily—they will recover! In general, new stems should be cut back to within 10 to 20 centimeters (4 to 8 inches) of their point of origin, with care taken that the top remaining axillary bud of each stem is pointing outward. This promotes growth that leaves the center open for better air circulation. Any dead or diseased canes should be removed and the number of remaining canes limited to three or four per plant.

Raspberries, Blackberries, and Their Relatives

Berry canes are biennial. They are produced from the base the first year, branch during the summer, and usually produce fruit on the branches the second year, although in milder climates they may also produce fruit on the first year's growth. The canes die after the second year.

Old, dead canes should be removed and all but three or four canes developing from each crown should be pulled out when the

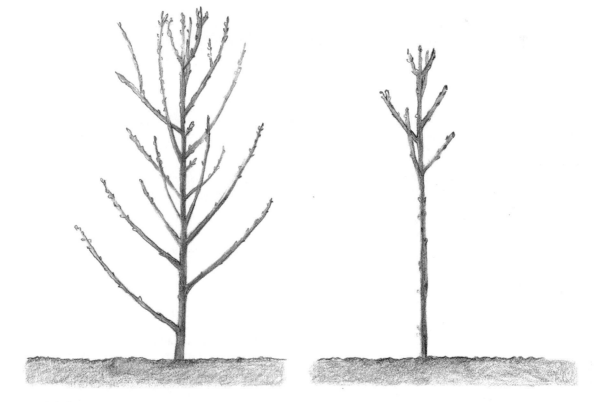

FIGURE A4.1   A young peach tree. *A.* Before pruning. *B.* After pruning.

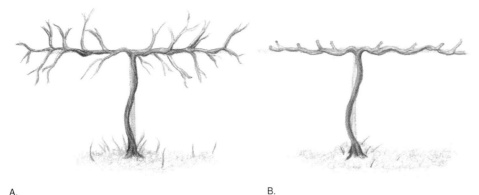

A.   B.

FIGURE A4.2   A grapevine. *A.* Before pruning. *B.* After pruning.

FIGURE A4.3   A bonsai plant. This Sitka spruce tree is little more than 30 centimeters (1 foot) tall and is over 40 years old.
(Courtesy Guy Downing)

ground is soft. New canes should be cut back to lengths of 1 meter (3 feet) or less in the spring. Branches of 1-year-old canes should be cut back in early spring to lengths of about 30 centimeters (12 inches) for larger fruits.

Bonsai

Container-grown trees that are dwarfed through the constant careful pruning of both roots and stems, the manipulation of soil mixtures, and the weighting of branches are called **bonsai** (Fig. A4.3). Some of these dwarfed trees attain ages of 50 to 75 years or more and may be less than 1 meter (3 feet) tall. Bonsai is an art that requires knowledge of the environmental requirements and tolerances of individual species.

In general, bonsai growers pinch out new growth above a bud every few days during a growing season but never prune when a plant is dormant. Refer to *Additional Reading* for more information on the subject.

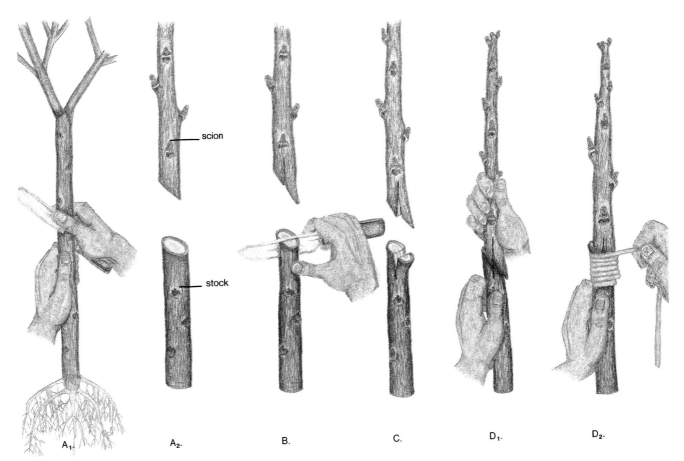

FIGURE A4.4 Stages in whip, or tongue, grafting. *A.* A smooth tangential cut is made at the bottom of the scion and at the top of the stock. *B.* Vertical cuts are made back into the centers of the stock and scion. *C.* The cuts are slightly widened to form a little tongue on each portion. *D.* The scion is inserted into the stock as tightly as possible without forcing a split. The graft then is bound with rubber strips and sealed with grafting wax.

MAJOR TYPES OF GRAFTING

Whip or Tongue Grafting

Whip, or *tongue grafting,* is widely used for relatively small material—that is, wood between 0.70 and 1.25 centimeters (0.25 and 0.50 inch) in diameter. The stock and scion (rooted portion and aerial portion, respectively) should be nearly identical in diameter to bring about maximum contact between the cambia. The scion should contain two or three buds, and the cuts on both the stock and scion should be made in an internode. As shown in Figure A4.4, a smooth tangential cut about 5 centimeters (2 inches) long is made with a sharp sterilized knife at the bottom of the scion and at the top of the stock. The angles of both cuts should also be as nearly identical as possible, and there should be no irregularities or undulations in the surfaces (such as those caused by a dull cutting instrument). A second cut is then made in both the stock and scion about one-third of the distance from the tip of the cut surfaces. This cut is made back into the wood, nearly parallel to the first cut so that it forms a little tongue. The scion is then inserted into the stock as tightly as possible, taking care not to force a split. In addition, the bottom edge of the scion should not protrude past the bottom of the

cut of the stock. The process is completed by binding the materials with plastic tape and adding grafting wax.

If the stock and scion are not identical in diameter, it is still possible to obtain a graft if care is taken to bring the cambia in close contact along one edge of the cuts (Fig. A4.5).

Splice Grafting

Splice grafting is sometimes used with plants in which the pith is extensive. It is essentially the same as whip grafting, except a second cut isn't made and no tongue is formed.

Cleft Grafting

Cleft grafting is used routinely when the diameter of the stock is considerably greater than that of the scion or scions. First, cut the stock branch or trunk at right angles, making sure the bark is not torn. If some bark is pulled loose by the saw, make a new cut. Commercial growers often minimize detachment of the bark by making a cut one-third of the way in on one side and then making a cut slightly lower on the opposite side. This usually leaves a surface with clean edges. Next, hammer a meat cleaver or heavy knife 5.0

CONVERSION TABLE

TO CONVERT	TO	MULTIPLY BY
Milliliters	Fluid ounces	0.03
Liters	Quarts	1.057
Liters	Gallons	0.2642
Fluid ounces	Milliliters	29.57
Quarts	Liters	0.9463
Gallons	Liters	3.785
Hectares	Acres	2.471
Square kilometers	Square miles	0.3861
Square inches	Square feet	6.944×10^{-3}
Square feet	Square inches	144.0
Acres	Hectares	0.4047
Square miles	Square kilometers	2.590

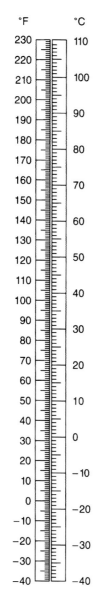

Temperature Conversion Scale

To convert Fahrenheit to Celsius use the following formula:

$C = \frac{5}{9}(°F - 32)$

To convert Celsius to Fahrenheit use the following formula:

$C = \frac{9}{5} °C + 32$

carbon-fixing reactions (kahr'bon fixing ree-ak'shunz) a cyclical series of chemical reactions that utilizes carbon dioxide and energy generated during the light reactions of photosynthesis. Sugars are produced, some of which are stored as insoluble carbohydrates while others are recycled; the reactions are independent of light and occur in the stroma of chloroplasts (pp. 163, 164)

carpel (kahr'pul) an ovule-bearing unit that is a part of a pistil (pp. 126, 402)

caryopsis (kare-ee-op'siss) a dry fruit in which the pericarp is tightly fused to the seed; does not split at maturity (p. 131)

Casparian strip (kass-pair'ee-un strip) a band of suberin around the radial and transverse walls of an endodermal cell (p. 65)

cell (sel) the basic structural and functional unit of living organisms; in plants it consists of protoplasm surrounded by a cell wall (p. 28)

cell biology (sel by-ol'uh-jee) the biological discipline involving the study of cells and their functions (p. 10)

cell cycle (sel sy'kul) a sequence of events involved in the division of a cell (p. 40)

cell membrane (sel mem'brayn) see *plasma membrane*

cell plate (sel playt) the precursor of the middle lamella; forms at the equator during telophase (p. 43)

cell sap (sel sap) the liquid contents of a vacuole (p. 39)

cell wall (sel wawl) the relatively rigid boundary of cells of plants and certain other organisms (p. 31)

centromere (sen'truh-meer) the dense constricted portion of a chromosome to which a spindle fiber is attached (also called *kinetochore*) (p. 41)

chemiosmosis (kem-ee-oz-moh'siss) a theory that energy is provided for phosphorylation by protons being "pumped" across inner mitochondrial and thylakoid membranes (p. 175)

chiasma (pl. **chiasmata**) (kyaz'mah; pl. ky-az'mah-tah) the X-shaped configuration formed by two chromatids of homologous chromosomes as they remain attached to each other during prophase I of meiosis (p. 205)

chlorenchyma (klor-en'kuh-mah) tissue composed of parenchyma cells that contain chloroplasts (p. 51)

chlorophyll (klor'uh-fil) green pigments essential to photosynthesis (p. 36)

chloroplast
containi
of most
(p. 36)

chromatid
strands
centrom

chromatin
staining
found in

chromoplast
containi
chloroph
yellow to

chromosom
consistin
and comp
chromoso
and appe
mitosis a

cilium (pl. **c**
short hai
on the ce
organism
and arran
common
of the cel

circadian rh
rith'um) a
growth ar
organism

cladistics (k
ification s
of shared

cladophyll (
that resen
phyllocla

class (klas) a
between a
(pp. 257,

climax veget
tay'shun)
perpetuat
culminatic
(p. 455)

coenocytic (s
nuclei not
by crossw
molds (p.

cohesion-ten
ten'shun t
explains tl
resulting f
cohesion c
and trache
columns b
(p. 150)

coleoptile (kc
sheath sur
of seedlin
(Poaceae)

A

abscisic acid (ab-siz'ik as'id) (**ABA**) a growth-inhibiting hormone of plants; involved with other hormones in dormancy (p. 186)

abscission (ab-sizh'un) the separation of leaves, flowers, and fruits from plants after the formation of an abscission zone at the base of their petioles, peduncles, and pedicels (p. 116)

achene (uh-keen') a single-seeded fruit in which the seed is attached to the pericarp only at its base (p. 130)

acid (as'id) a substance that dissociates in water, releasing hydrogen ions (p. 20)

active transport (ak'tiv trans'port) the expenditure of energy by a cell in moving a substance across a plasma membrane against a diffusion gradient (p. 148)

adventitious (ad-ven-tish'uss) said of buds developing in internodes or on roots or of roots developing along stems or on leaves (pp. 66, 88)

aerobic respiration (air-oh'bik res-puh-ray'shun) respiration that requires free oxygen (pp. 171, 172)

agar (ah'gur) a gelatinous substance produced by certain red algae and also a few brown algae; often used as a culture medium, particularly for bacteria (pp. 308, 310)

aggregate fruit (ag'gruh-git froot) a fruit derived from a single flower having several to many pistils (p. 127)

algin (al'jin) a gelatinous substance produced by certain brown algae; used in a wide variety of food substances and in pharmaceutical, industrial, and household products (pp. 304, 309)

allele (uh-leel') one of a pair of genes at the identical location (*locus*) on a pair of homologous chromosomes (p. 214)

Alternation of Generations (ol-tur-nay'shun uv jen-ur-ay'shunz) alternation between a haploid gametophyte phase and a diploid sporophyte phase in the life cycle of sexually reproducing organisms (p. 209)

amino acid (ah-mee'noh as'id) one of the organic, nitrogen-containing units from which proteins are synthesized; there are about 20 in all proteins (p. 22)

anaerobic respiration (an-air-oh'bik res-puh-ray'shun) respiration in which the hydrogen removed from the glucose during glycolysis is combined with an organic ion (instead of oxygen) (p. 171)

angiosperm (an'jee-oh-spurm) a plant whose seeds develop within ovaries that mature into fruits (p. 402)

annual (an'you-ul) a plant that completes its entire life cycle in a single growing season (pp. 83, 122)

annual ring (an'you-ul ring) a single season's production of xylem (wood) by the vascular cambium (p. 84)

annulus (an'yu-luss) a specialized layer of cells around a fern sporangium; aids in spore dispersal through a springlike action; also a membranous ring around the stipe of a mushroom (p. 372)

anther (an'thur) the pollen-bearing part of a stamen (p. 403)

antheridiophore (an-thur-id'ee-oh-for) a stalk that bears an antheridium (p. 348)

antheridium (pl. **antheridia**) (an-thur-id'ee-um; pl. an-thur-id'ee-ah) the male gametangium of certain algae, fungi, bryophytes, and vascular plants other than gymnosperms and angiosperms (pp. 301, 315, 348)

anthocyanin (an-thoh-sy'ah-nin) a water-soluble pigment found in cell sap; anthocyanins vary in color from red to blue (p. 39)

antibiotic (an-tee-by-ot'ik) a substance produced by a living organism that interferes with the normal metabolism of another living organism (p. 335)

apical dominance (ay'pi-kul dom'i-nunts) suppression of growth of lateral buds by hormones (p. 188)

apical meristem (ay'pi-kul mair'i-stem) a meristem at the tip of a shoot or root (pp. 50, 63)

apomixis (ap-uh-mik'sis) reproduction without fusion of gametes or meiosis in otherwise normal sexual structures (p. 248)

archegoniophore (ahr-kuh-goh'nee-oh-for) a stalk bearing an archegonium (p. 348)

archegonium (pl. **archegonia**) (ahr-kuh-goh'nee-um; pl. ahr-kuh-goh'nee-ah) the multicellular female gametangium of bryophytes and most vascular plants other than angiosperms (p. 348)

aril (air′il) an often brightly colored appendage surrounding the seed of certain plants (e.g., yew) (p. 388)

ascus (pl. **asci**) (as′kus; pl. as′eye) one of often numerous, frequently fingerlike hollow structures in which the fusion of two haploid nuclei is followed by meiosis; a row of ascospores (usually eight) is ultimately produced in each ascus on or within the sexually initiated reproductive bodies of cup (sac) fungi (p. 323)

asexual reproduction (ay-seksh′yule ree-proh-duk′shun) any form of reproduction not involving the union of gametes (p. 204)

assimilation (uh-sim-i-lay′shun) cellular conversion of raw materials into protoplasm and cell walls (p. 16)

atom (at′um) the smallest individual unit of an element that retains the properties of the element (p. 16)

ATP (ay-tee-pee) adenosine triphosphate, a molecule with three phosphate groups found in all living cells; the principal vehicle for energy storage and exchange in cell metabolism (p. 164)

autotrophic (aw-toh-troh′fik) descriptive of an organism capable of sustaining itself through conversion of inorganic substances to organic material (p. 273)

auxin (awk′sin) a growth-regulating substance produced either naturally by plants or synthetically (p. 182)

axil (ak′sil) the angle formed between a twig and the petiole of a leaf; normally the site of an *axillary bud* (also called *lateral bud*) (p. 80)

B

backcross (bak′kross) a cross involving a hybrid and one of its parents (p. 218)

bacteriophage (bak-teer′ee-oh-fayj) a virus whose host is a bacterium (p. 285)

bark (bahrk) tissues of a woody stem between the vascular cambium and the exterior (p. 85)

base (bayss) a substance that dissociates in water, releasing hydroxyl (OH⁻) ions (p. 20)

basidiospore (bah-sidd′ee-oh-spor) a spore produced on a basidium (p. 329)

basidium (pl. **basidi**‌ buh-sid′ee-ah) on‌ ous, frequently cl‌ structures in whic‌ haploid nuclei is f‌ the four resulting ‌ externally borne b‌ are produced on o‌ initiated reproduc‌ fungi (e.g., mushr‌ (pp. 328, 329)

berry (bair′ee) a thin-‌ usually develops f‌ ovary and commo‌ than one seed (p. 1‌

biennial (by-en′ee-ul)‌ normally requires ‌ complete its life cy‌ growth being stric‌ (p. 122)

biological controls (b‌ trohlz′) the use of ‌ inhibitors in comb‌ other destructive o‌

biome (by′ohm) simil‌ communities consi‌ worldwide scale (e‌ grassland biome) (‌

biotechnology (by-oh‌ manipulation of or‌ cells, or molecules‌ applications prima‌ human benefit (p. 2‌

biotic community (by‌ ee) an association ‌ and other organism‌ (p. 449)

blade (blayd) the cons‌ part of a leaf (also ‌ seaweed (pp. 103, 3‌

bond (bond) a force th‌ together (p. 19)

bonsai (bon-sy′) conta‌ (usually trees) that ‌ artificially through ‌ manipulation of the‌ (p. 528)

botany (bot′an-ee) sci‌ study of plants (p. 6‌

botulism (bot′yu-lizm)‌ consumption of foo‌ botulism bacteria (p‌

bract (brakt) a structur‌ leaflike and modifie‌ color (p. 112)

bryophyte (bry′oh-fyt)‌ terrestrial, aquatic, ‌ embryo-producing p‌ and phloem (e.g., m‌ hornworts) (p. 344)

co‌

co‌

co‌

cr‌

cu‌

cu‌

cu‌

cy‌

cy‌

c‌

c‌

c‌

c‌

c‌

c‌

c‌

c‌

gene (jeen) a unit of heredity; part of a linear sequence of such units occurring in the DNA of chromosomes (pp. 24, 214)

generative cell (jen′uh-ray-tiv sel) the cell of the male gametophyte of angiosperms that divides, producing two *sperms;* also, the cell of the male gametophyte of gymnosperms that divides, producing a *sterile cell* and a *spermatogenous cell* (p. 388)

genetic engineering (juh′net′ik en-juh-neer′ing) the introduction, by artificial means, of genes from one form of DNA into another form of DNA (p. 228)

genetics (juh-net′iks) the biological discipline involving the study of heredity (pp. 10, 213)

genotype (jeen′oh-typ) the genetic constitution of an organism; may or may not be visibly expressed, as contrasted with *phenotype* (p. 214)

genus (pl. **genera**) (jee′nus; pl. jen′er-ah) a category of classification between a family and a species (p. 256)

gibberellin (jib-uh-rel′in) one of a group of plant hormones that have a variety of effects on growth; they are particularly known for promoting elongation of stems (p. 185)

gill (gil) one of the flattened plates of compact mycelium that radiate out from the stalk on the underside of the caps of most mushrooms (p. 329)

girdling (gurd′ling) the removal of a band of tissues extending inward to the vascular cambium on the stem of a woody plant (p. 530)

gland (gland) a small body of variable shape and size that may secrete certain substances but that also may be functionless (pp. 57, 106)

glycolysis (gly-kol′uh-sis) the initial phase of all types of respiration in which glucose is converted to pyruvic acid without involving free oxygen (p. 171)

Golgi body (gohl′jee bod′ee) an organelle consisting of disc-shaped, often branching hollow tubules that apparently function in accumulating and packaging substances used in the synthesis of materials by the cell; also called *dictyosome;* collectively, the Golgi bodies of a cell may be referred to as the *Golgi apparatus* (p. 35)

graft (graft) the union of a segment of a plant, the *scion,* with a rooted portion, the *stock* (pp. 242, 529)

grain (grayn) see *caryopsis*

granum (pl. **grana**) (gra′num; pl. gra′nuh) a series of stacked thylakoids within a chloroplast (p. 36)

gravitational water (grav-uh-tay′shun-ul waw′tur) water that drains out of the pore spaces of a soil after a rain (p. 75)

gravitropism (grav-uh-troh′pism) growth response to gravity (p. 191)

ground meristem (grownd mair′i-stem) meristem that produces all the primary tissues other than the epidermis and stele (e.g., cortex, pith) (pp. 50, 63, 81)

growth (grohth) progressive increase in size and volume through natural development (pp. 15, 181)

guard cell (gahrd sel) one of a pair of specialized cells surrounding a stoma (pp. 57, 106)

guttation (guh-tay′shun) the exudation of water in liquid form from leaves due to root pressure (p. 153)

gymnosperm (jim′noh-spurm) a plant whose seeds are not enclosed within an ovary during their development (e.g., pine tree) (p. 383)

H

haploid (hap′loyd) having one set of chromosomes per cell, as in gametophytes; also referred to as having *n* chromosomes (as contrasted with 2*n* chromosomes in the *diploid* cells of sporophytes) (p. 209)

haustorium (pl. **haustoria**) (haw-stor′ee-um; pl. haw-stor′ee-uh) a protuberance of a fungal hypha or plant organ such as a root that functions as a penetrating and absorbing structure (p. 69)

heartwood (hahrt′wood) nonliving, usually darker-colored wood whose cells have ceased to function in water conduction (p. 85)

herbaceous (hur-bay′shuss or ur-bay′shuss) referring to nonwoody plants (p. 83)

herbal (hur′bul or ur′bul) a 16th- and 17th-century botany book that emphasized medicinal uses, edibility, and other utilitarian functions of plants (p. 7)

herbarium (pl. **herbaria**) (hur-bair′ee-um or ur-bair′ee-um; pl. hur-bair′ee-uh) a collection of dried pressed specimens, usually mounted on paper and provided with a label that gives collection information and an identification (p. 413)

heterocyst (het′uh-roh-sist) a transparent, thick-walled, slightly enlarged cell occurring in the filaments of certain blue-green bacteria (p. 279)

heterospory (het-uh-ross′por-ee) the production of both microspores and megaspores (p. 364)

heterotrophic (het-ur-oh-troh′fick) incapable of synthesizing food and therefore dependent on other organisms for it (p. 272)

heterozygous (het-uh-roh-zy′guss) having a pair of genes with contrasting characters at the same location on homologous chromosomes (p. 215)

holdfast (hold′fast) attachment organ or cell at the base of the thallus or filament of certain algae or blue-green bacteria (pp. 298, 303)

homologous chromosomes (hoh-mol′uh-guss kroh′muh-sohmz) pairs of chromosomes that associate together in prophase I of meiosis; each member of a pair is derived from a different parent (p. 205)

homozygous (hoh-moh-zy′guss) having a pair of genes with identical characters at the same location on a pair of homologous chromosomes (p. 214)

hormone (hor′mohn) an organic substance generally produced in minute amounts in one part of an organism and transported to another part of the organism where it controls or affects growth and development (p. 181)

hybrid (hy′brid) offspring of two parents that differ in one or more genes (pp. 216, 233)

hydathode (hy′duh-thohde) structure at the tip of a leaf vein through which water is forced by root pressures (p. 153)

hydrolysis (hy-drol′uh-sis) the breakdown of complex molecules to simpler ones as a result of the union of water with the compound; usually controlled by enzymes (p. 177)

hydrosere (hy′droh-sear) a primary succession that is initiated in a wet habitat (p. 456)

hygroscopic water (hy-gruh-skop′ik waw′tur) water that is chemically bound to soil particles and therefore unavailable to plants (p. 75)

hypha (pl. **hyphae**) (hy′fuh; pl. hy′fee) a single, usually tubular, threadlike filament of a fungus; *mycelium* is a collective term for hyphae (p. 320)

hypocotyl (hy-poh-kot′ul) the portion of an embryo or seedling between the radicle and the cotyledon(s) (p. 136)

hypodermis (hy-poh-dur′mis) a layer of cells immediately beneath the epidermis and distinct from the parenchyma cells of the cortex in certain plants (pp. 109, 384)

hypothesis (hy-poth'uh-sis) a postulated explanation for some observed facts that must be tested experimentally before it can be accepted as valid or discarded as incorrect (p. 6)

I

imbibition (im-buh-bish'un) adsorption of water and subsequent swelling of organic materials because of the adhesion of the water molecules to the internal surfaces (p. 146)

indusium (pl. **indusia**) (in-dew'zee-um; pl. in-dew'zee-uh) the small, membranous, sometimes umbrellalike covering of a developing fern *sorus* (p. 372)

inferior ovary (in-feer'ee-or oh'vuh-ree) an ovary to which parts of the calyx, corolla, and stamens have become more or less united so they appear to be attached at the top of it (pp. 124, 409)

inflorescence (in-fluh-res'ints) a collective term for a group of flowers attached to a common axis in a specific arrangement (p. 124)

inorganic (in-or-gan'ik) descriptive of compounds having no carbon atoms (p. 20)

integument (in-teg'yu-mint) the outermost layer of an ovule; usually develops into a seed coat; a gymnosperm ovule usually has a single integument, and an angiosperm ovule usually has two integuments (pp. 383, 386, 403)

intermediate-day plant (in-tur-me'dee-ut day plant) a plant that has two critical photoperiods; it will not flower if the days are either too short or too long (p. 197)

internode (in'tur-nohd) a stem region between nodes (p. 80)

ion (eye'on) a molecule or atom that has become electrically charged through the loss or gain of one or more electrons (p. 19)

isogamy (eye-sog'uh-me) sexual reproduction in certain algae and fungi having gametes that are alike in size (p. 299)

isotope (eye'suh-tohp) one of two or more forms of an element that have the same chemical properties but differ in the number of neutrons in the nuclei of their atoms (p. 17)

K

kinetochore (kuh-net'uh-kor) see *centromere*

kingdom (king'dum) the highest category of classification (e.g., Plant Kingdom, Animal Kingdom) (p. 258)

knot (not) a portion of the base of a branch enclosed within wood (p. 95)

Krebs cycle (krebz' sy'kul) a complex series of reactions following glycolysis in aerobic respiration that involves ATP, mitochondria, and enzymes and that results in the combining of free oxygen with protons and electrons from pyruvic acid to make water (p. 172)

L

lamina (lam'uh-nuh) see *blade*

lateral bud (lat'uh-rul bud) see *axil*

laticifer (luh-tis'uh-fur) specialized cells or ducts resembling vessels; they form branched networks of *latex*-secreting cells in the phloem and other parts of plants (p. 86)

leaf (leef) a flattened, usually photosynthetic structure arranged in various ways on a stem (pp. 50, 102)

leaf gap (leef gap) a parenchyma-filled interruption in a stem's cylinder of vascular tissue immediately above the point at which a branch of vascular tissue (*leaf trace*) leading to a leaf occurs (p. 81)

leaflet (leef'lit) one of the subdivisions of a compound leaf (p. 104)

leaf scar (leef skahr) the suberin-covered scar left on a twig when a leaf separates from it through abscission (p. 80)

leaf trace (leef trays) see *leaf gap*

legume (leg'yoom) a dry fruit that splits along two "seams," the seeds being attached along the edges (p. 128)

lenticel (lent'uh-sel) one of usually numerous, slightly raised, somewhat spongy groups of cells in the bark of woody plants; lenticels permit gas exchange between the interior of a plant and the external atmosphere (pp. 58, 83)

leucoplast (loo'kuh-plast) a colorless plastid commonly associated with starch accumulation (p. 37)

light reactions (lyt ree-ak'shunz) a series of chemical and physical reactions through which light energy is converted to chemical energy with the aid of chlorophyll molecules; in the process, water molecules are split, with hydrogen ions and electrons being produced and oxygen gas being released; ATP and $NADPH_2$ also are created (pp. 163, 164)

lignin (lig'nin) a polymer with which certain cell walls (e.g., those of wood) become impregnated (pp. 33, 52)

ligule (lig'yool) the tiny tonguelike appendage at the base of a spike moss (*Selaginella*) or quillwort (*Isoetes*) leaf; also, the outgrowth from the upper and inner side of a grass leaf at the point where it joins the sheath; also, the conspicuous straplike portion of the corolla of an outer floret in the flower head of a member of the Sunflower Family (Asteraceae) (p. 364)

linkage (link'ij) the tendency of two or more genes located on the same chromosome to be inherited together (p. 218)

lipid (lip'id) a general term for fats, fatty substances, and oils (p. 21)

locule (lok'yool) a cavity within an ovary or a sporangium (p. 128)

long-day plant (long day plant) a plant in which flowering is not initiated unless exposure to more than a critical day length occurs (p. 197)

M

mass flow hypothesis (mass flo hy-poth'uh-sus) see *pressure-flow hypothesis*

megaphyll (meg'uh-fill) a leaf having branching veins; it is associated with a leaf gap (p. 359)

megaspore (meg'uh-spor) a spore that develops into a female gametophyte (pp. 364, 386, 387)

megaspore mother cell (meg'uh-spor muth'ur sel) a diploid cell that produces megaspores upon undergoing meiosis (pp. 364, 386, 403)

meiocyte (my'oh-syt) see *spore mother cell*

meiosis (my-oh'sis) the process of two successive nuclear divisions through which segregation of genes occurs and a single diploid ($2n$) cell becomes four haploid (n) cells (p. 204)

mericloning (mair′i-kloh-ning) multi-plication of plants through cultur-ing and artificial dividing of shoot meristems (p. 235)

meristem (mair′i-stem) a region in which undifferentiated cells divide (pp. 41, 50)

mesocarp (mez′uh-karp) the middle region of the fruit wall that lies between the exocarp and the endocarp (p. 126)

mesophyll (mez′uh-fil) parenchyma (chlorenchyma) tissue between the upper and lower epidermis of a leaf (p. 107)

metabolism (muh-tab′uh-lizm) the sum of all the interrelated chemical processes occurring in a living organism (p. 16)

microfilament (my′kroh-fil′uh-mint) a protein filament involved with cytoplasmic streaming and with contraction and movement in animal cells (p. 38)

microphyll (my′kroh-fil) a leaf having a single unbranched vein not associated with a leaf gap (p. 359)

micropyle (my′kroh-pyl) a pore or opening in the integuments of an ovule through which a pollen tube gains access to an embryo sac or arche-gonium of a seed plant (pp. 386, 403)

microspore (my′kroh-spor) a spore that develops into a male gametophyte (pp. 364, 385, 403)

microspore mother cell (my′kroh-spor muth′ur sel) a diploid cell that produces microspores upon undergoing meiosis (pp. 364, 385, 403)

microsporophyll (my-kroh-spor′uh-fil) a leaf, usually reduced in size, on or within which microspores are produced (p. 364)

microtubule (my′kroh-t(y)oo-byul) an unbranched tubelike proteinaceous structure commonly found inside the plasma membrane where it apparently regulates the addition of cellulose to the cell wall (p. 38)

middle lamella (mid′ul luh-mel′uh) a layer of material, rich in pectin, that cements two adjacent cell walls together (p. 31)

midrib (mid′rib) the central (main) vein of a pinnately veined leaf or leaflet (p. 105)

mitochondrion (pl. **mitochondria**) (my-toh-kon′dree-un; pl. my-toh-kon′dree-uh) an organelle containing enzymes that function in the Krebs cycle and the electron transport chain of aerobic respiration (p. 36)

mitosis (my-toh′sis) nuclear division, usually accompanied by cytokinesis, during which the chromatids of the chromosomes separate and two genetically identical daughter nuclei are produced (p. 40)

mixture (miks′chur) a substance containing two or more ingredients, the atoms and molecules of which retain a separate identity and are not in a fixed proportion to one another (p. 19)

molecule (mol′uh-kyul) the smallest unit of an element or compound retaining its own identity; consists of two or more atoms (pp. 16, 18)

monocotyledon (mon-oh-kot-uh-lee′dun) a class of angiosperms whose seeds have a single cotyledon; the term is commonly abbreviated to *monocot* (pp. 83, 122)

monoecious (moh-nee′shuss) having unisexual male flowers or cones and unisexual female flowers or cones both on the same plant (p. 410)

monohybrid cross (mon-oh-hy′brid kross) a cross involving a single pair of genes with contrasting characters in the parents (p. 217)

monokaryotic (mon-oh-kair-ee-ot′ik) having a single nucleus in each cell or unit of the mycelium in club fungi (p. 328)

monomer (mon′oh-mur) a simple indi-vidual molecular unit of a polymer (p. 21)

motile (moh′tul) capable of independent movement (p. 268)

multiple fruit (mul′tuh-pul froot) a fruit derived from several to many individual flowers in a single inflorescence (p. 128)

mushroom (mush′room) a sexually initiated phase in the life cycle of a club fungus, usually consisting of an expanded *cap* and a *stalk* (*stipe*) (p. 329)

mutation (myu-tay′shun) an inheritable change in a gene or chromosome (pp. 234, 247)

mycelium (my-see′lee-um) a mass of fungal hyphae (p. 320)

mycorrhiza (pl. **mycorrhizae**) (my-kuh-ry′zuh; pl. my-kuh-ry′zee) a symbiotic association between fungal hyphae and a plant root (p. 69)

N

n (en) having one set of chromosomes per cell (*haploid*) (p. 209)

NAD (en-ay-dee) nicotinamide adenine dinucleotide, a molecule that during respiration temporarily accepts electrons whose negative charges are balanced by also accepting protons and thereby hydrogen atoms (pp. 172, 174)

NADP (en-ay-dee-pee) nicotinamide adenine dinucleotide phosphate, a high-energy storage molecule that temporarily accepts electrons from Photosystem I in the light reactions of photosynthesis (p. 167)

nastic movement (nass′tik moov′mint) a nondirected movement of a flat organ (e.g., petal, leaf) in which the organ alternately bends up and down (p. 190)

neutron (new′tron) an uncharged particle in the nucleus of an atom (p. 17)

node (nohd) region of a stem where one or more leaves are attached (pp. 50, 80)

nonmeristematic tissue (non-mair′i-stem-atic tish′yu) a tissue composed of cells that have assumed various shapes and sizes related to their functions as they matured following their production by a meristem (p. 51)

nucellus (new-sel′us) ovule tissue within which an embryo sac develops (pp. 383, 386)

nuclear envelope (new′klee-ur en′vuh-lohp) a porous double membrane enclosing a nucleus (p. 38)

nucleic acid (new-klay′ik as′id) see *DNA, RNA*

nucleolus (pl. **nucleoli**) (new-klee′oh-luss; pl. new-klee′oh-ly) a somewhat spherical body within a nucleus; contains primarily RNA and protein; there may be more than one nucleolus per nucleus (p. 38)

nucleotide (new′klee-oh-tyd) the struc-tural unit of DNA and RNA (p. 24)

nucleus (new′klee-uss) the organelle of a living cell that contains chromosomes and is essential to the regulation and control of all the cell's functions; also, the core of an atom (pp. 17, 38)

nut (nutt) one-seeded dry fruit with a hard, thick pericarp; develops with a cup or cluster of bracts at the base (p. 131)

O

oil (oyl) a fat in a liquid state (p. 22)

oogamy (oh-og′uh-mee) sexual reproduction in which the female gamete or egg is nonmotile and larger than the male gamete or sperm, which is motile (p. 301)

oogonium (pl. **oogonia**) (oh-oh-goh'nee-um; pl. oh-oh-goh'nee-ah) a female sex organ of certain algae and fungi; consists of a single cell that contains one to several eggs (pp. 301, 315)

operculum (oh-per'kyu-lum) the lid or cap that protects the peristome of a moss sporangium (p. 353)

orbital (or'buh-till) a volume of space in which a given electron occurs 90% of the time (p. 17)

order (or'dur) a category of classification between a class and a family (p. 258)

organelle (or-guh-nel') a membrane-bound body in the cytoplasm of a cell; there are several kinds, each with a specific function (e.g., mitochondrion, chloroplast)[1] (p. 31)

organic (or-gan'ik) pertaining to or derived from living organisms and pertaining to the chemistry of carbon-containing compounds (p. 20)

osmosis (oz-moh'sis) the diffusion of water or other solvents through a differentially permeable membrane from a region of higher concentration to a region of lower concentration (p. 145)

osmotic potential (oz-mot'ik puh-ten'shil) potential pressure that can be developed by a solution separated from pure water by a differentially permeable membrane (the pressure required to prevent osmosis from taking place) (p. 145)

osmotic pressure (oz-mot'ik presh'ur) see *osmotic potential*

ovary (oh'vuh-ree) the enlarged basal portion of a pistil that contains an ovule or ovules and usually develops into a fruit (p. 124)

ovule (oh'vyool) a structure of seed plants that contains a female gametophyte and has the potential to develop into a seed (pp. 124, 383, 386)

P

palisade mesophyll (pal-uh-sayd' mez'uh-fil) mesophyll having one or more relatively uniform rows of tightly packed, elongate, columnar parenchyma (chlorenchyma) cells beneath the upper epidermis of a leaf (p. 107)

palmately compound (pahl'mayt-lee kom'pownd) having leaflets or principal veins radiating out from a common point (p. 104)

1. Ribosomes, which are considered organelles, are an exception in that they are not bounded by a membrane.

palmately veined (pahl'mayt-lee vaynd) see *palmately compound*

papilla (pl. **papillae**) (puh-pil'uh; pl. puh-pill'ay) a small, usually rounded or conical protuberance (p. 299)

parenchyma (puh-ren'kuh-muh) thin-walled cells varying in size, shape, and function; the most common type of plant cell (p. 51)

parthenocarpic (par-thuh-noh-kar'pik) developing fruits from unfertilized ovaries; the resulting fruit is, therefore, usually seedless (p. 408)

passage cell (pas'ij sel) a thin-walled cell of an endodermis (p. 65)

pectin (pek'tin) a water-soluble organic compound occurring primarily in the middle lamella; becomes a jelly when combined with organic acids and sugar (p. 31)

pedicel (ped'i-sel) the individual stalk of a flower that is part of an inflorescence (p. 124)

peduncle (pee'dun-kul) the stalk of a solitary flower or the main stalk of an inflorescence (p. 124)

peptide bond (pep'tyd bond) the type of chemical bond formed when two amino acids link together in the synthesis of proteins (p. 22)

perennial (puh-ren'ee-ul) a plant that continues to live indefinitely after flowering (p. 122)

pericarp (per'uh-karp) collective term for all the layers of a fruit wall (p. 126)

pericycle (per'uh-sy-kul) tissue sandwiched between the endodermis and phloem of a root; often only one or two cells wide in transverse section; the site of origin of lateral roots (p. 65)

periderm (pair'uh-durm) outer bark; composed primarily of cork cells (p. 57)

peristome (per'uh-stohm) one or two series of flattened, often ornamented structures (teeth) arranged around the margin of the open end of a moss sporangium; the teeth are sensitive to changes in humidity and facilitate the release of spores (p. 353)

petal (pet'ul) a unit of a corolla; usually both flattened and colored (p. 124)

petiole (pet'ee-ohl) the stalk of a leaf (pp. 80, 103)

$P_{far-red}$ or P_{fr} (pee-far-red *or* pee-ef-ahr) a form of phytochrome (which see) (p. 198)

pH scale (pee-aitch) a symbol of hydrogen ion concentration indicating the degree of acidity or alkalinity (p. 20)

phage (fayj) see *bacteriophage*

phellogen (fel'uh-jun) see *cork cambium*

phenotype (fee'noh-typ) the physical appearance of an organism (p. 214)

pheromone (fer'uh-mohn) something produced by an organism that facilitates chemical communication with another organism (p. 489)

phloem (flohm) the food-conducting tissue of a vascular plant (p. 54)

photon (foh'ton) a unit of light energy (p. 165)

photoperiodism (foh-toh-pir'ee-ud-izm) the initiation of flowering and certain vegetative activities of plants in response to relative lengths of day and night (p. 197)

photosynthesis (foh-toh-sin'thuh-sis) the conversion of light energy to chemical energy; water, carbon dioxide, and chlorophyll are all essential to the process, which ultimately produces carbohydrate, with oxygen being released as a by-product (pp. 16, 161)

photosynthetic unit (foh-toh-sin-thet'ik yew'nit) one of two groups of about 250 to 4 pigment molecules each that function together in chloroplasts in the light reactions of photosynthesis; the units are exceedingly numerous in each chloroplast (p. 163)

photosystem (foh'toh-sis-tum) collective term for a specific functional aggregation of photosynthetic units (p. 165)

phytochrome (fy'tuh-krohm) protein pigment associated with the absorption of light; it is found in the cytoplasm of cells of green plants and occurs in interconvertible active and inactive forms ($P_{far\ red}$ and P_{red}); facilitates a plant's capacity to detect the presence (or absence) and duration of light (p. 198)

pilus (pl. **pili**) (py'lis; pl. py'lee) the equivalent of a conjugation tube in bacteria (p. 269)

pinna (pl. **pinnae**) (pin'uh; pl. pin'ee) a primary subdivision of a fern frond; the term is also applied to a leaflet of a compound leaf (p. 372)

pinnately compound (pin'ayt-lee kom'pownd) having leaflets or veins attached on both sides of a common axis (e.g., rachis, midrib) (p. 104)

pinnately veined (pin'ayt-lee vaynd) see *pinnately compound*

pistil (pis'tul) a female reproductive structure of a flower; composed of one or more carpels and consisting of an ovary, style, and stigma (p. 124)

pit (pit) a more or less round or elliptical thin area in a cell wall; pits occur in pairs opposite each other, with or without shallow, domelike borders (pp. 46, 54)

pith (pith) central tissue of a dicot stem and certain roots; consists of parenchyma cells that become proportionately less of the volume of woody plants as cambial activity increases the organ's girth (p. 81)

plankton (plank'ton) free-floating aquatic organisms that are mostly microscopic (p. 296)

plant anatomy (plant uh-nat'uh-mee) the botanical discipline that pertains to the internal structure of plants (p. 8)

plant community (plant kuh-myu'nuh-tee) an association of plants inhabiting a common environment and interacting with one another (p. 448)

plant ecology (plant ee-koll'uh-jee) the science that deals with the relationships and interactions between plants and their environment (p. 9)

plant geography (plant jee-og'ruh-fee) the botanical discipline that pertains to the broader aspects of the space relations of plants and their distribution over the surface of the earth (p. 9)

plant morphology (plant mor-fol'uh-jee) the botanical discipline that pertains to plant form and development (p. 9)

plant physiology (plant fiz-ee-ol'uh-jee) the botanical discipline that pertains to the metabolic activities and processes of plants (p. 8)

plant taxonomy (plant tak-son'uh-mee) the botanical discipline that pertains to the classification, naming, and identification of plants (p. 9)

plasma membrane (plaz'muh mem'brayn) the outer boundary of the protoplasm of a cell; also called *cell membrane,* particularly in animal cells (p. 33)

plasmid (plaz'mid) one of up to 30 or 40 small, circular DNA molecules usually present in a bacterial cell (p. 228)

plasmodesma (pl. **plasmodesmata**) (plaz-muh-dez'muh; pl. plaz-muh-dez'muh-tah) minute strands of cytoplasm that extend between adjacent cells through pores in the walls (p. 33)

plasmodium (pl. **plasmodia**) (plaz-moh'dee-um; pl. plaz-moh'dee-ah) the multinucleate, semiviscous liquid, active form of slime mold; moves in a "crawling-flowing" motion (p. 312)

plasmolysis (plaz-mol'uh-sis) the shrinking in volume of the protoplasm of a cell and the separation of the protoplasm from the cell wall due to loss of water via osmosis (p. 146)

plastid (plas'tid) an organelle associated primarily with the storage or manufacture of carbohydrates (e.g., *leucoplast, chloroplast*) (p. 36)

plumule (ploo'myool) the terminal bud of the embryo of a seed plant (p. 136)

polar nuclei (poh'lur new'klee-eye) nuclei, frequently two in number, that unite with a sperm in an embryo sac, forming a primary endosperm nucleus (p. 403)

pollen grain (pahl'un grayn) a structure derived from the microspore of seed plants that develops into a male gametophyte (pp. 124, 385, 403)

pollen tube (pahl'un t(y)oob) a tube that develops from a pollen grain and conveys the sperms to the female gametophyte (pp. 388, 407)

pollination (pahl-uh-nay'shun) the transfer of pollen from an anther to a stigma (p. 407)

pollinium (pl. **pollinia**) (pah-lin'ee-um; pl. pah-lin'ee-ah) a cohesive mass of pollen grains commonly found in members of the Orchid Family (Orchidaceae) and the Milkweed Family (Asclepiadaceae) (p. 412)

polymer (pahl'i-mur) a large molecule composed of many monomers (p. 21)

polyploidy (pahl'i-ploy-dee) having more than two complete sets of chromosomes per cell (p. 234)

pome (pohm) a simple fleshy fruit whose flesh is derived primarily from the receptacle (p. 127)

population (pop-yew-lay'shun) a group of organisms, usually of the same species, occupying a given area at the same time (pp. 448, 449)

P$_{red}$ or **P$_r$** (pee-red *or* pee-ahr) a form of phytochrome (which see) (p. 198)

pressure-flow hypothesis (presh'ur floh hy-poth'uh-sis) the theory that food substances in solution in plants flow along concentration gradients between the sources of the food and sinks (places where the food is utilized) (p. 151)

primary consumer (pry'mer-ree kon-soo'mur) organism that feeds directly on producers (p. 451)

primary tissue (pry'mer-ee tish'yu) a tissue produced by an apical meristem (e.g., epidermis, cortex, primary xylem and phloem, pith) (p. 50)

primordium (pry-mord'ee-um) an organ or structure (e.g., leaf, bud) at its earliest stage of development (p. 81)

procambium (proh-kam'bee-um) a tissue produced by the primary meristem that differentiates into primary xylem and phloem (pp. 50, 63, 81)

producer (pruh-dew'sur) an organism that manufactures food through the process of photosynthesis (p. 451)

prokaryotic (proh-kair-ee-ot'ik) having a cell or cells that lack a distinct nucleus and other membrane-bound organelles (e.g., bacteria) (pp. 30, 267)

proplastid (proh-plas'tid) a tiny, undifferentiated organelle that can duplicate itself and that may develop into a chloroplast, leucoplast, or other type of plastid (p. 37)

protein (proht'ee-in *or* proh'teen) a polymer composed of many amino acids linked together by peptide bonds (p. 22)

prothallus (pl. **prothalli**) (proh-thal'us; pl. proh-thal'eye) the gametophyte of ferns and their relatives; also called *prothallium* (p. 372)

protoderm (proh'tuh-durm) the primary meristem that gives rise to the epidermis (pp. 50, 63, 81)

proton (proh'ton) a positively charged particle in the nucleus of an atom (p. 17)

protonema (proh-tuh-nee'muh) a green, usually branched, threadlike or sometimes platelike growth from a bryophyte spore; gives rise to "leafy" gametophytes (p. 349)

protoplasm (proh'tuh-plazm) the living part of a cell (includes the cytoplasm and nucleus) (p. 20)

pruning (proon'ing) removal of portions of plants for aesthetic purposes, for improving quality and size of fruits or flowers, or for elimination of diseased tissues (p. 527)

pyrenoid (py'ruh-noyd) a small body found on the chloroplasts of certain green algae and hornworts; associated with starch accumulation; may occur singly on a chloroplast or may be numerous (p. 297)

pyruvic acid (py-roo'vik as'id) the organic compound that is the end product of the glycolysis phase of respiration (p. 172)

Q

quiescence (kwy′ess-ens) a state in which a seed or other plant part will not germinate or grow unless environmental conditions normally required for growth are present (p. 200)

R

rachis (ray′kiss) the axis of a pinnately compound leaf or frond extending between the lowermost leaflets or pinnae and the terminal leaflet or pinna (corresponds with the midrib of a simple leaf) (p. 104)

radicle (rad′i-kuhl) the part of an embryo in a seed that develops into a root (pp. 61, 136)

ray (ray) radially oriented tiers of parenchyma cells that conduct food, water, and other materials laterally in the stems and roots of woody plants; generally continuous across the vascular cambium between the xylem and the phloem; the portion within the wood is called a *xylem ray,* while the extension of the same ray in the phloem is called a *phloem ray* (pp. 54, 85)

receptacle (ree-sep′tuh-kul) the commonly expanded tip of a peduncle or pedicel to which the various parts of a flower (e.g., calyx, corolla) are attached (p. 124)

recessive (ree-ses′iv) descriptive of a member of a pair of genes whose phenotypic expression is masked or suppressed by the dominant gene (p. 214)

red tide (red tyd) the marine phenomenon that results in the water becoming temporarily tinged with red due to the sudden proliferation of certain dino-flagellates that produce substances poisonous to animal life and humans (p. 294)

reproduction (ree-proh-duk′shun) the development of new individual organisms through either sexual or asexual means (p. 15)

resin canal (rez′in kuh-nal′) a tubular duct of many conifers and some angio-sperms that is lined with resin-secreting cells (pp. 85, 384)

respiration (res-puh-ray′shun) the cellular breakdown of sugar and other foods, accompanied by release of energy; in aerobic respiration, oxygen is utilized (pp. 16, 171, 174)

rhizoid (ry′zoyd) delicate root- or root-hairlike structures of algae, fungi, the gametophytes of bryophytes, and certain structures of a few vascular plants; function in anchorage and absorption but have no xylem or phloem (p. 315)

rhizome (ry′zohm) an underground stem, usually horizontally oriented, that may be superficially rootlike in appearance but that has definite nodes and internodes (p. 88)

ribosome (ry′boh-sohm) a granular particle composed of two subunits consisting of RNA and proteins; ribosomes lack membranes, are the sites of protein synthesis, and are very numerous in living cells (p. 35)

RNA (ar-en-ay) the standard abbreviation for *ribonucleic acid,* an important cellular molecule that occurs in three forms, all involved in communication between the nucleus and the cytoplasm and in the synthesis of proteins (pp. 24, 222)

root (root) a plant organ that functions in anchorage and absorption; most roots are produced below ground (p. 50)

root cap (root kap) a thimble-shaped mass of cells at the tip of a growing root; functions primarily in protection (p. 62)

root hair (root hair) a delicate protu-berance that is part of an epidermal cell of a root; root hairs occur in a zone behind the growing tip (p. 64)

runner (run′ur) a stem that grows horizontally along the surface of the ground; typically has long internodes; see also *stolon* (p. 88)

S

salt (salt) a substance produced by the bonding of ions that remain after hydrogen and hydroxyl ions of an acid and a base combine to form water (p. 20)

samara (sah-mair′uh) a dry fruit whose pericarp extends around the seed in the form of a wing (p. 132)

saprobe (sap′rohb) an organism that obtains its food directly from nonliving organic matter (p. 272)

sapwood (sap′wood) outer layers of wood that transport water and minerals in a tree trunk; sapwood is usually lighter in color than heartwood (p. 85)

science (sy′ints) a branch of study involved with the systematic observation, recording, organization, and classification of facts from which natural laws are derived and used predictively (p. 6)

scion (sy′un) a segment of plant that is grafted onto a stock (p. 529)

sclereid (sklair′id) a sclerenchyma cell that usually has one axis not conspicuously longer than the other; may vary in shape and is heavily lignified (p. 52)

sclerenchyma (skluh-ren′kuh-muh) tissue composed of lignified cells with thick walls; functions primarily in strengthening and support (p. 52)

secondary consumer (sek′on-dair-ee kon-soo′mer) an organism that feeds on other consumers (p. 451)

secondary tissue (sek′un-der-ee tish′yu) a tissue produced by the vascular cambium or the cork cambium (e.g., virtually all the xylem and phloem in a tree trunk) (p. 64)

secretory cell (see′kruh-tor-ee sel) cell (or tissue) producing a substance or sub-stances that are moved outside the cells (p. 53)

secretory tissue (see′kruh-tor-ee tish′yu) see *secretory cell*

seed (seed) a mature ovule containing an embryo and bound by a protective seed coat (pp. 124, 204)

seed coat (seed′ koht) the outer boundary layer of a seed; developed from the integument(s) (pp. 383, 386, 403)

semipermeable membrane (sem-ee-pur-me-uh-bil mem-brayn) see *differentially permeable membrane*

senescence (suh-ness′ints) the breakdown of cell components and membranes that leads to the death of the cell (p. 188)

sepal (see′puhl) a unit of the calyx that frequently resembles a reduced leaf; sepals often function in protecting the unopened flower bud (p. 124)

sessile (sess′uhl) without petiole or pedicel; attached directly by the base (p. 103)

seta (see′tuh) the stalk of a bryophyte sporophyte (p. 348)

sexual reproduction (seksh′yule ree-proh-duk′shun) reproduction involving the union of gametes (p. 204)

short-day plant (short day plant) a plant in which flowering is initiated when the days are shorter than its critical photoperiod (p. 197)

sieve plate (siv playt) an area of the wall of a sieve-tube member that contains several to many perforations that permit cytoplasmic connections between similar adjacent cells, the cytoplasmic strands being larger than plasmodesmata (p. 54)

sieve tube (siv t(y)oob) a column of sieve-tube members arranged end to end; food is conducted from cell to cell through sieve plates (p. 54)

sieve-tube member (siv t(y)oob mem′bur) a single cell of a sieve tube (p. 54)

silique (suh-leek′) a dry fruit that splits along two "seams," with the seeds borne on a central partition (p. 128)

simple leaf (sim′pul leef) a leaf with the blade undivided into leaflets (p. 104)

solvent (sol′vent) a substance (usually liquid) capable of dissolving another substance (p. 145)

sorus (pl. **sori**) (sor′uss; pl. sor′eye) a cluster of sporangia; the term is most frequently applied to clusters of fern sporangia (p. 372)

species (spee′seez; *species* is spelled and pronounced the same way in either singular or plural form; there is no such thing as a *specie*) the basic unit of classification; a population of individuals capable of interbreeding freely with one another but, which, because of geographic, reproductive, or other barriers, do not in nature interbreed with members of other species (p. 256)

sperm (spurm) a male gamete; except for those of red algae and angiosperms, sperms are frequently motile and are usually smaller than the corresponding female gametes (pp. 204, 301)

spice (spyss) an aromatic organic plant product used to season or flavor food or drink (p. 506)

spindle (spin′dul) an aggregation of fiberlike threads (microtubules) that appears in cells during mitosis and meiosis; some threads are attached to the centromeres of chromosomes, whereas other threads extend directly or in arcs between two invisible points designated as poles (p. 42)

spine (spyn) a relatively strong, sharp-pointed, woody structure usually located on a stem; usually a modified leaf or stipule (p. 110)

spongy mesophyll (spun′jee mez′uh-fil) mesophyll having loosely arranged cells and numerous air spaces; generally confined to the lower part of the interior of a leaf just above the lower epidermis (p. 107)

sporangiophore (spuh-ran′jee-uh-for) the stalk on which a sporangium is produced (p. 321)

sporangium (pl. **sporangia**) (spuh-ran′jee-um; pl. spuh-ran′jee-uh) a structure in which spores are produced; may be either unicellular or multicellular (pp. 313, 321)

spore (spor) a reproductive cell or aggregation of cells capable of developing directly into a gametophyte or other body without uniting with another cell (*note:* a bacterial spore is not a reproductive cell but is an inactive phase that enables the cell to survive under adverse conditions); *sexual spores* formed as a result of meiosis are often called *meiospores;* spores produced by mitosis may be referred to as *vegetative spores* (pp. 210, 313)

spore mother cell (spor muth′ur sel) a diploid cell that becomes four haploid spores or nuclei as a result of undergoing meiosis (pp. 210, 348)

sporophyll (spor′uh-fil) a modified leaf that bears a sporangium or sporangia (p. 361)

sporophyte (spor′uh-fyt) the diploid (2n) spore-producing phase of the life cycle of an organism exhibiting Alternation of Generations (pp. 209, 210)

stamen (stay′min) a pollen-producing structure of a flower; consists of an anther and usually also a filament (p. 124)

stele (steel) the central cylinder of tissues in a stem or root; usually consists primarily of xylem and phloem (p. 83)

stem (stem) a plant axis with leaves or enations (p. 50)

stigma (stig′muh) the pollen receptive area of a pistil; also, the eyespot of certain motile algae (p. 124)

stipe (styp) the supporting stalk of sea-weeds, mushrooms, and certain other stationary organisms (p. 329)

stipule (stip′yool) one of a pair of appendages of varying size, shape, and texture present at the base of the leaves of some plants (pp. 80, 103)

stock (stok) the rooted portion of a plant to which a scion is grafted (p. 242)

stolon (stoh′lun) a stem that grows vertically below the surface of the ground; typically has relatively long internodes; see also *runner* (p. 90)

stoma (pl. **stomata**) (stoh′muh; pl. stoh′mah-tuh) a minute pore or opening in the epidermis of leaves, herbaceous stems, and the sporophytes of hornworts (*Anthoceros*); flanked by two guard cells that regulate its opening and closing and thus regulate gas exchange and transpiration; the guard cells and pore are also collectively referred to as a *stoma* (pp. 36, 57, 106)

strobilus (pl. **strobili**) (stroh′buh-luss; pl. stroh′buh-leye) an aggregation of sporophylls on a common axis; usually resembles a cone or is somewhat conelike in appearance (pp. 361, 368)

stroma (stroh′muh) a region constituting the bulk of the volume of a chloroplast or other plastid; contains enzymes that in chloroplasts play a key role in carbon fixation, carbohydrate synthesis, and other photosynthetic reactions (p. 36)

style (styl) the structure that connects a stigma and an ovary (p. 124)

suberin (soo′buh-rin) a fatty substance found primarily in the cell walls of cork and the Casparian strips of endodermal cells (pp. 58, 82)

succession (suk-sesh′un) an orderly progression of changes in the composition of a community from the initial development of vegetation to the establishment of a climax community (p. 455)

sucrose (soo′krohs) a disaccharide composed of glucose and fructose; the primary form in which sugar produced by photosynthesis is transported throughout a plant (p. 21)

superior ovary (soo-peer′ee-or oh′vuh-ree) an ovary that is free from the calyx, corolla, and other floral parts, so the sepals and petals appear to be attached at its base (pp. 124, 409)

symbiosis (sim-by-oh′siss) an intimate association between two dissimilar organisms that benefits both of them (mutualism) or is harmful to one of them (parasitism) (p. 267)

syngamy (sin′gam-mee) a union of gametes; fertilization (p. 210)

T

2n (too-en) having two sets of chromosomes; diploid (p. 209)

3n (three-en) having three sets of chromosomes; triploid (p. 408)

tendril (ten′dril) a slender structure that coils on contact with a support of suitable diameter; usually is a modified leaf or leaflet, and aids the plant in climbing (p. 109)

thallus (pl. **thalli**) (thal′uss; pl. thal′eye) a multicellular plant body that is usually flattened and not organized into roots, stems, or leaves (pp. 304, 338, 347)

thylakoid (thy′luh-koyd) coin-shaped membranes whose contents include chlorophyll and that are arranged in stacks that form the *grana* of chloroplasts (p. 36)

tissue (tish′yu) an aggregation of cells having a common function (p. 50)

tissue culture (tish′yu kult′yur) the culture of isolated living tissue on an artificial medium (p. 234)

tracheid (tray′kee-id) a xylem cell that is tapered at the ends and has thick walls containing pits (p. 54)

transpiration (trans-puh-ray′shun) loss of water in vapor form; most transpiration takes place through the stomata (p. 103)

tropism (troh′pizm) the response of a plant organ or part to an external stimulus, usually in the direction of the stimulus (p. 190)

tuber (t(y)oo′bur) a swollen, fleshy underground stem (e.g., white potato) (p. 90)

turgid (tur′jid) firm or swollen because of internal water pressures resulting from osmosis (p. 146)

turgor movement (turr′gor moov′mint) the movement that results from changes in internal water pressures in a plant part (p. 192)

turgor pressure (tur′gur presh′ur) pressure within a cell resulting from the uptake of water (p. 146)

U

unisexual (yu-nih-seksh′yu-ul) a term usually applied to a flower lacking either stamens or a pistil (p. 410)

V

vacuolar membrane (vak-yu-oh′lur mem′brayn) the delimiting membrane of a cell vacuole; also called *tonoplast* (p. 39)

vacuole (vak′yu-ohl) a pocket of fluid that is separated from the cytoplasm of a cell by a membrane; may occupy more than 99% of a cell′s volume in plants; also, food-storage or contractile pockets within the cytoplasm of unicellular organisms (p. 39)

vascular bundle (vas′kyu-lur bun′dul) a strand of tissue composed mostly of xylem and phloem and usually enveloped by a bundle sheath (p. 83)

vascular cambium (vas′kyu-lur kam′bee-um) a narrow cylindrical sheath of cells that produces secondary xylem and phloem in stems and roots (pp. 41, 50)

vascular plant (vas′kyu-lur plant) a plant having xylem and phloem (p. 344)

vein (vayn) a term applied to any of the vascular bundles that form a branching network within leaves (p. 107)

venter (ven′tur) the site of the egg in the enlarged basal portion of an archegonium (p. 351)

vessel (ves′uhl) one of usually very numerous cylindrical "tubes" whose cells have lost their cytoplasm; occur in the xylem of most angiosperms and a few other vascular plants; each vessel is composed of *vessel members* laid end to end; the perforated or open-ended walls of the vessel members permit water to pass through freely (p. 54)

vessel member (ves′uhl mem′bur) a single cell of a vessel (p. 54)

virus (vy′riss) a minute particle consisting of a core of nucleic acid, usually surrounded by a protein coat; incapable of growth alone and can reproduce only within, and at the expense of, a living cell (p. 283)

vitamin (vyt′uh-min) a complex organic compound produced primarily by photosynthetic organisms; various vitamins are essential in minute amounts to facilitate enzyme reactions in living cells (p. 182)

W

whorled (wirld) having three or more leaves or other structures at a node (p. 103)

X

xerosere (zer′roh-sear) a primary succession that initiates with bare rock (p. 455)

xylem (zy′lim) the tissue through which most of the water and dissolved minerals utilized by a plant are conducted; consists of several types of cells (p. 54)

Z

zoospore (zoh′uh-spor) a motile spore occurring in algae and fungi (p. 297)

zygote (zy′goht) the product of the union of two gametes (pp. 204, 407)

Index

X

Xanthophyceae, 292
Xanthophyll, 115, 116, 308
Xerophyte, 109, 449
Xerosere, **455–56**
X-radiation, 117, 221, 336, 370, 440
Xylem, 53, **54,** 65, 66, 69, *71, 84, 85, 91,* 107, *109,* 116, 146, 150, 359, 403
   annual ring of, 66, *84–85, 86–87,* 98, 99, 384, 385
   components, 54, 82
   function of, 54, 82
   parenchyma, 54
   primary, 63, 65, 66, 81, 83, 84, *86–87*
   ray cells, 54, *86–87,* 384
   secondary, 66, 82, *83,* 84, *86–87*

Y

Yam, 66, 72, 421
Yareta, 117
Yarrow, American, 438
Yarrow, European, 438
Yeast, 29, 232, 234, 286, 323, 326
Yeast infection, 278
Yellow fever, 284, 287
Yellow-green algae (Xanthophyceae), 292
Yellow pine, 394
Yellowstone National Park, 174, 279
Yellow vetchling, 109
Yellow water weed, 67
Yew, *104, 388,* 397
   English, 397
   Pacific, 398
Yogurt bacteria, 278
Yosemite National Park, 469
Yucca, 411

Z

Zebra, 28
Zinc, 76, 154, 327
Zinnia, 28, 246, 416
Zoologist, 46
Zoosporangium, *316*
Zoospore, *297, 298, 300. See also individual organisms in Chapters 18–19.*
Zosterophyllum, 360
Zuni Indian, 428
Zygomycete, 321–22. *See also specific organisms in Chapters 18–23.*
Zygomycota, 262, **321–23**
Zygospore, 321, *322*
Zygote, 140, 204, 209, 215, 344